Advances in
QUANTUM CHEMISTRY

VOLUME **57**

Advances in QUANTUM CHEMISTRY

THEORY OF CONFINED QUANTUM SYSTEMS: PART ONE

VOLUME **57**

Editors

JOHN R. SABIN
Quantum Theory Project
University of Florida
Gainesville, Florida

ERKKI BRÄNDAS
Department of Quantum Chemistry
Uppsala University
Uppsala, Sweden

Special Editor

SALVADOR A. CRUZ
Physics Department
Universidad Autonoma Metropolitana-Iztapalapa
AP 55 534 09340
Mexico City
Mexico

ELSEVIER

Amsterdam • Boston • Heidelberg • London • New York • Oxford
Paris • San Diego • San Francisco • Singapore • Sydney • Tokyo

Academic Press is an imprint of Elsevier

Academic Press is an imprint of Elsevier
Linacre House, Jordan Hill, Oxford OX2 8DP, UK
32 Jamestown Road, London NW1 7BY, UK
Radarweg 29, PO Box 211, 1000 AE Amsterdam, The Netherlands
30 Corporate Drive, Suite 400, Burlington, MA 01803, USA
525 B Street, Suite 1900, San Diego, CA 92101-4495, USA

First edition 2009

ISBN: 978-0-12-374764-8
ISSN: 0065-3276

For information on all Academic Press publications visit our website at elsevierdirect.com

Printed and bound in USA
09 10 11 12   10 9 8 7 6 5 4 3 2 1

CONTENTS

PREFACE

The study of confined quantum systems has attracted increasing attention from several research groups in the world due to the unusual physical and chemical properties exhibited by such systems when subject to spatial limitation. Such novel properties, not present in conventional materials, have marked a new era for the synthesis of modern materials – structured at the nanoscale – and leading to what is now called nanotechnology.

As is the case with any scientific development with important technological consequences, basic research plays a fundamental role whereby appropriate models are designed to explore and predict the physical and chemical behavior of a system. A confined quantum system is a clear example where theory constitutes a cornerstone for explanation and prediction of new properties of spatially limited atoms, molecules, electrons, excitons, etc. Theoretical study of possible confined structures might also suggest and stimulate further experimental investigations. In essence, the design of novel materials with exceptional properties requires proper theoretical modeling.

A collection of contributions from leading scientists in the field dealing with different theoretical approaches for tackling a wide variety of quantum systems under different confinement conditions is presented in this and its companion volume: Vol. 58. These volumes also provide an updated view of recent developments and open problems. The type of confined systems studied ranges from the hydrogen, helium and many-electron atoms within geometrical boundaries, atoms and molecules encaged inside fullerene and fullerene-like cages, plasma effects on confined many-electron atoms up to low-dimensional multi-electron quantum dots, to engineering of quantum confined silicon nanostructures.

We hope that these volumes will inspire both new and senior researchers to participate in this exciting and timely field. Although the topics covered in these volumes are by no means exhaustive, and although not all the experts are included, we are certain that the wide span of topics given here constitutes an important reference for the state-of-the-art development in the field of confined quantum systems.

We are grateful to all contributors for devoting their valuable time and dedication in contributing to these volumes.

Salvador A. Cruz
Special Editor

CONTRIBUTORS

Numbers in parentheses indicate the pages where the authors' contributions can be found.

N. Aquino (123)
Departamento de Física, Universidad Autónoma Metropolitana-Iztapalapa, Apartado Postal 55-534, 09340 México D. F
e-mail: naa@xanum.uam.mx

Richard F.W. Bader (285)
Department of Chemistry, McMaster University, Hamilton ON, L8S 4M1, Canada
e-mail: bader@mcmaster.ca

B.L. Burrows (173)
Mathematics Section, Faculty of Computing, Engineering and Technology, Staffordshire University, Beaconside, Stafford, ST18, 0DP UK
e-mail: B.L.Burrows@staffs.ac.uk, brian.burrows2@btopenworld.com

M. Cohen (173)
Department of Physical Chemistry, The Hebrew University of Jerusalem, Jerusalem 91904, Israel
e-mail: maurice@fh.huji.ac.il

Salvador A. Cruz (255)
Departamento de Física, Universidad Autónoma Metropolitana-Iztapalapa, Apdo. Postal 55 534, 09340, México, D. F., México
e-mail: cruz@xanum.uam.mx

J. Garza (241)
Departamento de Química, División de Ciencias Básicas e Ingeniería, Universidad Autónoma Metropolitana-Iztapalapa, San Rafael Atlixco 186, Col. Vicentina. Iztapalapa. C. P. 09340. México D. F., México
e-mail: jgo@xanum.uam.mx

Cecil Laughlin (203)
School of Mathematical Sciences, University of Nottingham, Nottingham NG7 2RD, UK
e-mail: Cecil.Laughlin@nottingham.ac.uk

Eugenio Ley-Koo (79)
Instituto de Física, Universidad Nacional Autónoma de México Apartado Postal 20-364, 01000 México, D. F., México
e-mail: eleykoo@fisica.unam.mx

H.E. Montgomery Jr (25)
Chemistry Program, Centre College, 600 West Walnut Street, Danville, KY 40422, USA
e-mail: montgomery2@alltel.net

S.H. Patil (1)
Department of Physics, Indian Institute of Technology, Bombay 400 076, India
e-mail: sharad@phy.iitb.ac.in

V.I. Pupyshev (25)
Laboratory of Molecular Structure and Quantum Mechanics, Chemistry Department, Moscow State University, Moscow, 119991, Russia
e-mail: vip@moleq.chem.msu.su, vipupyshev@mail.ru

K.D. Sen (25)
School of Chemistry, University of Hyderabad, Hyderabad-500 046, India
e-mail: sensc@uohyd.ernet.in, kalidas.sen@gmail.com

R. Vargas (241, 311)
Departamento de Química, División de Ciencias Básicas e Ingeniería, Universidad Autónoma Metropolitana-Iztapalapa, San Rafael Atlixco 186, Col. Vicentina. Iztapalapa. C. P. 09340. México D. F., México
e-mail: ruvf@xanum.uam.mx

Y.P. Varshni (1)
Department of Physics, University of Ottawa, Ottawa, Canada KIN 6N5
e-mail: ypvsj@uottawa.ca

CHAPTER 1

Properties of Confined Hydrogen and Helium Atoms

S.H. Patil[a] and **Y.P. Varshni**[b]

Contents

1. INTRODUCTION

Recently, considerable attention has been given to the analysis of the properties of confined atoms and molecules [1,2]. They have been used in many fields, to analyse the effect of pressure on the energy levels and polarizabilities of atoms [3–9], as a cell model in liquid state [10,11], in semiconductor dots [12–17], and atoms encapsulated in fullerene [18]. It is also of importance in astrophysics [19] for the analysis of mass-radius relation of white

[a] Department of Physics, Indian Institute of Technology, Bombay 400 076, India
[b] Department of Physics, University of Ottawa, Ottawa, Canada KIN 6N5

Advances in Quantum Chemistry, Vol. 57
ISSN: 0065-3276, DOI: 10.1016/S0065-3276(09)00605-4

dwarfs and for the analysis of ionised plasma properties. In particular, the properties of the confined hydrogen atom have been analysed using different techniques [5–9,20–26]. The confined helium atom has also received some attention [27–33]. In short, confined hydrogen and helium atoms are very interesting and important in understanding the properties of confined quantum systems, and have many significant practical implications.

Here, we present a brief review of the analysis of the hydrogen atom with different forms of confinement, using different approaches. The main emphasis will be on developing a simple and clear understanding of the consequences of confinement on the energies, polarizabilities, and other properties of the atom. Some of the results will be extended to the description of the confined helium atom. We will begin with a discussion of some general properties of the wave functions of the confined hydrogen atom, the cusp and inflexion properties, and virial relation. Then the properties of the hydrogen atom confined to a spherical box will be analysed using a numerical approach and model wave functions, leading to a simple description of the energies and polarizabilities of the confined atom. Some of these considerations will be extended to the helium atom. After this, the properties of the hydrogen atom in an oscillator potential will be considered, for the case where the nucleus is at the centre of the potential, leading to simple expressions for the energies and polarizabilities. The analysis can be extended to the case of off-centre confining potentials, and to confinement in a circular disc. Finally, we will conclude with a summary of the results. Atomic units $e = \hbar = m_e = 1$ will be used unless stated otherwise.

2. SOME GENERAL PROPERTIES

Here, we begin by analysing some general properties of the energy eigenfunctions of a confined hydrogenic system, the cusp and inflexion properties, and virial relation.

2.1. The Schrödinger equation

The Schrödinger equation for the energy eigenfunctions of a confined hydrogenic system is

$$-\frac{1}{2}\nabla^2\psi - \frac{Z}{r} + V_c(r)\psi = E\psi, \tag{2.1}$$

where Z is the charge of the nucleus and $V_c(r)$ is the confining potential. A very simple and useful model for the confining potential is that of a spherical box with the nucleus at the centre, for which the potential is

$$\begin{aligned} V_c(r) &= 0 \quad \text{for } r < R \\ &= \infty \quad \text{for } r > R. \end{aligned} \tag{2.2}$$

Alternatively, we can use the oscillator potential for confinement,

$$V_c(r) = \frac{1}{2}kr^2, \tag{2.3}$$

which in a way is more appropriate since it increases more gradually. Using spherical coordinates, the solutions for these spherically symmetric potentials can be written in the form

$$\psi(r,\theta,\phi) = R_l(r)Y_l^m(\theta,\phi), \tag{2.4}$$

where $R_l(r)$ satisfies the radial equation

$$-\frac{1}{2}\frac{\mathrm{d}^2}{\mathrm{d}r^2}[rR_l(r)] + \frac{l(l+1)}{2r^2}[rR_l(r)] + \left[-\frac{Z}{r} + V_c(r)\right][rR_l(r)] = E[rR_l(r)]. \tag{2.5}$$

The radial wave function has some interesting and useful cusp and inflexion properties [25].

2.2. Cusp and inflexion properties

When $r \to 0$, the angular momentum and Coulomb potential terms dominate, leading to

$$\frac{\mathrm{d}^2}{\mathrm{d}r^2}[rR_l(r)] - \frac{l(l+1)}{r^2}[rR_l(r)] + \frac{2Z}{r}[rR_l(r)] = O[rR_l(r)] \quad \text{for } r \to 0. \tag{2.6}$$

Using a power series expansion,

$$R_l(r) = r^l(c_0 + c_1 r + \cdots) \tag{2.7}$$

and equating coefficients of power terms r^l, one obtains

$$2(l+1)c_1 + 2Zc_0 = 0 \Rightarrow c_1 = -\frac{Z}{l+1}c_0. \tag{2.8}$$

Therefore, one has for the threshold behaviour

$$R_l(r) = c_0 r^l\left(1 - \frac{Z}{l+1}r + \cdots\right) \quad \text{for } r \to 0, \tag{2.9}$$

which for $l = 0$ is related to the Kato [34] cusp property of Coulombic wave functions. It is a very useful property in the development of model wave functions.

At the points r_i, where

$$R_l(r_i) = 0, \tag{2.10}$$

one has from Equation (2.5),

$$\frac{\mathrm{d}^2}{\mathrm{d}r^2}[rR_l(r)] = 0 \quad \text{at } r = r_i, \tag{2.11}$$

which implies that the derivative has an extremum at $r = r_i$, and r_i are the inflexion points of $rR_l(r)$. This simple property can be quite useful in the development of model wave functions.

2.3. Virial relation

The energy eigenstates satisfy the relation

$$\frac{1}{2}\langle p^2 \rangle = \frac{1}{2}\langle \vec{r} \cdot \vec{\nabla} V \rangle \tag{2.12}$$

for the expectation values, which follows from the expectation value of the commutator of Hamiltonian H and $\vec{r} \cdot \vec{p}$. For the confining potential in Equation (2.2), one can take

$$V(r) = -\frac{Z}{r} + V_0 \theta(r - R), \tag{2.13}$$

where V_0 is a large positive quantity and θ is the Heaviside step function. This leads to

$$\langle \vec{r} \cdot \vec{\nabla} V \rangle = \left\langle \frac{Z}{r} \right\rangle + V_0 R^3 |R_l(R)|^2. \tag{2.14}$$

The relation

$$\frac{1}{R_l(r)} \frac{\mathrm{d}R_l(r)}{\mathrm{d}r} \approx -(2V_0)^{1/2} \quad \text{for } r > R \tag{2.15}$$

then leads to the virial relation

$$\frac{1}{2}\langle p^2 \rangle = \frac{1}{2}\left\langle \frac{Z}{r} \right\rangle + \frac{1}{4} R^3 \left[\frac{\mathrm{d}R_l}{\mathrm{d}r} \right]^2_{r=R} \tag{2.16}$$

for confinement in a sphere. This relation is particularly useful for developing model wave functions for excited states.

3. H IN A SPHERICAL BOX

Here we consider [25] the properties of H at the centre of a spherical box of radius R, using a numerical approach to obtain the energies and polarizabilities. We also develop some model wave functions, simple expressions for the energies and polarizability, deduce the critical radius R for which $E = 0$, and extend the analysis to the confined helium atom with effective screening.

3.1. Numerical solutions

To obtain numerical solutions for the hydrogenic system with spherical confinement, as in Equation (2.2), we start with the radial equation

$$-\frac{1}{2}\left(\frac{\mathrm{d}^2}{\mathrm{d}r^2}+\frac{2}{r}\frac{\mathrm{d}}{\mathrm{d}r}\right)R_l+\frac{l(l+1)}{2r^2}R_l-\frac{Z}{r}R_l=ER_l \quad \text{for } r<R. \tag{3.1}$$

Substituting the series expansion

$$R_l(r)=\sum_{i=0} c_i r^{l+i} \tag{3.2}$$

and equating terms of the same order, one obtains

$$c_1=-\frac{Z}{l+1}c_0, \qquad c_{i+2}=-2\frac{Zc_{i+1}+Ec_i}{(i+2)(2l+i+3)}. \tag{3.3}$$

The energy eigenvalues E are then deduced from the condition that the wave function vanishes at the boundary $r = R$,

$$\sum_{i=0} c_i R^{l+i}=0. \tag{3.4}$$

The constant c_0 is determined from the normalization condition

$$\int_0^R [R_l(r)]^2 r^2 \mathrm{d}r=1. \tag{3.5}$$

The energy eigenvalues for the confined hydrogen atom, for some states and some values of R are given in Table 1. A large range of eigenvalues is available in the literature [8,23]. The expectation values of some related dynamical quantities can be deduced [25] for these states.

The response of an atom or ion to an external potential

$$V_l(\vec{r})=r^l P_l(\cos\theta) \tag{3.6}$$

Table 1 The energies of hydrogen atom confined to a box of radius R, obtained from numerical calculations, model wave functions, and the simple expression in Equation (3.39). For each state and R, the first row is from numerical calculation, the second row in brackets is from model wave function, and the third row in brackets is from the simple expression in Equation (3.39). The last column is the dipole polarizability for the ground state

R	$E(1s)$	$E(2s)$	$E(2p)$	$E(3p)$	$E(3d)$	α_1
10.0	−0.5000	−0.1128	−0.1189	0.0492	−0.0071	–
	(−0.5000)	(−0.1128)	(−0.1188)	(0.0492)	(−0.0071)	–
	(−0.500)	(−0.1072)	(−0.1159)	(0.0250)	(−0.0106)	(4.445)
5.0	−0.4964	–	0.0076	–	0.3291	3.422
	(−0.4960)	(0.1420)	(0.0076)	(0.7077)	(0.3292)	(3.411)
	(−0.4933)	(0.1432)	(0.012)	(0.5953)	(0.3074)	(3.756)
3.0	−0.4240	1.1116	0.4813	–	1.2928	1.192
	(−0.4240)	(1.1077)	(0.4813)	(2.5203)	(1.2933)	(1.189)
	(−0.4138)	(1.0611)	(0.4809)	(2.277)	(1.245)	(1.780)
2.0	−0.1250	3.3275	1.576	6.269	3.3275	0.3425
	(−0.1250)	(3.3088)	(1.576)	(6.2868)	(3.3293)	(0.3422)
	(−0.1138)	(3.1884)	(1.570)	(5.869)	(3.248)	(0.515)
1.0	2.3740	16.57	8.223	27.47	14.968	0.0288
	(2.3741)	(16.48)	(8.226)	(27.59)	(14.977)	(0.0287)
	(2.3719)	(16.14)	(8.185)	(26.59)	(14.807)	(0.0360)
0.6	9.528	–	24.94	–	–	4.12×10^{-3}
	(9.5287)	–	(24.95)	–	(43.485)	(4.11×10^{-3})
	(9.5042)	(48.77)	(24.85)	(77.49)	(43.162)	(4.70×10^{-3})
0.5	14.748	72.67	–	114.64	–	–
	–	(72.64)	–	(115.35)	–	–
	(14.714)	(71.66)	–	(112.85)	–	–
0.4	24.634	–	–	–	–	8.53×10^{-4}
	(24.638)	–	(58.48)	–	–	(8.48×10^{-4})
	(24.584)	–	(58.49)	–	–	(9.30×10^{-4})
0.2	111.07	–	–	–	–	–
	(111.09)	–	–	–	–	(5.54×10^{-5})
	(110.94)	–	–	–	–	(5.81×10^{-5})

is described in terms of the polarizabilities. For the hydrogenic system in the ground state, the second order energy shift is given by

$$\delta E^{(2)} = -\frac{1}{2}\alpha_l \tag{3.7}$$

with

$$\alpha_l = 2\sum_{i\neq 0} \frac{|\langle\psi_i|V_l|\psi_0\rangle|^2}{E_i - E_0}. \tag{3.8}$$

The polarizability α_l can be written as

$$\alpha_l = 2\langle\psi^{(1)}|V_l|\psi_0\rangle. \tag{3.9}$$

Here $|\psi^{(1)}\rangle$ is the first order perturbation in the ground state which satisfies the equation

$$(H_0 - E_0)|\psi^{(1)}\rangle = V_l|\psi_0\rangle, \tag{3.10}$$

where H_0 is the unperturbed Hamiltonian and E_0 is the corresponding ground state energy. For the hydrogen atom with spherical confinement, one can take

$$\psi^{(1)}(\vec{r}) = P_l(\cos\theta)R_l^{(1)}(r), \tag{3.11}$$

which leads to

$$-\frac{1}{2}\left(\frac{\mathrm{d}^2}{\mathrm{d}r^2} + \frac{2}{r}\frac{d}{dr}\right)R_l^{(1)} + \frac{l(l+1)}{2r^2}R_l^{(1)} - \frac{Z}{r}R_l^{(1)} - E_0R_l^{(1)} = r^l R_0 \quad \text{for } r < R, \tag{3.12}$$

where R_0 is the ground state solution to Equation (3.1). Taking

$$R_l^{(1)}(r) = \sum_{i=0} d_i r^{l+i} \tag{3.13}$$

and R_0 in Equation (3.2), one obtains

$$d_1 = -\frac{Z}{l+1}d_0, \qquad d_{i+2} = -2\frac{Zd_{i+1} + E_0 d_i + c_i}{(i+2)(2l+i+3)} \tag{3.14}$$

with c_i given in Equation (3.3). The constant d_0 is determined by requiring that the wave function vanishes at $r = R$,

$$\sum_{i=0} d_i R^{l+i} = 0. \tag{3.15}$$

With the unperturbed and perturbed wave functions in Equations (3.2) and (3.13), the polarizability expression in Equation (3.9) leads to

$$\alpha_l = \frac{2}{2l+1}\int_0^R \left(\sum c_i r^i\right)\left(\sum d_j r^j\right) r^{2l+2}\mathrm{d}r. \tag{3.16}$$

The calculated values [25] of α_1 for the hydrogen atom, for some values of R are given in Table 1. It is interesting to observe that the polarizability decreases rapidly with decreasing radius of confinement. This is expected since the smaller domain of confinement decreases the flexibility of the electron.

3.2. Model wave functions

One can develop simple but accurate model wave functions for a confined hydrogen atom, based on the general cusp and inflexion properties discussed in Section 2.2 and the virial relation in Equation (2.16).

For the lowest energy state with a given l, one can consider a wave function of the form [25]

$$R_l(r) = Ar^l(r-R)(1+b_1r+b_2r^2)\mathrm{e}^{-ar}. \tag{3.17}$$

Carrying out an expansion for small r, and comparing with the cusp condition in Equation (2.8), one obtains for $Z=1$,

$$\frac{1}{R} - b_1 + a = \frac{1}{l+1}. \tag{3.18}$$

The inflexion property in Equation (2.11) at $r=R$ leads to

$$l+1+b_1(l+2)R+b_2(l+3)R^2 - aR(1+b_1R+b_2R^2) = 0. \tag{3.19}$$

The remaining parameter is determined by requiring that the virial relation in Equation (2.16) is satisfied. For simplicity we choose a value for a, determine b_1 from Equation (3.18), and b_2 from Equation (3.19). Then a is varied to obtain the value of a for which the virial relation in Equation (2.16) is satisfied. The total energy is then deduced as a sum of the expectation values of the kinetic energy and the potential energy,

$$E = \langle T\rangle + \langle V\rangle. \tag{3.20}$$

It is to be noted that with the wave function written in the form

$$R_l(r) = A\sum C_i r^{l+i}\mathrm{e}^{-ar}, \tag{3.21}$$

the various quantities can be written in terms of integrals of the type

$$I(n) = \int_0^R r^n e^{-2ar}\,dr = \frac{n!}{(2a)^{n+1}}\left[1 - \sum_{i=0}^{n} \frac{(2aR)^i}{i!} e^{-2aR}\right]. \tag{3.22}$$

The energies obtained [25] for $l = 0, 1, 2$, for some values of R, are given in Table 1. It may be noted that these values are close to the accurate, numerically obtained values. This demonstrates the importance of the general cusp and inflexion properties, and virial relation, for the energy eigenfunctions. The values of the parameter a, normalization constant A and some related expectation values are given in Ref. [25].

For the first excited state with a given l, and a node at $r = r_1$, we consider a wave function of the form

$$R_l(r) = Ar^l(r - R)(r - r_1)(1 + b_1 r + b_2 r^2)e^{-ar}. \tag{3.23}$$

Carrying out an expansion for small r, and comparing with the cusp condition in Equation (2.8), one obtains for $Z = 1$,

$$\frac{1}{R} + \frac{1}{r_1} - b_1 + a = \frac{1}{l+1}. \tag{3.24}$$

The inflexion property in Equation (2.11) leads to, at $r = R$,

$$\begin{aligned}&[(l+1)(R - r_1) + R - aR(R - r_1)](1 + b_1 R + b_2 R^2)\\&\quad + (b_1 + 2b_2 R)R(R - r_1) = 0,\end{aligned} \tag{3.25}$$

and at $r = r_1$,

$$\begin{aligned}&[(l+1)(r_1 - R) + r_1 - ar_1(r_1 - R)](1 + b_1 r_1 + b_2 r_1^2)\\&\quad + (b_1 + 2b_2 r_1)r_1(r_1 - R) = 0.\end{aligned} \tag{3.26}$$

The remaining parameter is determined by requiring that the virial relation in Equation (2.16) is satisfied. For simplicity, we choose values of a and of r_1, then determine b_1 from Equation (3.24), and b_2 from Equation (3.25). Use these values in Equation (3.26) and vary r_1 so that Equation (3.26) is satisfied. Finally, the parameter a is determined by requiring that the virial relation in Equation (2.16) is satisfied. The total energy is then deduced as a sum of the expectation values of the kinetic and potential energies, as in Equation (3.20). Here the calculations are also simplified by expressing the wave function in the form in Equation (3.21) and using the integrals in Equation (3.22). The energies for the $2s$ and $3p$ states are given in Table 1 for some values of R. These values are observed to be close to the numerically deduced accurate values.

One can develop a model wave function for the first order perturbed wave function in Equation (3.10) and use it to evaluate the polarizability in Equation (3.9). Here it is interesting to observe [35] that

$$F = \langle \psi^{(1)} | (H_0 - E_0) | \psi^{(1)} \rangle - \langle \psi^{(1)} | V | \psi_0 \rangle - \langle \psi_0 | V | \psi^{(1)} \rangle \tag{3.27}$$

has an extremum when $|\psi^{(1)}\rangle$ is a solution to Equation (3.10), i.e. $\delta F = 0$ for small variations in $|\psi^{(1)}\rangle$. Specifically, if we take

$$|\psi^{(1)}\rangle = D|\eta_l\rangle, \tag{3.28}$$

where $|\eta_l\rangle$ is a specific choice and D is a variational parameter, the extremum condition $\partial F/\partial D = 0$ leads to

$$D = \frac{\langle \eta_l | V_l | \psi_0 \rangle}{\langle \eta_l | (H_0 - E_0) | \eta_l \rangle}. \tag{3.29}$$

Using this expression in Equations (3.28) and (3.9),one obtains

$$\alpha_l = 2 \frac{|\langle \eta_l | V_l | \psi_0 \rangle|^2}{\langle \eta_l | (H_0 - E_0) | \eta_l \rangle}. \tag{3.30}$$

For the unperturbed ground state, we take the wave function

$$\psi_0(\vec{r}) = Y_0^0(\theta, \phi) R_0(r), \tag{3.31}$$

where $R_0(r)$ is given in Equation (3.17), and

$$\begin{aligned} \psi^{(1)} &= P_l(\cos\theta) R_l^{(1)}(r), \\ R_l^{(1)}(r) &= D r^l (r - R)(1 + d_1 r + d_2 r^2) \mathrm{e}^{-ar} \end{aligned} \tag{3.32}$$

with D being the variational parameter. With the expansion of $R_l^{(1)}(r)$ in powers of r, the cusp property in Equation (2.8) implies

$$d_1 = \frac{1}{R} + a - \frac{1}{l+1}. \tag{3.33}$$

The inflexion property in Equation (2.11) at $r = R$ leads to

$$d_2 = -\frac{(l+1) + d_1 R(l+2) - aR(1 + d_1 R)}{(l+3)R^2 - aR^3}. \tag{3.34}$$

Then the model wave function $R_0(r)$ in Equation (3.17) and the perturbed wave function in Equation (3.32) allow us to obtain the polarizabilities in Equation (3.30). The predicted values [25] of α_1 for some values of R are given in Table 1, and are observed to be close to the numerically calculated accurate values.

3.3. Simple expressions for energies

Here, some general properties of the spectrum of the confined hydrogen atom are considered, leading to a simple expression for the energies. It may be noted that since the confining model potentials are qualitative simulations of complicated situations, a qualitative description of the spectrum is very useful.

For $R \to \infty$, the hydrogenic system with the confining potential in Equation (2.2) tends to the free hydrogenic system, and the energies tend to

$$E(n, R) \to -\frac{1}{2n^2}, \quad n = 1, 2 \ldots, \text{for } R \to \infty, \tag{3.35}$$

where n is the principal quantum number. For $R \to 0$, the system tends to a particle in a spherical box for which the solutions are given in terms of spherical Bessel functions,

$$R_l(r) = A j_l(kr), \quad k = (2E)^{1/2}, \quad j_l(u) = u^l \left(-\frac{1}{u}\frac{\mathrm{d}}{\mathrm{d}u}\right)^l \left(\frac{\sin u}{u}\right). \tag{3.36}$$

The energies are determined by the boundary condition

$$j_l(u_{l,j}) = 0, \qquad u_{l,j} = k_{l,j} R \tag{3.37}$$

which leads to the energies

$$E_l(n.R) \to \frac{u_{l,j}^2}{2R^2}, \quad \text{for } R \to 0. \tag{3.38}$$

We now combine the expressions in Equations (3.35) and (3.38), and consider a simple interpolation

$$E_l(n, R) = -\frac{1}{2n^2} + \frac{u_{l.j}^2 b}{2R(\mathrm{e}^{bR} - 1)}, \quad j = 1, 2, \ldots, \tag{3.39}$$

where the radial quantum number $n_r = j - 1$ and $n = n_r + l + 1$. Now, one has accurate values [36] for the zeros of the Bessel functions of order m. Since the spherical Bessel functions correspond to Bessel functions of half integral order, one can obtain the zeros of spherical Bessel functions by replacing m by $l + 1/2$. Therefore one gets

$$u_{l,j} = \beta - \frac{l(l+1)}{2\beta} - \frac{l(l+1)(7l^2+7l-6)}{3(2\beta)^3},$$
$$\beta = \left(j + \frac{1}{2}l\right)\pi, \quad j = 1, 2, \ldots. \tag{3.40}$$

Specifically, one gets

$$u_{0,j} = j\pi, \qquad u_{1,j} = (j+1/2)\pi - \frac{1}{(j+1/2)\pi} - \frac{2}{3(j+1/2)^3\pi^3},$$
$$j = 1, 2, \ldots. \tag{3.41}$$

For the determination of the parameter b in the expression for E in Equation (3.39), we note that the asymptotic behaviour of the hydrogenic wave function for large r is

$$\psi \to \mathrm{e}^{-r/n} \quad \text{for } r \to \infty \tag{3.42}$$

and therefore, for large values of R, one expects the corrections to the energies to be of the order

$$\delta E \sim \mathrm{e}^{-R/n}. \tag{3.43}$$

Comparing this with the corrections to the hydrogenic energies in Equation (3.39) for large R,

$$\delta E \sim \mathrm{e}^{-bR}, \tag{3.44}$$

one may consider b to be of order $1/n$. It may also be noted that b has the dimensions of $1/r$, and the Bohr radius is of order n^2. We take b to be the average

$$b = \frac{1}{2}\left(\frac{1}{n} + \frac{1}{n^2}\right). \tag{3.45}$$

The predictions of the simple expression in Equation (3.39), with $u_{l,j}$ given in Equation (3.40) and b in Equation (3.45), are given in Table 1 for some states and some values of R. They are observed to be in good agreement with the accurate, numerically deduced values. A very important observation is that the degeneracy of the hydrogenic energies is removed by the effect of confinement, and the separation between the energies increases rapidly as the confining radius R decreases.

One can also deduce a simple expression for the dipole polarizability of the hydrogenic system confined to a sphere. For this, one observes that for large values of R, the polarizability tends to the hydrogenic value

$$\alpha_1 \to \frac{9}{2Z^4} \quad \text{for } R \to \infty. \tag{3.46}$$

For $R \to 0$, the polarizability will tend to that of a particle in a sphere. For deducing the dipole polarizability in this case, we use the expression in Equation (3.9). The equation for the first order perturbation in the wave function for the ground state can be written as

$$\begin{aligned} -\frac{1}{2}\left[\frac{\mathrm{d}^2}{\mathrm{d}r^2} + \frac{2}{r}\frac{d}{dr} - \frac{2}{r^2}\right] R^{(1)} - E_0 R^{(1)} &= r R^{(0)}, \\ E_0 = \frac{\pi^2}{2R^2}, \qquad R^{(0)} &= (2/R)^{1/2}\frac{1}{r}\sin(\pi r/R). \end{aligned} \tag{3.47}$$

The solution can be written as a linear combination of the solutions to the homogeneous and inhomogeneous equations, leading to

$$\begin{aligned} R^{(1)}(r) = {} & (2/R)^{1/2}\left(\frac{R}{2\pi}\right) r\cos(kr) \\ & + C\left[\frac{\sin(kr)}{k^2r^2} - \frac{\cos(kr)}{kr}\right], \qquad k = \pi/R. \end{aligned} \tag{3.48}$$

Imposing the boundary condition that it vanishes at $r = R$, one gets

$$C = \frac{1}{2}(2/R)^{1/2}R^2. \tag{3.49}$$

The expression for α_1 in Equation (3.9) then leads to

$$\alpha_1 = 2(1/3)\int_0^R R^{(0)}(r) r R^{(1)}(r) r^2 \mathrm{d}r = \frac{R^4}{3\pi^4}[3/4 + \pi^2]. \tag{3.50}$$

Combining the expressions in Equations (3.46) and (3.50), we consider a simple interpolation

$$\alpha_1 = \frac{1}{\frac{2Z^4}{9} + \frac{3\pi^4}{R^4(\pi^2+3/4)}} \tag{3.51}$$

which tends to the appropriate limiting values for $R \to \infty$ and for $R \to 0$. This simple expression provides a very useful description of the variation of the dipole polarizability for different values of the confining radius R. Some of the predicted values are given in Table 1.

3.4. Critical values of R

An important property of a confined system is the critical value of R for which the energy E is zero. For values of R less than the critical value, the energy E is positive and the electron can tunnel through the confining barrier. One can deduce a simple but accurate expression for the critical radius R by using the Wentzel–Kramers–Brillouin approach [37,38]. Taking into account the Langer correction, the WKB energies are given by

$$\int_{r_1}^{r_2} \mathrm{d}r \left[2\left(E + \frac{Z}{r} - \frac{(l+1/2)^2}{2r^2}\right)\right]^{1/2} = (n_r + 1 + c_1 + c_2)\pi, n_r = 0, 1, \ldots, \tag{3.52}$$

where r_1, r_2 are the turning points, $c_i = -1/4$ for the case where the potential varies linearly, and $c_i = 0$ for the case where the potential shoots up to infinity, at the turning point. For the case of $E = 0$, the outer turning point is at the confining radius, so that

$$\int_{r_1}^{R} \mathrm{d}r \frac{1}{r}\left[2\left(r - \frac{1}{2}(l+1/2)^2\right)\right]^{1/2} = (n_r + 3/4)\pi, \tag{3.53}$$

where we have taken $Z = 1$ for the case of hydrogen. Carrying out the integrations, this leads to

$$2\left[\left(2R - (l+1/2)^2\right)^{1/2} - (l+1/2)\tan^{-1}\left(\frac{2R}{(l+1/2)^2} - 1\right)^{1/2}\right] = (n_r + 3/4)\pi. \tag{3.54}$$

Carrying out an expansion of $\tan^{-1}$ for large R, one obtains

$$\left(2R - (l+1/2)^2\right)^{1/2} - (l+1/2)\left(\frac{\pi}{2} - \frac{(l+1/2)}{(2R)^{1/2}}\right) = \frac{1}{2}(n_r + 3/4)\pi, \quad n_r = 0, 1, \ldots. \tag{3.55}$$

To leading order, this is equivalent to

$$\left(2R + (l + 1/2)^2\right)^{1/2} = \frac{1}{2}\,(n_r + l + 5/4)\,\pi, \tag{3.56}$$

so that one has

$$R = \frac{1}{8}(n + 1/4)^2\pi^2 - \frac{1}{2}(l + 1/2)^2, \quad n = n_r + l + 1, \tag{3.57}$$

where n is the principal quantum number. The predicted values for some states are

$$\begin{aligned} &R(1\text{s}) = 1.803\ (1.83), \qquad R(2\text{p}) = 5.12\ (5.1), \qquad R(3\text{d}) = 9.91\ (9.8), \\ &R(2\text{s}) = 6.12\ (6.2), \qquad R(3\text{p}) = 11.91 \end{aligned} \tag{3.58}$$

where the numbers in the brackets are from numerical calculations. The predictions of the simple expression in Equation (3.57) are accurate which indicates the usefulness of the WKB approach in the present context.

3.5. Energies of confined helium and isoelectronic ions

We now consider the extension [33] of our results to helium and isoelectronic ions, in terms of effective screening. We begin by considering the scaling properties of the hydrogenic system.

The Schrödinger equation for the hydrogenic system confined to a sphere can be written in the form

$$-\frac{1}{2}\nabla^2\psi - \frac{Z}{r}\psi + V_0\theta(r - R)\psi = E(Z, R)\psi, \tag{3.59}$$

where V_0 is a large positive number, and θ is the Heaviside step function. A scale transformation

$$\vec{r} = \lambda\vec{r}\,' \tag{3.60}$$

leads to

$$-\frac{1}{2}\nabla'^2\psi - \frac{Z\lambda}{r'}\psi + V_0\lambda^2\theta(r' - R/\lambda)\psi = \lambda^2 E(Z, R)\psi. \tag{3.61}$$

Taking $\lambda = 1/Z$, one obtains

$$-\frac{1}{2}\nabla'^2\psi - \frac{1}{r'}\psi + \frac{V_0}{Z^2}\theta(r' - ZR)\psi = \frac{1}{Z^2}E(Z, R)\psi. \tag{3.62}$$

Comparison with Equation (3.59) leads to

$$E(Z, R) = Z^2 E(1, ZR). \tag{3.63}$$

Thus, the energies for the general case of nuclear charge Z are related to the energies of a hydrogen atom with scaled radius ZR.

The two-electron atom or ions confined to a sphere of radius R, are described by the Hamiltonian H,

$$H = -\frac{1}{2}(\nabla_1^2 + \nabla_2^2) + V(\vec{r}_1, \vec{r}_2), \tag{3.64}$$

$$\begin{aligned} V(\vec{r}_1, \vec{r}_2) &= -Z\left(\frac{1}{r_1} + \frac{1}{r_2}\right) + \frac{1}{r_{12}} \quad \text{for } r_1 < R, r_2 < R, \\ &= \infty \qquad \text{for } r_1 > R \text{ or } r_2 > R. \end{aligned}$$

The complications due to the $1/r_{12}$ term are greatly simplified by simulating it with a screening of the nuclear charge.

For the ground state, we take the effective Hamiltonian inside the box to be

$$\begin{aligned} H = &-\frac{1}{2}(\nabla_1^2 + \nabla_2^2) - Z\left(\frac{1}{r_1} + \frac{1}{r_2}\right) \\ &+ S(R)\left(\frac{1}{r_1} + \frac{1}{r_2}\right) \quad \text{for } r_1 < R, r_2 < R, \end{aligned} \tag{3.65}$$

where $S(R)$ represents the amount of screening as a function of the confining radius. The ground state energy of the system is then given by

$$E(R) = 2Z'^2 E_0(1, Z'R), \qquad Z' = Z - S(R), \tag{3.66}$$

where $E_0(1, Z'R)$ is the ground state energy of the hydrogen atom with confinement radius $Z'R$. For $R \to \infty$, the energy tends to the energy of the free two electron system, which implies that

$$-[Z - S(\infty)]^2 = E(\infty) \Rightarrow S(\infty) = Z - [-E(\infty)]^{1/2}. \tag{3.67}$$

For $R \to 0$, one can treat the system as two electrons in a spherical box with the Coulomb interaction as a perturbation. For the ground state wave function, one has

$$\psi = A\frac{1}{r_1 r_2}\sin\left(\frac{\pi r_1}{R}\right)\sin\left(\frac{\pi r_2}{R}\right). \tag{3.68}$$

Calculating the expectation values of $1/r_i$ and $1/r_{12}$ with this wave function, one obtains [33]

$$\left\langle \frac{1}{r_{12}} \right\rangle = 0.367 \left(\left\langle \frac{1}{r_1} \right\rangle + \left\langle \frac{1}{r_2} \right\rangle \right) \Rightarrow S(0) = 0.367 \tag{3.69}$$

for the effective screening for $R \to 0$. Combining the results in Equations (3.67) and (3.69), we take our interpolated screening as

$$S(R) = S(\infty) + [0.367 - S(\infty)]\mathrm{e}^{-(Z-5/16)R}, \tag{3.70}$$

which tends to the appropriate limits for $R \to 0$ and $R \to \infty$. The exponential factor ensures that the screening tends to the free-atom screening when R is greater than the effective radius of the atom. For He, one has

$$\begin{aligned} \text{He:} \quad & Z = 2, \qquad E(\infty) = -2.9037, \qquad S(\infty) = 0.296, \\ & \Rightarrow S(R) = 0.296 + 0.071\mathrm{e}^{-(2-5/16)R}. \end{aligned} \tag{3.71}$$

Then the ground state energies of the confined helium atom are given by

$$E_{\mathrm{He}}((1\mathrm{s})(1\mathrm{s}), R) = 2Z'^2 E_0(1, Z'R), \qquad Z' = Z - S(R), \tag{3.72}$$

with $S(R)$ given in Equation (3.71), and E_0 is the hydrogen ground state energy obtained from Equation (3.39) with $n = 1$, $u_{01} = \pi$, $b = 1$,

$$E_0(1, Z'R) = -\frac{1}{2} + \frac{\pi^2}{2(Z'R)(\mathrm{e}^{Z'R} - 1)}. \tag{3.73}$$

Some of the predicted values [33] are given in Table 2, for some values of R, along with some variationally deduced values [29,39].

For the excited states $(1\mathrm{s})(nl)$, we take the effective Hamiltonian inside the box to be

$$H = -\frac{1}{2}\nabla_1^2 - \frac{Z}{r_1} - \frac{1}{2}\nabla_2^2 - \frac{Z'}{r_2} \qquad \text{for } r_1 < R, r_2 < R, \tag{3.74}$$

where Z' is the screened charge seen by the outer electron. Then the energies of the confined helium are given by

$$E_{\mathrm{He}}((1\mathrm{s})(nl), R) = Z^2 E_0(1, ZR) + Z'^2 E_l(n, Z'R), \tag{3.75}$$

where $Z = 2$, and $E_l(n, R)$ are the hydrogenic energies in Equation (3.39). The screened charge is determined by requiring that it leads to the correct

Table 2 Energies of He confined to a sphere of radius R, for the ground state $(1s)^{2\,1}S$ from Equation (3.72), for the $(1s2s)^1S$ and $(1s2s)^3S$ states from Equation (3.77), and $(1s2p)^1P$ and $(1s2p)^3P$ from Equation (3.79). The energies in the brackets are from Refs. [29] and [39] obtained with model wave functions

R	$E((1s)^{2\,1}S)$	$E((1s2s)^1S)$	$E((1s2s)^3S)$	$E((1s2p)^1P)$	$E((1s2p)^3P)$
6.0	−2.9036	−2.0175	−2.0658	−2.0489	−2.0624
	(−2.9035)	–	(−2.1171)	(−2.0547)	(−2.0684)
5.0	−2.9030	−1.9030	−1.9615	−1.9857	−2.0012
	(−2.9032)	–	(−2.0473)	(−1.9928)	(−2.0126)
4.0	−2.8988	−1.6524	−1.7277	−1.8499	−1.8686
	(02.9003)	–	(−1.8734)	(−1.8598)	(−1.8907)
3.0	−2.8682	−1.0141	−1.1193	−1.5069	−1.5312
	(−2.8727)	–	(−1.3679)	(−1.5217)	(−1.5758)
2.0	−2.6065	1.1482	0.9809	−0.3334	−0.3692
	(−2.6065)	–	(−0.5862)	(−0.3414)	(−0.4907)
1.0	0.9124	15.407	15.050	7.751	7.680
	(1.0142)	–	(15.5451)	(8.0312)	(7.5265)

unconfined atom energy for $R \to \infty$,

$$E_{\rm He}(\infty) = -\frac{Z^2}{2} - \frac{Z'^2}{2n^2} \Rightarrow Z' = n[-2E_{\rm He}(\infty) - Z^2]^{1/2}. \tag{3.76}$$

Specifically, one has

$$\begin{aligned}
&(1s)(2s)^1S: \quad E_{\rm He}(\infty) = -2.1460, \qquad Z' = 1.0807,\\
&(1s)(2s)^3S: \quad E_{\rm He}(\infty) = -2.1752, \qquad Z' = 1.1837,\\
&E_{\rm He}((1s)(2s), R) = Z^2 E_0(1, ZR) + Z'^2 E_0(2, Z'R)
\end{aligned} \tag{3.77}$$

with

$$E_0(n, R) = -\frac{1}{2n^2} + \frac{\pi^2 n^2 b}{2R(e^{bR} - 1)}, \qquad b = \frac{1}{2}\left(\frac{1}{n} + \frac{1}{n^2}\right). \tag{3.78}$$

For the $(1s)(2p)$ states, one has

$$\begin{aligned}
&(1s)(2p)^1P: \quad E_{\rm He}(\infty) = -2.1238, \qquad Z' = 0.9952,\\
&(1s)(2p)^3P: \quad E_{\rm He}(\infty) = -2.1332, \qquad Z' = 1.0323,\\
&E_{\rm He}((1s)(2p), R) = Z^2 E_0(1, ZR) + Z'^2 E_1(2, Z'R)
\end{aligned} \tag{3.79}$$

with

$$E_1(n,R) = -\frac{1}{2n^2} + \frac{u_{1j}b}{2R(e^{bR}-1)}, \quad j = n-1, \tag{3.80}$$

$$u_{1j} = (j+1/2)\pi - \frac{1}{(j+1/2)\pi} - \frac{2}{3(j+1/2)^3\pi^3},$$

given in Equations (3.39) and (3.40), and b given in Equation (3.78). Some of the predicted values [33] are given in Table 2, for some values of R, along with some variationally deduced values [29,39].

4. H IN CONFINING OSCILLATOR POTENTIAL

Another useful model for confinement is in terms of a simple harmonic oscillator potential, which in a sense is more appropriate since its increase is gradual and analytical. We will consider a hydrogen atom in an oscillator potential, and a simple, analytical description of its properties.

4.1. Energy spectrum

The Schrödinger equation for a hydrogen atom at the centre of a confining oscillator potential is

$$-\frac{1}{2}\nabla^2\psi - \frac{Z}{r}\psi + \frac{1}{2}kr^2\psi = E\psi. \tag{4.1}$$

It is separable in terms of spherical coordinates and one can obtain the energy spectrum by solving the radial equation numerically, for example by converting [26], the differential equation, into a difference equation. Here, we will develop simple, analytical solutions which provide an insight into the general properties.

For $k \to 0$, one has hydrogenic spectrum,

$$E_{l,n} = -\frac{Z^2}{2n^2} \quad \text{for } k \to 0. \tag{4.2}$$

One also has a simple expression [40] for the expectation value of r^2, which then leads to

$$E_{l,n} = -\frac{Z^2}{2n^2} + \frac{k}{4Z^2}n^2[5n^2+1-3l(l+1)] \quad \text{for } k \to 0, \tag{4.3}$$

correct to order k. For $k \to \infty$, the oscillator potential dominates and one gets

$$E_{l,n} \to k^{1/2}(2n_r + l + 3/2) = k^{1/2}(2n - l - 1/2) \quad \text{for } k \to \infty, \tag{4.4}$$

Table 3 The energies of hydrogen atom at the centre of the oscillator potential in Equation (4.1), obtained from the expression in Equation (4.5), and the dipole polarizabilities from Equation (4.9). The values in brackets are the numerically calculated values for the energies, and the variationally deduced polarizabilities from Ref. [26]

k	$E(1s)$	$E(2s)$	$E(2p)$	$E(3p)$	$E(3d)$	α_1
0.001	−0.4985 (−0.4984)	−0.107 (−0.106)	−0.112 (−0.118)	−0.0004 (−0.008)	−0.0154 (−0.0164)	3.952 (3.733)
0.01	−0.4864 (−0.4865)	0.006 (0.002)	−0.0313 (−0.0344)	0.244 (0.284)	0.1694 (0.1420)	3.063 (2.629)
0.05	−0.4387 (−0.437)	0.3234 (0.284)	0.195 (0.161)	0.767 (0.664)	0.571 (0.481)	2.070 (1.598)
0.1	−0.3864 (−0.384)	0.5998 (0.522)	0.393 (0.326)	1.173 (1.021)	0.886 (0.752)	1.612 (1.186)
0.2	−0.2927 (−0.296)	1.015 (0.879)	0.689 (0.575)	1.755 (1.538)	1.337 (1.146)	1.185 (0.8357)
0.5	−0.0607 (−0.088)	1.878 1.625	1.306 (1.096)	2.916 (2.591)	2.239 (1.953)	0.720 (0.4874)
1.0	0.250 (0.180)	2.875 (2.500)	2.018 (1.709)	4.230 (3.802)	3.260 (2.882)	0.462 (0.3078)
2.0	0.743 (0.594)	4.303 (3.771)	3.038 (2.603)	6.091 (5.538)	4.707 (4.218)	0.281 (0.1871)

with $n = n_r + l + 1$. For a simple interpolation, we consider an expression

$$E_{l,n} = -\frac{Z^2}{2n^2} + \frac{kc_0}{1 + k^{1/2}c_0/(2n - l - 1/2)},$$
$$c_0 = \frac{1}{4Z^2}n^2[5n^2 + 1 - 3l(l + 1)]. \tag{4.5}$$

This has the correct limiting forms in Equation (4.3) for $k \to 0$, and in Equation (4.4) for $k \to \infty$. The predicted values of the energy for some states, and some values of k are given in Table 3. They are fairly close to the numerically evaluated values [26].

4.2. Dipole polarizability

One can develop simple, analytical expressions for the dipole polarizability α_1 based on some general properties. For $k \to 0$, one has

$$\alpha_1 \to \frac{9}{2Z^4} \quad \text{for } k \to 0, \tag{4.6}$$

corresponding to the free hydrogenic system. For $k \to \infty$, one has for the electron in an oscillator potential and an external electric field $\vec{F}$

$$\begin{aligned} V &\to \frac{1}{2}kr^2 + \vec{F}\cdot\vec{r} \quad \text{for } k \to \infty, \\ &= \frac{1}{2}k\left(\vec{r} + \frac{1}{k}\vec{F}\right)^2 - \frac{1}{2k}F^2. \end{aligned} \tag{4.7}$$

This corresponds to

$$\alpha_1 \to \frac{1}{k} \quad \text{for } k \to \infty. \tag{4.8}$$

One can then consider a simple interpolation

$$\alpha_1 = \frac{(9/2)}{(Z^2 + 3(k/2)^{1/2})^2}, \tag{4.9}$$

which tends to the appropriate expressions in Equation (4.6) for $k \to 0$ and in Equation (4.8) for $k \to \infty$. The predictions of this simple expression for the hydrogen atom in an oscillator potential are given in Table 3, for some values of k, along with the predictions of a model variational wave function from Ref. [26]. They are in qualitative agreement with each other.

5. OFF-CENTRE CONFINEMENT

In some cases, the nucleus may not be at the centre of the confining potential. If there is rotational symmetry about the line of nuclear displacement, the elliptic coordinates may lead to separable solutions. For example, in the case of a hydrogen atom with the nucleus at the focus $(0, 0, -R/2)$ of a confining ellipsoid, one can take the elliptic coordinates of the electron as

$$u = \frac{r_1 + r_2}{R}, \quad v = \frac{r_1 - r_2}{R}, \quad \phi; \qquad r_1 = \left|\vec{r} + \frac{1}{2}\vec{R}\right|, \quad r_2 = \left|\vec{r} - \frac{1}{2}\vec{R}\right|, \tag{5.1}$$

with $\vec{r}$ being the vectorial distance of the electron from the centre. The Schrödinger equation for the hydrogen atom in elliptic coordinates is

$$-\frac{1}{2}\left(\frac{2}{R}\right)^2 \frac{1}{u^2-v^2}\left[\frac{\partial}{\partial u}(u^2-1)\frac{\partial}{\partial u}+\frac{\partial}{\partial v}(1-v^2)\frac{\partial}{\partial v}+\left(\frac{1}{u^2-1}\right.\right.$$
$$\left.\left.+\frac{1}{1-v^2}\right)\frac{\partial^2}{\partial\phi^2}\right]\psi-\frac{1}{R(u+v)}\psi=E\psi \tag{5.2}$$

inside the ellipsoid. Taking

$$\psi(u,v,\phi)=A(u^2-1)^{|m|/2}(1-v^2)^{|m|/2}F(u)G(v)\mathrm{e}^{\mathrm{i}m\phi}, \tag{5.3}$$

one obtains

$$\begin{aligned}(u^2-1)\frac{\mathrm{d}^2F}{\mathrm{d}u^2}+2(|m|+1)u\frac{\mathrm{d}F}{\mathrm{d}u}+RuF+\frac{1}{2}R^2E(u^2-1)F&\\ =\beta F,\quad u\geq 1,&\\ (v^2-1)\frac{\mathrm{d}^2G}{\mathrm{d}v^2}+2(|m|+1)v\frac{\mathrm{d}G}{\mathrm{d}v}+RvG+\frac{1}{2}R^2E(v^2-1)G&\\ =\beta G,\quad -1\leq v\leq 1,&\end{aligned} \tag{5.4}$$

where β is the separation constant. The confinement to an ellipsoid of semi-major axis R_e implies that $F(u)$ must vanish at $u=u_0$,

$$F(u_0)=0,\quad u_0=2R_e/R. \tag{5.5}$$

The solutions have been obtained [24] by solving these equations numerically, and by using model wave functions.

It is interesting to observe that a hydrogen atom off-centre in an oscillator potential with

$$V(\vec{r})=-\frac{1}{|\vec{r}+\vec{R}/2|}+\frac{1}{2}kr^2 \tag{5.6}$$

also leads to separable solutions in elliptic coordinates. In elliptic coordinates, one has

$$r^2=\frac{1}{2}\left[r_1^2+r_2^2-\frac{1}{2}R^2\right]=\frac{1}{4}R^2(u^2+v^2-1). \tag{5.7}$$

Use of the factorised form in Equation (5.3) leads to

$$(u^2-1)\frac{d^2F}{du^2}+2(|m|+1)u\frac{dF}{du}+RuF-\frac{1}{16}kR^4u^4F$$
$$+\frac{1}{2}R^2\left(E+\frac{1}{8}kR^2\right)(u^2-1)F=\beta F, \quad u\geq 1, \tag{5.8}$$

and

$$(v^2-1)\frac{d^2G}{dv^2}+2(|m|+1)v\frac{dG}{dv}+RvG-\frac{1}{16}kR^4v^4G$$
$$+\frac{1}{2}R^2\left(E+\frac{1}{8}kR^2\right)(v^2-1)G=\beta G, \quad -1\leq v\leq 1, \tag{5.9}$$

with β being the separation constant. What is striking is that the equations for $F(u)$ and $G(v)$ are identical, but are relevant in different domains. The solutions for the energies and some other properties have been obtained [26] by solving these equations numerically and by using model wave functions. It is observed that shifting the nucleus from the centre generally increases the energies, as well as removing the spherical symmetry of the states.

6. SUMMARY

We have reviewed the properties of a hydrogen atom confined to a sphere, which provides a very useful model for the analysis of the implications of confinement in different situations. Apart from the different numerical approaches for deducing its properties, one can obtain simple, analytical descriptions of the energy spectrum and polarizabilities. It implies that the energy separation in the higher excited states increases very rapidly with decreasing radius of confinement, and one has a simple expression for the critical radius at which the energy is zero, which is important in the tunneling of the electrons. The polarizability also decreases as the confinement radius decreases. The approach is used to describe the properties of confined helium in terms of scaling and screening effects in the two electron atom. The analysis has been extended to confinement with the nucleus at the centre of an oscillator potential. For this case, the separation between the excited state energies increases with the strength of the confining potential, but more gently. The spectrum can also be analysed with off-centre confinement. These two models of confinement, inside a sphere, and in oscillator potential, illustrate the general properties of quantum confinement. Some of these considerations can be extended to deduce the properties of a hydrogen or helium atom confined to planar disc [38], leading to similar results. Overall, it is a topic of great interest and practical importance for the analysis of atoms influenced by external pressure, confined to cavities, encapsulated inside fullerene, and other effects. It deserves considerable attention.

The authors acknowledge support from the Natural Science and Engineering Research Council of Canada to one of the authors (Y.P.V), and from A.I.C.T.E to other author (S.H.P.) as an emeritus fellow.

REFERENCES

[1] W. Jaskolski, Phys. Rep. 271 (1996) 1.
[2] V.K. Dolmatov, A.S. Baltenkov, J.-P. Connerade, Radiat. Phys. Chem. 70 (2004) 417.
[3] G.A. Artego, F.M. Fernandez, E.A. Castro, J. Chem. Phys. 80 (1984) 1569.
[4] P.O. Fröman, S. Yngve, N. Fröman, J. Math. Phys. 28 (1987) 1818.
[5] J. Gorecki, W.B. Brown, J. Phys. B 22 (1989) 2659.
[6] P.L. Goodfriend, J. Phys. B 23 (1990) 1373.
[7] J.L. Marin, S.A. Cruz, J. Phys. B 24 (1991) 2899.
[8] S. Goldman, C. Joslin, J. Phys. Chem. 96 (1992) 6021.
[9] K.R. Brownstein, Phys. Rev. Lett. 71 (1993) 1427.
[10] J.M.H. Levett, R.P. Hurst, J. Chem. Phys. 32 (1960) 96.
[11] I.H. Hillier, J. Walkley, J. Chem. Phys. 41 (1964) 3205.
[12] J.L. Zhu, J.J. Xiong, B.L. Gu, Phys. Rev. B 41 (1990) 6001.
[13] D.S. Chuu, C.M. Hsiao, W.N. Mei, Phys. Rev. B 46 (1992) 3898.
[14] J.L. Zhu, J.H. Zhao, W.H. Duan, B.L. Gu, Phys. Rev. B 46 (1992) 7546.
[15] N. Porras-Montenegro, S.T. Perez-Merchancano, Phys. Rev. 46 (1992) 9780.
[16] J.L. Zhu, X. Chen, Phys. Rev. B 50 (1994) 4497.
[17] Z.Y. Deng, J.K. Guo, T.R. Lai, Phys. Rev. B 50 (1994) 5736.
[18] J.-P. Connerade, V.K. Dolmatov, S.T. Manson, J. Phys. B 32 (1999) L395.
[19] G.M. Harris, J.E. Roberts, J.G. Trulio, Phys. Rev. 119 (1960) 1832;
F.C. Auluck, Proc. Natl. Inst. Sci. India 8 (1942) 147.
[20] Y.P. Varshni, J. Phys. B 31 (1998) 2849.
[21] M.E. Changa, A.V. Scherbinin, V.I. Pupyshev, J. Phys B 33 (2000) 421.
[22] S. Ting-yun, Q. Hao-xue, L. Bai-wen, J. Phys. B 33 (2000) L349.
[23] Y.P. Varshni, J. Phys. B 30 (1997) L589;
R. Dutt, A. Mukherjee, Y.P. Varshni, Phys. Lett A 280 (2001) 318.
[24] S.H. Patil, J. Phys. B 34 (2001) 1049.
[25] S.H. Patil, J. Phys. B 35 (2002) 255.
[26] S.H. Patil, Y.P. Varshni, Can. J. Phys. 82 (2004) 917.
[27] E.V. Ludena, J. Chem. Phys. 69 (1978) 1770.
[28] E.V. Ludena, M. Gregori, J. Chem. Phys. 71 (1979) 2235.
[29] C. Joslin, S. Goldman, J. Phys. B 25 (1992) 1965.
[30] J.L. Marin, S.A. Cruz, J. Phys. B 25 (1992) 4365.
[31] S.A. Cruz, J. Soullard, Int. J. Quant. Chem. 83 (2001) 271.
[32] Y.P. Varshni, Eur. Phys. J. D 22 (2003) 229.
[33] S.H. Patil, Y.P. Varshni, Can. J. Phys. 82 (2004) 647.
[34] T. Kato, Commun. Pure Appl. Math. 10 (1954) 151.
[35] S.H. Patil, K.T. Tang, Asymptotic Methods in Quantum Mechanics, Springer, Berlin, 2000.
[36] M. Abramowitz, I.A. Stegun, Handbook of Mathematical Functions, Dover Publications, New York, 1965.
[37] A. Sinha, R. Roychoudhury, Y.P. Varshni, Physica B 325 (2003) 214.
[38] S.H. Patil, Y.P. Varshni, Can. J. Phys. 84 (2006) 165.
[39] A. Banerjee, C. Kamal, A. Choudhury, Phys. Lett. A 350 (2006) 121.
[40] L.D. Landau, E.M. Lifshitz, Quantum Mechanics, Pergamon Press, Oxford, 1989.

CHAPTER 2

Exact Relations for Confined One-Electron Systems

K.D. Sen[a], **V.I. Pupyshev**[b] and **H.E. Montgomery Jr**[c]

Contents

[a] School of Chemistry, University of Hyderabad, Hyderabad-500 046, India

[b] Laboratory of Molecular Structure and Quantum Mechanics, Chemistry Department, Moscow State University, Moscow, 119991, Russia

[c] Chemistry Program, Centre College, 600 West Walnut Street, Danville, KY 40422, USA

Advances in Quantum Chemistry, Vol. 57
ISSN: 0065-3276, DOI: 10.1016/S0065-3276(09)00606-6

1. INTRODUCTION: SOME NOTES ON HISTORY

A particle in an infinite square well is usually the first problem solved by quantum chemistry students, but most of the attention in quantum chemistry is devoted to "free" systems. That is, to quantum problems solved for a system in a whole space. Analysis of problems for some subregion Ω of the space is important mainly when one tries to model systems in highly inhomogeneous media or in intense external fields. The attention to confined models has been recently renewed due to interest in the analysis of a large number of new types of molecular systems and materials. For a wide range of physical situations, one may consider the Schrödinger equation for some subsystem and use the nontrivial boundary conditions on the boundary $\partial\Omega$ of the region Ω. In traditional mathematical physics, one considers the following self-adjoint boundary conditions on the wavefunctions:

$$\psi(\mathbf{r})|_{\mathbf{r}\in\partial\Omega} = 0 \quad \text{(Dirichlet boundary condition)}, \tag{1.1a}$$

$$\partial_n\psi(\mathbf{r})|_{\mathbf{r}\in\partial\Omega} = 0 \quad \text{(Neumann boundary condition)}, \tag{1.1b}$$

$$\partial_n\psi(\mathbf{r}) - P(\mathbf{r})\psi(\mathbf{r})|_{\mathbf{r}\in\partial\Omega} = 0 \quad \text{(General boundary condition)}, \tag{1.1c}$$

where ∂_n is the normal derivative, that is, $\partial_n\psi = (\mathbf{n}, \nabla\psi)$ for the unit vector $\mathbf{n}$ externally normal to the surface $\partial\Omega$ at a given point and $\nabla\psi$ is the gradient of the wavefunction ψ, $P(\mathbf{r})$ is here some real-valued function. The Neumann boundary condition means $P \equiv 0$, while for Dirichlet boundaries one may suppose formally $|P| \to \infty$.

Probably one of the first physics paper that analyzed an atomic system in a bounded region was the work of Wigner and Seitz [1,2] on the theory of periodic structures, published in 1933–1934. The Schrödinger equation for an atom in a lattice was studied with the Neumann boundary conditions (named in honor of Carl Neumann, who had studied the differential equations and potential theory for such problems at the end of 19th century).

In this work Ω was a sphere of radius R and the nucleus was placed at the center of the sphere. This reduced the problem to that of the radial function only. In 1911, H. Weyl solved some vibrational problems [3], which now may be interpreted as describing the structure of the highly excited part of the spectrum of a free particle in a bounded region Ω with Dirichlet boundary conditions. Weyl's famous asymptotic formulae for the density of states in a region of large volume, that depends on the volume but not on the form of the region Ω (see e.g. Sect. VI.4. in [4], or Sect. XIII.15 in [5]), are usually used in physical chemistry when the partition function is calculated for translational motion of an ideal gas. Nowadays the next term in this asymptotic expression is usually studied in the theory of chaos (see e.g. Sect. 7.2 of [6]).

This example illustrates why we do not intend to give a complete review of confined systems problems. Sometimes we can only mention a few papers from a large literature on a given topic. As a rule, we consider the proofs of the statements only when it is difficult to find the corresponding result (or its generalized form, as used here) in the literature. It is impossible to reference all the texts which have considered confined systems, as sometimes it is much simpler to derive something than to find it in literature. The following short prehistory illustrates this statement.

In the context of the Wigner–Seitz theory, in 1937 Brillouin [7] gave a formal analysis of the atomic energy variations under boundary deformations using contact coordinate transformations that transform the boundary modifications into Hamiltonian transformations for a fixed region and generate the associated commutation relations. This allows application of the usual form of perturbation theory for the problem.

In 1936, Fröhlich [8] used the Wigner–Seitz theory to calculate the properties of alkali metals and used some nontrivial constructions to solve the boundary value problems. This work initiated a detailed analysis of the problem by J. Bardeen [9], who expressed the energy derivative on R, $\partial_R E(R)$, through the radial wave function values on the boundary (note that Bardeen thanks E. Wigner for discussions of this part of the paper). Bardeen's method of calculation was based on differentiation and integration by parts. Fröhlich, in his short paper of 1938 [10], described the evaluation of $\partial_R E(R)$, for both the Dirichlet and Neumann boundary value problems, in a general way using Green's formulae. In particular, the problem was solved for spherically symmetric problems.

The independent work of Michels et al. [11] at the end of 1937 considered the hydrogen atom in an impenetrable sphere as a physical model for hydrogen under high pressure and studied the effects of confinement on the polarizability of hydrogen. Usually this paper is referred to as the first work on the confined atom problem, where the Dirichlet boundary conditions were used for quantum-mechanical problems. This work was followed by the work of Sommerfeld et al. [12,13] with the introduction of confluent hypergeometric functions, and from that time the number of publications is too large for our short review. We note that it is not the usual practice to mention the earlier works [1–3,7–10] in the context of quantum chemical boundary value problems. For detailed analysis of the later literature on confined systems see e.g. [14–20] and references therein, and the other papers of this volume. In the work presented, we restrict our attention to "exact" statements on confined one-electron systems and use elementary methods for the proofs.

2. SOME GENERAL CONSTRUCTIONS AND NOTATIONS

2.1. The energy functional

In this text we consider one-particle systems with the Hamiltonian $\hat{H} = -\frac{1}{2}\Delta + V(\mathbf{r})$, where V is a potential with a possible singularity of the Coulomb type. The remaining part of V is supposed here to be a non-oscillatory, smooth function bounded from below (a general description of potentials that define the self-adjoined Hamiltonians is a nontrivial problem, see Sect. X.1, X.4 in [5]). When the Schrödinger equation is studied in the region $\Omega = R^3$, that is the "free" system is studied, the traditional boundary condition for the wavefunctions $\psi \in L_2(R^3)$ is used. Otherwise, one of the boundary conditions of the type (1.1) must be used on the boundary of the region Ω. It is difficult to give exact descriptions of the regions Ω that are important from the physical point of view, except for the case $\Omega = R^3$ and regions with smooth boundaries. Usually one considers Ω be a bounded region of R^3 with a sufficiently simple boundary $\partial\Omega$. For example, one may use a diffeomorphic image of a polyhedron as Ω. A detailed mathematical description of possible regions is not simple; see [21–24] for the theory of differential operators for our problems. The mathematical aspects of the theory are well studied in classical textbooks [5] (see chapter XIII), [21] (see especially Sect. 2.5), [22–25], though the basic principles of the theory are nowadays far from intuitively clear constructions. Crudely speaking, for the usual quantum chemical Hamiltonians this is a theory of Sobolev's spaces [21,22]. For example, when the closure of the region Ω is a *compact* set, the spectrum of the Hamiltonian consists of the discrete energy levels $\{E_j\}$ of finite degeneracy, bounded from below and with the only limiting point $E = +\infty$. This follows from the so called Rellich–Kondrachov embedding theorems (see [23] or Sect. VI.1 of [22]).

The formal definition of the Hamiltonian holds at least for infinitely differentiable functions with the boundary condition desired, but this description is insufficient and one must use a more complex construction. For example, for smooth functions, square integrable in Ω simultaneously with their gradients, one may define the form $\hat{H}(\varphi, \psi)$ by the relation

$$\hat{H}(\varphi, \psi) = \frac{1}{2}\langle\nabla\varphi|\nabla\psi\rangle_\Omega + \langle\varphi|V\psi\rangle_\Omega - \frac{1}{2}\int_{\partial\Omega} P\varphi^*\psi d\sigma, \tag{2.1}$$

where $\langle *|*\rangle_\Omega$ is the usual notation for the scalar product in $L_2(\Omega)$. When the definition of Ω is clear, we use the simplified notation $\langle *|*\rangle$. The surface integral in this relation corresponds to the boundary conditions. This term ensures the symmetry of the form $\hat{H}(\varphi, \psi)$ for the functions φ, ψ that obey the boundary conditions. In particular, this term is absent for functions that obey Dirichlet or Neumann conditions. Note, that when ψ satisfies boundary conditions (1.1), one may write $\hat{H}(\varphi, \psi) = (\varphi, \hat{H}\psi)$.

For sufficiently smooth functions φ normalized to 1 that obey the boundary conditions, the energy functional $\hat{H}(\varphi, \varphi)$ is bounded from below. This allows construction of the so called Friedrichs extension (see e.g. Sect. X.3 of [5]) of the formal operator $\hat{H}$ to find the self-adjoint operator in $L_2(\Omega)$. As a result, the Hamiltonian $\hat{H}$ may be defined for the set D_H, the domain of definition of $\hat{H}$, being a dense set in $L_2(\Omega)$. Different boundary conditions generate different self-adjoint operators $\hat{H}$ with different domains.

The form (2.1) may be used to formulate the variational principle for the systems under consideration. It is important to note that one may search for the minimum of the energy functional in a much wider class of functions than D_H, but the minimum is achieved on a function from D_H, being the solution of the usual differential Schrödinger equation

$$\hat{H}\psi_j = E_j\psi_j, \tag{2.2}$$

for the ground state ($j = 0$). For excited states, a similar statement holds due to the minimax principle (see Sect. XIII.1 in [5]). To mark the region, we use the symbol $E_j(\Omega)$ for the jth eigenvalue of H. It is possible to show that when the potential or boundary conditions vary smoothly with some parameter, one may differentiate the wavefunctions and energies. These statements are nontrivial results of a general theory (see e.g. [25], Sect. VI, Section 6 of VII, and [5], Sect. XII).

We intend to discuss physical problems here and the usual "physical level" of accuracy is used to avoid discussion of important, but formal problems.

2.2. Spherically symmetric problems

For the spherically symmetric system, one supposes Ω to be a sphere of radius R with its center at the coordinate origin. In this case, the potential is a function of $r = |\mathbf{r}|$. Here, we use wavefunctions of the form $\psi(\mathbf{r}) =$

$R_{n,\ell}(r)Y_{\ell m}(\theta, \phi)$ with spherical harmonics $Y_{\ell m}(\theta, \phi)$ and radial functions $R_{n,\ell}(r)$, normalized to 1 on the interval $[0, R]$ with the weight r^2. Much simpler are the functions $\varphi_{n,\ell}(r) = rR_{n,\ell}(r)$, normalized to 1 with the weight 1. At the point $r = 0$ the radial functions $\varphi_{n,\ell}(r)$ obey the boundary condition $\varphi_{n,\ell}(0) = 0$. It should be noted that for boundary conditions of the type (1.1c) one must modify the spherically symmetric boundary constant $P = p(R)$. The equivalent conditions at the point $r = R$ are:

$$\partial R_{n,\ell}(R) = PR_{n,\ell}(R), \qquad \partial\varphi_{n,\ell}(R) = \left(P + \frac{1}{R}\right)\varphi_{n,\ell}(R) = p\varphi_{n,\ell}(R). \quad (2.3)$$

In practice, both types of wavefunctions $R_{n,\ell}(r)$ and $\varphi_{n,\ell}(r)$ are used, just as are both boundary constants P and p. The conditions stated in Equation (2.3) have interesting consequences on the uncertainty like relationships for the free and confined systems. We shall discuss them later, in Section 8.

2.3. Local properties of wavefunctions

In practice, the situation is essentially simplified, as a nontrivial study of eigenfunctions $\psi_j(r)$ of Hamiltonians by embedding theorems gives rise to consideration of the usual infinitely differentiable functions, except at the points where the same is not true for the potential. For example, it is not true for the points where the Coulomb singularity is placed, but Kato's cusp condition holds for these points (see [26,27], or Sect. XIII.11 of [5] for differential properties of wavefunctions). In addition, the generalized form of cusp-conditions may be proved [28]. For the model confined systems of hydrogen-like atoms and the isotropic harmonic oscillator, the density and its derivatives at the location of the origin defined by the nuclear position and the equilibrium position, respectively, can be used to estimate extremely accurate values of the energy for all states. We shall return to this topic in Section 6.

2.4. Extension of the region and nodal properties of wavefunctions

The simplest example of the use of the above mentioned statements is the following. Let us consider the Dirichlet problem for systems in regions Ω and $\Omega_1 \supset \Omega$. Let ψ be the ground state wavefunction for the Dirichlet problem in Ω, that is, $\psi|_{\partial\Omega} = 0$. One may consider the continuation ψ^1 of this function into Ω_1 by zero value. That is $\psi^1(\mathbf{r}) = \psi(\mathbf{r})$ for $\mathbf{r} \in \Omega$ and $\psi^1(\mathbf{r}) = 0$ for all $\mathbf{r} \in \Omega_1 \setminus \Omega$. As ψ^1 fulfills the Dirichlet boundary conditions for Ω_1, one can see that the energy functional has the same values for ψ and ψ^1. Hence, its minimal value is *lower* for Ω_1. This is a strong inequality (otherwise ψ^1 minimizes the energy functional and ψ^1 is a solution of the differential equation (see Section 2.1); but the relation $\psi^1 = 0$ for a region $\Omega_1 \setminus \Omega$ means

$\psi^1 \equiv 0$ due to uniqueness of the solution *of the differential equations)*. Hence,

$$E(\Omega) > E(\Omega_1) \quad \text{for } \Omega \subset \Omega_1. \tag{2.4}$$

This important relation holds for all the states for the Dirichlet problem (see e.g. [5], Sect. XIII.15). However, the analogous relation is wrong for the other types of boundaries [29]. For example, the constant function in a sphere (that is, the wavefunction $\psi(\mathbf{r}) = \text{const}$) satisfies the Neumann boundary conditions and has a zero value for the kinetic energy. Hence the energy functional is the mean value for the potential, that is, it equals $-3/(2R)$ for the hydrogen atom. When the sphere radius R goes to zero, the energy value decreases, in contrast to Equation (2.4).

The other useful property of the wavefunctions for Dirichlet boundary problems that follows from the variational principle is the famous property of the nodal points. One may consider for any function ψ the maximal connective open regions, $\omega(\psi)$, where the function possesses a constant sign. It is easy to show from relation (2.4), that if ψ' and ψ are eigenfunctions of the differential operator $\hat{H}$ and there are regions $\omega(\psi')$ and $\omega(\psi)$ with the property $\omega(\psi') \subset \omega(\psi) \subset \Omega$, then the energy value $E_{\psi'}$ is greater than E_{ψ}. In addition, the behavior outside $\omega(\psi')$ (in particular, the boundary conditions on $\partial\Omega$) are unessential details for this statement [30].

In particular, the ground state function has a constant sign (and it is a non-degenerate state). When all the states (including degenerate ones) are numbered from 0 and energy E_k increases with increasing k, the kth stationary wavefunction has no more than $k+1$ regions of constant sign (see [31]). For the use of these statements in molecular orbitals and symmetry analysis see [30]. Analogous statements hold for other types of boundary conditions, but here we use only the ground state properties of one-dimensional problems.

A simple result of the properties described is the following. If one finds an eigenfunction ψ of the Hamiltonian and a region $\omega(\psi)$ of constant sign, then this is a ground state function for the Dirichlet problem in $\Omega = \omega(\psi)$. For example, it is clear, that the 2p-atomic orbital of the free hydrogen atom defines the ground state wavefunction (and ground state energy $-1/8$ au) for the Dirichlet boundary problem in the half-space of R^3 for a hydrogen atom at the boundary. In [32] this idea was used in the qualitative consideration of the energy variations for a number of Dirichlet problems for the hydrogen atom.

Another example of the use of this statement in boundary value problems is given by the wavefunction $\psi(\mathbf{r}) = Ce^{-r}$ for a hydrogen atom in a sphere. It is clear that the constant energy value $E(R) = -1/2$ au for any R value corresponds to the ground state of the spherically symmetric problem with the boundary condition $\partial_n\psi = -\psi$ [29].

2.5. Comparison theorems

There exist a number of useful statements, known as comparison theorems, that allow comparison of the wavefunctions $\psi_{1,2}$ for two Schrödinger equations with potentials $V_{1,2}$ and corresponding energies $E_{1,2}$. For example, for some region ω, such that for any $\mathbf{r} \in \omega$ one may write $V_2 - E_2 \geq V_1 - E_1$ and $\psi_1 \geq 0$, then the inequality $\psi_1 \geq \psi_2$ for the points on the boundary $\partial\omega$ means $\psi_1 \geq \psi_2$ in ω. For proofs of different variants and generalizations of these statements and their application to quantum chemical problems see [33].

One may use such statements to find the optimal position of the one-electron atom in a given external potential $U(\mathbf{r})$ or in a cavity with Dirichlet boundaries. For example, let the nucleus be placed in the plane Σ and the potential $U(\mathbf{r})$ in one of the corresponding half-spaces, Σ^+ or Σ^-, be greater than in the other in the following sense. For $\mathbf{r} \in \Sigma^+$ one may put $\mathbf{r} = \mathbf{a} + \alpha\mathbf{n}$, where $\mathbf{a} \in \Sigma$, $\mathbf{n}$ is normal to Σ, and $\alpha \geq 0$. Reflection σ with respect to Σ transforms Σ^+ to Σ^- and vice versa. Let us suppose that for $\mathbf{r} \in \Sigma^+$ we have

$$U(\mathbf{r}) = U(\mathbf{a} + \alpha\mathbf{n}) \leq U(\mathbf{a} - \alpha\mathbf{n}) = U(\sigma\mathbf{r}). \tag{2.5}$$

In this case one may use the symmetry of the Coulomb potential and apply the comparison theorem for the ground state wavefunction $\psi(\mathbf{r})$ and the "reflected" function $\varphi(\mathbf{r}) = \psi(\sigma\mathbf{r})$, that corresponds to the problem in Σ^+ with external potential $U_1(\mathbf{r}) = U(\sigma\mathbf{r})$. As $\varphi(\mathbf{r})$ and $\psi(\mathbf{r})$ are identical on Σ and at infinity, by the comparison theorem one finds $\psi(\mathbf{r}) \geq \varphi(\mathbf{r}) = \psi(\sigma\mathbf{r})$. That is, the wavefunction ψ is greater in the region Σ^+, than in Σ^-, in the sense of Equation (2.5). It is clear, that in this case the Hellman–Feynman force is oriented into Σ^+, that is, its $\mathbf{n}$-projection is positive. Practically the same discussion was used in [34] to prove the monotone nature of the adiabatic potential for the ground state of the one-electron diatomic molecule (in the absence of the internuclear repulsion term). This statement is also easily modified for the Dirichlet boundary value problem for some region Ω. One may formally require the external potential U to be infinite out of Ω (for the analysis of this statement see [35]).

From this discussion, it follows that the optimum position of the hydrogen atom in a spherical cavity with impenetrable walls is at the sphere center, at least for the ground electronic state. Obviously, the analogous statement is true if Ω is a cube or cylinder.

To some extent the statement holds for excited states that are the lowest states of a given symmetry type, but only when a small shift along the normal $\mathbf{n}$ to the plane Σ does not change the symmetry type [35]. For example, for an atom on the symmetry axis of the cylinder, one may state the optimal position is at the symmetry center for the lowest π- or δ-states, but one cannot require u- or g-symmetry for the state. Numerical experiments confirm this statement [36]. To analyze the shifts perpendicular to the main axes we have to use alternative methods [37].

These statements hold not only for atoms, but for any system with a central field potential subject to some additional potential U.

3. COMMUTATION RELATIONS AND HYPERVIRIAL THEOREMS

It is usually considered that boundary value problems do not differ essentially from free systems. To some extent this is true, and the discussion above confirms this statement. However, even a quick consideration demonstrates that existence of nontrivial boundary conditions requires special attention to some details that may be ignored, or supposed to be almost trivial, for the case when the region Ω is the whole space. In this section we describe some situations, where the details mentioned essentially change the physical picture and physical intuition.

3.1. Integrals of motion and boundary value problems

It is natural to exploit the commutation relations for quantum chemical considerations, as they are the traditional background of the algebraic structure of quantum theory. There are many formal relations like

$$\hat{H}\hat{A} - \hat{A}\hat{H} = \hat{B} \tag{3.1}$$

with appropriate operators $\hat{A}$ and $\hat{B}$, that may be checked for the infinitely differentiable functions that differ from zero in some compact internal subregion of Ω only.

Let us consider the case $\hat{B} = 0$, when the Hermitian operator $\hat{A}$ corresponds to some integral of motion from the usual point of view. When the operator relation (3.1) is written for the Hamiltonian eigenfunction ψ with eigenvalue E, one may write,

$$\hat{H}\hat{A}\psi = E\hat{A}\psi \tag{3.2}$$

and it only *seems* to be clear, that $\hat{A}\psi$ is the eigenfunction of the Hamiltonian $\hat{H}$ with energy E. This statement is wrong for the boundary value problem, as the function $\phi = \hat{A}\psi$, in general, does not obey the same boundary conditions as ψ.

The statement is true only for the case, usually supposed to be realized, when it follows for any function ψ from D_H, the domain of definition of $\hat{H}$, that $\hat{A}\psi \in D_H$. For example, this is the case for the spherically symmetric systems in a sphere with spherically symmetric boundary conditions that may be formulated as relations for radial functions. The rotation operators

$\hat{A}$ do not remove ψ from D_H. Hence, one may use angular momentum conservation for spherically symmetric problems.

In general, the relation $\hat{A}\psi \in D_H$ is not realized. As a result, even when, for a free problem, $\hat{A}$ is an integral of motion, this will not be the case for a problem in a bounded region. An excellent example is given by the hydrogen atom, where an additional integral of motion exists, the so called Lenz vector $\hat{\mathbf{A}} = \frac{\mathbf{r}}{|\mathbf{r}|} - \frac{1}{2}\left(\hat{\mathbf{p}} \times \hat{\mathbf{L}} + \hat{\mathbf{L}} \times \hat{\mathbf{p}}\right)$, where $\hat{\mathbf{p}}$ and $\hat{\mathbf{L}}$ are momentum and angular momentum vectors of the particle. For the classical atomic system, existence of this additional integral results in elliptic orbits with constant orientation. From the quantum point of view the result is additional degeneracy of the hydrogen energy levels [38].

Note that for a hydrogen atom at the center of an impenetrable spherical cavity of radius R, the classical orbit consists of a system of elliptical fragments and only for the case $R = |\hat{\mathbf{L}}|^2$ may one find that for any energy E and a given angular momentum, all of the orbits are closed (they contain exactly two elliptical fragments); see [39] for details. The quantum analog of this situation is also nontrivial. Really, the Dirichlet boundary conditions both for $\hat{A}\psi$ and ψ may not be realized, but it is possible for ψ and $\left(\hat{A}_x + \mathrm{i}\hat{A}_y\right)^2 \psi$. In this case, one may write some relation for the radial function and its first and second derivative on $\partial\Omega$. The Schrödinger equation allows us to exclude the second derivative. As a result, the boundary condition for the radial function $\varphi(r)$ (see Section 2.2) means that some expression of the type $c(R, E)\partial\varphi|_{\partial\Omega} = 0$ vanishes. As it is impossible for both the functions $\partial\varphi$ and φ to vanish at the boundary, this means $c(R, E) = 0$. These relations are derived in [40]. In particular, for some R values, one may note that the function $c(R, E)$ does not depend on the value of E. This gives an interesting regular degeneracy of *all* the pairs of the states $(n, \ell + 2)$ and $(n + 1, \ell)$ at $R = (\ell + 1)(\ell + 2)$. Formally, this degeneracy is associated with the Gauss relations for the confluent hypergeometric function, that allow us to analyze not only the hydrogen problem (see [41]), but also the confined harmonic oscillator (for details see [42]).

It is strange that the analogous regular effects for other types of boundary conditions for the hydrogen atom have not been attempted to date. Similar trends for the confined isotropic harmonic oscillator will be discussed in Section 8.

3.2. Hypervirial theorems

Let us calculate the mean value of operators of the form presented by relations (3.1) on the eigenfunction ψ of the Hamiltonian. It is easy to see that

$$\left\langle \psi \left| \hat{B}\psi \right.\right\rangle = \left\langle \psi \left| \left(\hat{H}\hat{A} - \hat{A}\hat{H}\right)\psi \right.\right\rangle = E\left\langle \psi \left| \hat{A}\psi \right.\right\rangle - \left\langle \psi \left| \hat{H}\hat{A}\psi \right.\right\rangle. \qquad (3.3)$$

The exponential decrease of the wavefunctions for the stationary states of free systems usually ensures that $\hat{A}\psi \in D_H$. As a result, one may write the relation

$$\left\langle \psi \left| \hat{H}\hat{A}\psi \right.\right\rangle = \left\langle \hat{H}\psi \left| \hat{A}\psi \right.\right\rangle = E\left\langle \psi \left| \hat{A}\psi \right.\right\rangle, \tag{3.4}$$

and, hence, $\langle \psi | \hat{B}\psi \rangle = 0$. But this "hypervirial" relation is only true for the case when the first relation of Equation (3.4) is true, that is when $\hat{A}\psi \in D_H$.

The traditional "movement" of $\hat{H}$ from the right hand side of the scalar product to the left in order to express it as $\langle \hat{H}\psi | \hat{A}\psi \rangle$ is not trivial here. It requires some integral transformations, namely, integration "by parts" or using Green's formulae. As a result, for a boundary value problem the commutation relation described expresses the mean value of $\hat{B}$ as some surface integral over $\partial\Omega$. It is interesting that this is a common situation, that these additional terms can be expressed through $\partial_R E(R)$ (see e.g. Equation (4.7)).

There exist a large number of different hypervirial relations (see one method of generation of such relations in [43]), but the best known is the virial theorem. We will consider this relation in the next subsection.

It is worthwhile to write some of the useful relations that can be derived in the way described by analysis of commutators of the type $[\hat{H}, [\hat{H}, r^k]]$ for the spherically symmetric system with the potential $V(r)$. For a state with a given angular momentum, one can use the symbol $\langle C \rangle$ for mean values. When the Dirichlet problem is studied for a sphere in D-dimensional space R^D we find that for $q = \ell + (D-3)/2$ (one may use any $D \geq 2$, but s-states for $D = 2$ are excluded)

$$\begin{aligned} \partial E(R) R^{n+1} = {} & n\left[q(q+1) - \frac{n^2-1}{4} \right] \langle r^{n-2} \rangle - 2(n+1)\langle r^n \rangle E \\ & + 2(n+1)\langle r^n V \rangle + \langle r^{n+1} \partial V \rangle. \end{aligned} \tag{3.5}$$

This relation holds only if all the integrals used converge. For the case $D = 3$ it was studied for model confined systems in [44–46]. For a free system in D dimensions it was used in [43].

3.3. Scaling

It is worth noting the mutual connections between the commutation relations such as Equation (3.1) and exponential transformations, for example, in the context of Hausdorf's relations. Note that the coordinate transformation is a standard method for the analysis of boundary value problems (see e.g. [25]). An important type of commutation relation is naturally connected with the scaling procedure, that is, a transformation of the type $r \to \alpha r$ (in this context, again see Ref. [7]). Here we use this method to derive the virial theorem. Fock exploited a similar variational approach to derive the virial theorem [47] for approximate wavefunctions.

Let us write the Schrödinger equation for a set $\{Z_a, R_a\}$ of nuclei with charges Z_a placed at the points R_a to study a molecule placed in a cavity $\Omega(\lambda)$ with Dirichlet boundary conditions. Here λ is some linear parameter used to describe the region modifications by scaling. The eigenvalue problem may be written as

$$\begin{aligned}&\hat{H}(\mathbf{r}|m, \{Z_a, R_a\}, \Omega(\lambda))\psi(\mathbf{r}|m, \{Z_a, R_a\}, \Omega(\lambda))\\&\quad = E(m, \{Z_a, R_a\}, \Omega(\lambda))\psi(\mathbf{r}|m, \{Z_a, R_a\}, \Omega(\lambda)),\end{aligned} \tag{3.6}$$

where $\mathbf{r}$ is the vector of variables and m is the particle mass. When we use a function of the form $u(\mathbf{r}, \alpha) = \alpha^{3/2}\psi(\alpha\mathbf{r}|m, \{Z_a, R_a\}, \Omega(\lambda))$ where $\alpha^{3/2}$ is the normalization factor, this function corresponds to the Dirichlet boundaries for $\Omega(\alpha^{-1}\lambda)$. The trivial transformations show clearly that $u(\mathbf{r}, \alpha)$ is the solution of the equation for the Hamiltonian

$$\begin{aligned}\hat{H}(\alpha\mathbf{r}|m, \{Z_a, R_a\}, \Omega(\lambda)) &= \hat{H}(\mathbf{r}|m\alpha^2, \{\alpha^{-1}Z_a, \alpha^{-1}R_a\}, \Omega(\alpha^{-1}\lambda))\\&= \alpha^{-2}\hat{H}(\mathbf{r}|m, \{\alpha Z_a, \alpha^{-1}R_a\}, \Omega(\alpha^{-1}\lambda)).\end{aligned} \tag{3.7}$$

But this means, that the following relations hold

$$\begin{aligned}E(m, R_a, Z_a, \Omega(\lambda)) &= E(m\alpha^2, \{\alpha^{-1}Z_a, \alpha^{-1}R_a\}, \Omega(\alpha^{-1}\lambda))\\&= \alpha^{-2}E(m, \{\alpha Z_a, \alpha^{-1}R_a\}, \Omega(\alpha^{-1}\lambda)),\\u(\mathbf{r}, \alpha) &= \alpha^{3/2}\psi(\alpha\mathbf{r}|m, \{Z_a, R_a\}, \Omega(\lambda))\\&= \psi(\mathbf{r}|m, \{\alpha Z_a, \alpha^{-1}R_a\}, \Omega(\alpha^{-1}\lambda)).\end{aligned} \tag{3.8}$$

The derivative of relation (3.8) for energy with respect to α at the point $\alpha = 1$ immediately gives the virial theorem for confined Coulomb systems:

$$2\langle T\rangle + \langle V\rangle + \sum_a R_a\partial_{R_a}E + \lambda\partial_\lambda E = 0, \tag{3.9}$$

where $\langle T\rangle$ and $\langle V\rangle$ are the mean values of the kinetic and potential energy.

For the case of harmonic forces, associated with the system of centers with radius-vectors $\{S_a\}$, one may write by the same arguments

$$2\langle T\rangle - 2\langle V\rangle + \sum_a S_a\partial_{S_a}E + \lambda\partial_\lambda E = 0. \tag{3.10}$$

The only difference from the free system in these relations is the term $\lambda\partial_\lambda E$, generated by the boundary effects.

Sometimes analytical relations such as the one in Equation (3.8) are useful for qualitative discussions. For example, one may note, that

$$2E + \lambda\partial_\lambda E = \lambda^{-1}\partial_\lambda[\lambda^2 E]. \tag{3.11}$$

Hence, for a harmonic oscillator placed in the center of an impenetrable cavity (that is, when $S_a = 0$) it follows from the virial relation (3.10) that

$$\lambda^{-1}\partial_\lambda[\lambda^2 E] = 4\langle V\rangle. \tag{3.12}$$

In particular, for a harmonic oscillator, the values of $\lambda^2 E(\Omega(\lambda))$ increase monotonically with the cavity size λ enlargement for the confined oscillator, but these values decrease when the "inverted" oscillator is considered (i.e. the oscillator with negative force constant). In the same manner, one may demonstrate that $\lambda^2 E(\Omega(\lambda))$ increases monotonically for a hydrogen atom at the center of a cavity $\Omega(\lambda)$. This is useful for the case of small cavities (see Section 4.2).

For the case of general type conditions, the $P(\mathbf{r})$ function in Equation (1.1c) can also be varied under scaling. For example, for Neuman's boundary conditions, one can use direct evaluation of the mean values for integral relations with analysis of the surface integrals. It is difficult to formulate a general statement on the basis of the scaling procedure, as the vector $\mathbf{n}$ for the normal to $\partial\Omega$ changes for scaled regions. The only case where $\mathbf{n}$ is not changed under scaling is the case where Ω is a sphere. This is the case, when the form of the virial theorem is the same for Neumann's conditions, as for Dirichlet conditions (see also [44,45]).

The possibility of using the virial/hypervirial relations for approximate wavefunctions depends on the form and nature of the classes of approximate functions. The general theory is presented in [48]. For the use of the virial theorem for confined systems, see reference [49].

For a free particle in a cavity $\Omega(\lambda)$ and for a linear scale λ one may put $\langle V\rangle = 0$ and $\langle T\rangle = E(\lambda)$. Hence it follows from Equation (3.9) or Equation (3.12), that $E(\lambda) = c\lambda^{-2}$ for some constant that depends on the state and the form of the region. For this case, the volume of the region $\Omega(\lambda)$ may be estimated as $V(\lambda) = A\lambda^3$ and one may write for the partition function,

$$\sum_j \exp(-E_j(\lambda)/kT) = \Phi(V^{2/3}T, \Omega(1)) = \Phi(V^{2/3}T) \tag{3.13}$$

for some function $\Phi(z)$ of one variable. Note, that the pressure P is a derivative of the free energy ($\sim T\ln\Phi$) with respect to V, and, hence,

$P \sim (T/V)(TV^{2/3})(2\Phi'/3\Phi)$. But for an ideal gas one has $P \sim T/V$, that is, for any fixed Ω one can write $\Phi(z) \sim z^{3/2}$. This relation gives Weyl's relations [3], as the Laplace transformation of $\Phi(z)$ defines the density of states, being proportional to the volume, V.

Another application of scaling is useful for the case of the Dirichlet problem for a homogeneous potential of degree n. In this case, the analog of relation (3.8) is followed by

$$E(m, \lambda) = m^{-\frac{n}{n+2}} E(1, \lambda m^{\frac{1}{n+2}}). \tag{3.14}$$

One may use such relations to consider nontrivial experimental isotopic shifts [50] in the spectra of confined hydrogen molecules in fullerene cavities, as was done in [51] to analyze the results of numerical experiments.

3.4. The Kirkwood–Buckingham relation

Note that one may use a pair of relations similar to Equations (3.5) and (3.9) to exclude the energy derivatives. This gives an integral relation, valid both for free and confined systems. One of the useful relations of such a kind was derived in an approximate way by Kirkwood [52] and Buckingham [53] to estimate the polarizability. For a simple proof and the history of these relations, widely used nowadays, see [54].

The formal proof of the relations mentioned is almost trivial. Let ψ be the real-valued eigenfunction of $\hat{H}$ with eigenvalue E. Let $f(\mathbf{r})$ be a real valued, continuous function with square-integrable gradient, equal to zero outside the region Ω. One may consider the values of the quadratic form given by Equation (2.1) for the function $f\psi$. It is clear that

$$\begin{aligned} \hat{H}(f\psi, f\psi) &= \frac{1}{2}\langle\nabla\psi|f^2|\nabla\psi\rangle + \langle f\psi|V|f\psi\rangle + \frac{1}{4}\langle\nabla\psi^2|\nabla f^2\rangle \\ &\quad + \frac{1}{2}\langle\psi|(\nabla f)^2|\psi\rangle. \end{aligned} \tag{3.15}$$

Hence one may use Green's formula to write

$$\begin{aligned} \langle\nabla\psi^2|\nabla f^2\rangle &= \int_\Omega (\nabla\psi^2, \nabla f^2)\mathbf{dr} = \int_\Omega \{\mathrm{div}(f^2\nabla\psi^2) - f^2\mathrm{div}\nabla\psi^2\}\mathbf{dr} \\ &= \int_{\partial\Omega} f^2\partial_n\psi^2 \mathrm{d}\sigma - \int_\Omega f^2\Delta\psi^2\mathbf{dr} = -\int_\Omega f^2\Delta\psi^2\mathbf{dr} \\ &= -2\langle\nabla\psi|f^2|\nabla\psi\rangle - 2\langle\psi|f^2|\Delta\psi\rangle, \end{aligned} \tag{3.16}$$

as f vanishes on $\partial\Omega$. Hence (3.15) may be rewritten as

$$\hat{H}(f\psi, f\psi) = \langle f^2\psi|\hat{H}\psi\rangle + \frac{1}{2}\langle\psi|(\nabla f)^2|\psi\rangle$$
$$= E\langle\psi|f^2|\psi\rangle + \frac{1}{2}\langle\psi|(\nabla f^2)|\psi\rangle. \tag{3.17}$$

This is the Kirkwood–Buckingham relation useful for analysis of boundary value problems. For example, Equation (2.4) follows from Equation (3.17) and the variational inequality for the ground state wavefunctions for Dirichlet problems in regions Ω and Ω_1. One can find some generalizations of this relation in [54,55].

In practice, Equation (3.17) may be exploited in the following way. Let $f(\mathbf{r})$ differ from 0 in the region Ω and for the points of some $\Omega_1 \subset \Omega$ one may suppose $f(\mathbf{r}) = 1$. Then for some constant C, that estimates $|\nabla f|^2$, the ground state energy $E(\Omega)$ for the Dirichlet problem in Ω may be estimated as [54]

$$E(\Omega) \le E + \frac{\langle\psi|(\nabla f^2)|\psi\rangle}{2\langle\psi|f^2|\psi\rangle} \le E + \frac{C\langle\psi|\psi\rangle_{\Omega_1^c}}{1 - \langle\psi|\psi\rangle_{\Omega_1^c}} \tag{3.18}$$

where Ω_1^c means a region, additional to $\Omega_1 : \Omega_1^c = R^3 \setminus \Omega_1$. We use similar relations later for estimations.

To illustrate one of the ways to use relation (3.17), let us choose some values $A > 0$ and γ, such that $0 < \gamma < 1$ and put $B = \gamma A$. One may define the spherically symmetric function $f(\mathbf{r}) = f(r)$ by

$$f(r) = 1 \quad \text{for } r > A, \qquad f(r) = 0 \quad \text{for } r < B, \qquad \text{and}$$
$$f(r) = (B^{-1} - A^{-1})^{-1}(B^{-1} - r^{-1}) \quad \text{for } B \le r \le A. \tag{3.19}$$

Note, that $f(r)$ is a continuous function and its gradient is a well-defined function out of the sets $r = A$ or $r = B$. The gradient of $f(r)$ is a square integrable function in R^3 and it is clear that, for the region $\omega(A)$, where ∇f differs from zero, one may write

$$\int_{\omega(A)} |\nabla f|^2 \mathrm{d}\mathbf{r} = 4\pi(B^{-1} - A^{-1})^{-2} \int_B^A r^{-4} r^2 \mathrm{d}r$$
$$= 4\pi(B^{-1} - A^{-1})^{-1}$$
$$= 4\pi(\gamma^{-1} - 1)^{-1} A. \tag{3.20}$$

Note, that the wavefunction $\psi(\mathbf{r})$ is a bounded function (see, e.g. [26]). Hence it follows from Equation (3.17), that when $A \to 0$, one may write

$\langle f\psi \mid f\psi\rangle \rightarrow 1$ and $\hat{H}(f\psi, f\psi) \rightarrow E$. That is, the condition that the wavefunction vanishes for some small sphere has a small effect on energy values at least for the ground state. The same is true for excited states too.

It should be stressed that the wavefunctions of the form

$$\varphi = f(\mathbf{r})\psi(\mathbf{r}) \tag{3.21}$$

with exact, approximate or basic functions ψ and some cut-off function f that ensures that the boundary conditions are met have been widely used in the theory of confined systems from the earliest works (see e.g. [10,11]).

4. ENERGY AND REGION MODIFICATIONS

Let us consider the one-parameter family of regions $\Omega(\lambda)$ and the corresponding boundary value problems. In this section we describe the main results concerning the energy changes with region modifications. Of course, the general result is given in Equation (2.4), but one needs a more accurate form to use for physical calculations. As a historical review one may refer to the text of the Section 1.

4.1. Integral relations for energy derivative

Let us consider Dirichlet problems for a modified system of regions $\Omega(\lambda)$. Let the value $\lambda = 0$ correspond to the region $\Omega = \Omega(0)$. For normalized eigenfunctions $\psi_\lambda = \psi(\mathbf{r}, \lambda)$ with eigenvalues $E(\lambda)$ for a given state of the system, one may write the relation

$$\left\langle \psi_0 \middle| \left[\hat{H} - E(\lambda)\right] \psi_\lambda \right\rangle = 0. \tag{4.1}$$

The derivative with respect to λ means

$$\left\langle \psi_0 \middle| \left[\hat{H} - E(\lambda)\right] \partial_\lambda \psi_\lambda \right\rangle_{\Omega(\lambda)} - \partial_\lambda E(\lambda) \langle \psi_0 \mid \psi_\lambda \rangle_{\Omega(\lambda)} = 0 \tag{4.2}$$

and using Green's formula at the point $\lambda = 0$ immediately gives

$$\begin{aligned}
\langle \psi_0 | [H - E(0)] \partial_\lambda \psi_0 \rangle_\Omega &= \int_\Omega \psi_0^* [\hat{H} - E(0)] \partial_\lambda \psi_0 \mathrm{d}x \\
&= \frac{1}{2} \int_{\partial\Omega} [\partial_\lambda \psi_0 \partial_n \psi_0^* - \psi_0^* \partial_n (\partial_\lambda \psi_0)] \mathrm{d}\sigma \\
&\quad + \langle [\hat{H} - E(0)] \psi_0 | \partial_\lambda \psi_0 \rangle_\Omega \\
&= \frac{1}{2} \int_{\partial\Omega} [\partial_\lambda \psi_0 \partial_n \psi_0^* - \psi_0^* \partial_n (\partial_\lambda \psi_0)] \mathrm{d}\sigma
\end{aligned} \tag{4.3}$$

as ψ_0 is the eigenfunction of $\hat{H}$ with the eigenvalue $E(0)$. The Dirichlet boundary condition means

$$\langle\psi_0|[\hat{H} - E(0)]\partial_\lambda\psi_0\rangle_\Omega = \frac{1}{2}\int_{\partial\Omega} \partial_\lambda\psi_0\partial_n\psi_0^* d\sigma. \tag{4.4}$$

In order to exclude the wavefunction derivatives with respect to λ, we note that for small λ the point $\mathbf{x} \in \partial\Omega$ transforms to the point $\mathbf{x}(\lambda) \in \partial\Omega(\lambda)$ as

$$\mathbf{x}(\lambda) = \mathbf{x} + \lambda\rho(\mathbf{x})\mathbf{n} + o(\lambda) \tag{4.5}$$

for some weight function $\rho(\mathbf{x})$ that defines the modification of the boundary for the family $\{\Omega(\lambda)\}$ in the direction of the outside normal $\mathbf{n}$ to $\partial\Omega$. The correction $o(\lambda)$ goes to zero as $\lambda \to 0$ not too slowly: $o(\lambda)/\lambda \to 0$. In this case, the Dirichlet boundary condition $\psi(\mathbf{x}(\lambda), \lambda) = 0$ for small λ means

$$0 = \partial_\lambda\psi(\mathbf{x}(\lambda), \lambda) + \rho(\mathbf{x}, \lambda)(\mathbf{n}, \nabla\psi(\mathbf{x}(\lambda), \lambda)) + \mathrm{o}(\lambda)/\lambda. \tag{4.6}$$

Hence, at the point $\lambda = 0$ one may write $\partial_\lambda\psi_0 = -\rho\partial_n\psi_0$ on $\partial\Omega$. Due to the normalization condition $\langle\psi_0|\psi_0\rangle_\Omega = 1$ the following relation follows from Equations (4.2) and (4.4),

$$\partial_\lambda E(0) = -\frac{1}{2}\int_{\partial\Omega} |\partial_n\psi_0|^2 \rho d\sigma. \tag{4.7}$$

A similar approach to derive Equation (4.7) was used in [9]. Practically the same construction was used in [10,56], but it used finite differences instead of derivatives and the corresponding limits. Note, that due to the Dirichlet boundary conditions for ψ_0 one may use the region $\Omega(\lambda)$ in Equations (4.1), (4.2) for any method of continuation of ψ_0 out of Ω: by zero value, or with continuous derivatives, as in both cases corrections to the integral are $\sim o(\lambda)$. The same result may be found by differentiation of the energy functional values. It is also worth noting that the normal derivative of ψ_0 does not vanish at $\partial\Omega$ simultaneously with ψ_0, if the wavefunction is not zero in Ω. Hence, the energy derivative is negative for extended regions, as ρ is nonnegative. This statement confirms Equation (2.4).

One may replace $|\partial_n\psi_0|$ at the boundary of the region by $|\nabla\psi_0|$, as $\nabla\psi_0$ is proportional to the normal $\mathbf{n}$. It is also easy to show that when the surface $\partial\Omega(\lambda)$ is described by the relation $S(x, \lambda) = 0$ and where the gradient ∇S is oriented along the normal $\mathbf{n}$, the ρ values may be calculated as $-\|\nabla S(x, 0)\|^{-1}\partial_\lambda S(x, 0)$. In particular, for a spherically symmetric potential and a sphere of radius R we have the equation for the surface $(x^2 + y^2 + z^2) -$

$R^2 = 0$. In this case $\rho = -\|\nabla S\|^{-1}\partial_\lambda S = 1$ and

$$\partial_R E(R) = -\frac{R^2}{2}(\partial R_{n\ell}(R))^2, \tag{4.8}$$

where $R_{n\ell}(r)$ is the radial part of wavefunction (see Section 2.2). This is the relation used first by Fröhlich [10]. He also used the result of integration by parts of a relation similar to (4.1), where ψ_0 was replaced with any solution of the Schrödinger equation ψ with energy E in the form

$$(E(\lambda) - E)\langle\psi|\psi_\lambda\rangle_{\Omega(\lambda)} = -\frac{1}{2}\int_{\partial\Omega((\lambda))} \psi^*\partial_n\psi_\lambda \mathrm{d}\sigma. \tag{4.9}$$

Fröhlich's relations were also used in a slightly modified form of the perturbation theory power series in the dimensionless parameter $(R-R_0)/R_0$ [57]. In its current form, derivatives like $\partial_R E(R)$ are usually derived on the basis of commutation relations as a consequence of some hypervirial relations in R^3 (see [44–46]). It is clear that these derivatives appear as the boundary terms in the usual integral relations (see also Sections 3.2 and 3.3).

4.2. Small cavities

The case of small cavities is simple, one may use a crude estimate (see [5], vol. II, Sect. X.2) for the functions, that obeys Dirichlet boundary conditions:

$$\langle\psi|\mathrm{r}^{-2}|\psi\rangle \leq 4\langle\nabla\psi|\nabla\psi\rangle. \tag{4.10}$$

For a bounded potential, an analogous relation holds due to Friedrichs's inequality (see [22,23]).

As the kinetic energy for a small cavity Ω of scale λ is of the order λ^{-2} (see Section 3.3), this relation means that the mean value of the Coulomb potential (being of the order not more than $\sim\lambda^{-1}$) or of any bounded potential is much less than the kinetic energy. Hence, for the systems studied, one may consider a simple "particle in a box" problem and the potential may be considered as a perturbation, being small enough in comparison to the energy level differences ($\sim\lambda^{-2}$). This is why, for any real potential, one may consider the free particle instead of the more complex problems for small cavities. The estimates of Section 3.3 give alternative arguments for such constructions.

For example, for small enough sphere radius R, the lowest states of the hydrogen atom may coincide in their order with the order of free particle levels in a D-dimensional spherical box. For $D = 2$ or $D = 3$ the ordering of the states is as follows, when not more than h-states ($\ell = 5$) are considered:

$D = 2$: 1s 1p 1d 2s 1f 2p 1g 2d 3s 1h 2f 3p 2g 3d 4s 2h 3f 4p 3g 4d
5s 3h 4f 5p 4g 5d 4h 5f 5g 5h;

$D = 3$: 1s 1p 1d 2s 1f 2p 1g 2d 1h 3s 2f 3p 2g 3d 4s 2h 3f 4p 3g 4d
5s 3h 4f 5p 4g 5d 4h 5f 5g 5h.

Note the state ordering is slightly different for R^2 and R^3 (see the 3s and 1h states).

Perturbation theory can also be used to describe the splitting of electron energy levels for a hydrogen atom under small shifts from the center of an impenetrable spherical cavity. For example, it is demonstrated in [58] that first-order perturbation theory shows that for any s-state, the position of the atom in the center corresponds to a minimum. For the states with $\ell \geq 1$ the situation is more complex. For the system with shifted nuclei, only m is a normal quantum number. In this case when $|m|$ is small, namely for $3m^2 < \ell(\ell + 1)$, the center corresponds to a maximum (this is the case of $2p_\sigma$-state, for example), while for large $|m|$ values the center corresponds to a minimum (as for $2p_\pi$-states). Similar constructions are used in [37] for hydrogen in a cylinder.

4.3. Dirichlet problem for large cavities

The Dirichlet boundary value problem for a large scale region $\Omega(\lambda)$ has special interest for our discussions, as it seems to be a reasonable approximation to the free problem. To some extent, this is a simple problem, when some properties of $\Omega(\lambda)$ are supposed to be satisfied. We suppose that enlargement of λ means extension of $\Omega(\lambda)$ and any point of R^3 belongs to some $\Omega(\lambda)$ for a large enough λ values. One may also suppose that for any λ the distance between boundaries of $\Omega(\lambda)$ and $\Omega(\lambda + \delta)$ is not less then $K\delta$ for any $\delta > 0$ and some constant K.

Here we use the standard Weyl constructions, described for the spectral theory of molecules, for example in [59]. For more modern constructions see [5,55]. For example, one may use, for the free eigenfunction ψ_j of the state with the energy E_j, the modified function $f\psi_j$ with the cutoff function $f(\mathbf{r})$, that is equal to 1 for some large region $\Omega(\lambda)$ and vanishes outside a neighborhood of this region, $\Omega(\lambda + \delta)$. In this case

$$\begin{aligned} \|(\hat{H} - E_j) f \psi_j\| &= \|f(\hat{H} - E_j)\psi_j - \psi_j \frac{1}{2}\Delta f - (\nabla\psi_j, \nabla f)\| \\ &\leq \frac{1}{2}\|\psi_j \Delta f\| + \|(\nabla\psi_j, \nabla f)\|. \end{aligned} \tag{4.11}$$

As derivatives of the cutoff function f differ from zero only in the external neighborhood of $\partial\Omega(\lambda)$, one may choose, for large enough δ values, the function f with small derivatives. As $\|\psi_j\| = 1$ and $\|\nabla\psi_j\|$ is estimated by the kinetic energy, for any $\varepsilon > 0$ one may choose δ, such that $\|(\hat{H} - E_j)\varphi_j\| \leq \varepsilon$, where φ_j is a normalized function $f\psi_j$. In this case, there exists some eigenvalue $E(\lambda + \delta)$, for the Dirichlet problem in the region $\Omega(\lambda + \delta)$, being close to $E_{j.}$. That is, $|E_j - E(\lambda+\delta)| \leq \varepsilon$ (otherwise, it follows from the spectral decomposition for the Dirichlet problem in $\Omega(\lambda + \delta)$ that $\|(\hat{H} - E_j)\varphi_j\| > \varepsilon$). Hence, the eigenspectrum for a free problem corresponds to some of the limiting points of the spectrum for a boundary value problem in a large region. One may find analyses of similar problems not only for energy, but for some other properties, in [60–63] and the references therein.

The discussion presented also proves that the wavefunctions and their gradients converge in the limit $\lambda \to \infty$ to the free functions and gradients in L_2-sense.

Of course, one can prove the same statement for other types of boundaries (1.1). Really, it is sufficient to modify $f(\mathbf{r})$ locally, near the boundary, in the region where ψ is almost constant. But, though it seems strange, for some types of boundary conditions (1.1) there exist energy levels $E(\lambda)$ that converge for $\lambda \to \infty$ to some limiting value E_∞, such that this is not the energy of the free system. The nature of this strange situation will clarified in Section 5.3.

Nevertheless, this is impossible for Dirichlet boundaries. Really, the energy functional, Equation (2.1), has the same form both for the free and the Dirichlet problem and its value for a function with support in $\Omega(\lambda)$ estimates the functional value for a free problem (one shall continue these functions to a whole space with zero values). Hence, the minimax principle (see [5], Sect. XIII.1) excludes the situation when there is any other limiting point for the sequence of energies $E(\lambda)$ for $\lambda \to \infty$. The same is evident by the use of Katriel's trick [64].

4.4. The case of a large sphere. An upper bound

It is possible to estimate the rate of convergence of the $E(\lambda)$ values for Dirichlet or other problems to the free system limits when one considers a spherically symmetric problem. It is interesting, as for any region one may estimate the energy by Equation (2.4) for inscribed and circumscribed spheres. In this case one may use as λ the radius R of the cavity. For crude estimates, one may use the Kirkwood–Buckingham relation in the form of Equation (3.18) to find the energy-shift $\Delta E(R) = E(\lambda) - E$ [65]. For example, for the exact eigenfunction ψ of a free problem and a cutoff function $f(\mathbf{r})$, that changes from 1 to 0 in the region $\Omega(R) \setminus \Omega(R_0)$ only for some large fixed R_0, one can see from Equation (3.18), that as ψ decreases at infinity at least exponentially, the $\Delta E(R)$ value also decreases exponentially and one of the upper bounds for the main part of asymptotic value is expressed through the radial part of the free wavefunction

$$\Delta E_{n\ell}(R) \leq \frac{1}{2}\langle\psi|(\nabla f^2)|\psi\rangle = \frac{1}{2}\int_{R_0}^{R} R_{n\ell}(r)^2|\partial f(r)|^2 r^2 \mathrm{d}r. \tag{4.12}$$

This is evident, at least, for the ground, ($n = \ell + 1$) state of a given symmetry type, but one may state it for the general case too (see [65]). The Euler equation that minimizes the integral in relation (4.12) with respect to variation in f immediately gives for some constant D

$$r^2 R_{n\ell}(r)^2 \partial f(r) = -D, \tag{4.13}$$

and the boundary conditions $f(R_0) = 1$ and $f(R) = 0$ means (one may suppose radial functions do not change their sign for $r > R_0$ and large enough R_0)

$$D = \left(\int_{R_0}^{R} [s^2 R_{n\ell}(s)^2]^{-1} \mathrm{d}s\right)^{-1}. \tag{4.14}$$

It is evident in this case that the optimal form of the inequality (4.12) is

$$\Delta E(R) \leq \frac{1}{2}\left(\int_{R_0}^{R} [s^2 R_{n\ell}(s)^2]^{-1} \mathrm{d}s\right)^{-1} = \frac{1}{2}D. \tag{4.15}$$

If the asymptotic value is considered, one can see that for the radial functions $\varphi(r) = rR_{n\ell}(r)$ (see Section 2.2) some simple relations hold. For example, for the usual potential functions $V(r)$ (in particular, the harmonic oscillator or hydrogen atom may be considered) one may suppose $(V - E)^{-1}\partial V \to 0$ for $r \to \infty$, while $|\varphi/\partial\varphi|$ is bounded. In this case the use of l'Hôpital's rule shows, that $\varphi\partial^2\varphi/(\partial\varphi)^2 \to 1$ as $r \to \infty$. Once more, the use of l'Hôpital's rule gives $\varphi(R)\partial\varphi(R)/D(R) \to -1/2$ for $R \to \infty$ and one may prove this relation be asymptotically correct. This gives the asymptotic estimate

$$\Delta E(R) \leq -\varphi(R)\partial\varphi(R) \tag{4.16}$$

for the Dirichlet problem [65].

4.5. The case of a large sphere

Now one may consider a more general situation by the method, described for one-dimensional problems by Hull and Julius [66]. Let $\varphi(r)$ be a normalized eigenfunction of the free Hamiltonian, with the eigenvalue E, that fulfills the condition $\varphi(0) = 0$ and is square integrable on $[0, \infty]$. The second solution of the differential Schrödinger equation for the same energy is denoted as $u(r)$. One may suppose $\varphi\partial u - u\partial\varphi = 1$ and this first order equation with respect to u has an evident solution for any known $\varphi(r)$, at least for the region where φ has no zero value. For large enough r values, one may suppose φ to be positive. Hence, for some constant c and some $a > 0$ sufficiently large, one may put

$$u(r) = \varphi(r)\left[c + \int_a^r \varphi(s)^{-2}\mathrm{d}s\right]. \tag{4.17}$$

Note that the integral in this relation is similar to the integral in Equation (4.14). It means, that for large r values, one may use the estimates from the end of Section 5.4, and asymptotically $u(r)$ behaves like $[\partial\varphi(r)]^{-1}$, that is, it increases monotonically, at least exponentially (see the end of preceding section).

The solution of the Schrödinger equation is reduced to solution of the integral equation, well known in the theory of Green's functions:

$$\chi(r) = \varphi(r) - 2\Delta E \int_0^r [u(r)\varphi(s) - \varphi(r)u(s)]\chi(s)\mathrm{d}s, \tag{4.18}$$

where ΔE is some parameter. Note, that the constant c from relation (4.17) is unessential for this equation and one may put $c = 0$. Direct calculation demonstrates χ is a regular solution of the Schrödinger equation with energy $E + \Delta E$, such that $\chi(0) = 0$ and $\partial\chi(0) = \partial\varphi(0)$. Of course, one may employ any boundary condition for χ at the point $r = R$ by the choice of the ΔE value. The original paper by Hull and Julius [66] solved a number of Dirichlet problems, but one may also consider a general boundary condition of the type $\partial\chi(R) = p\chi(R)$ (the Dirichlet boundary corresponds formally to the limiting case $p \to \infty$) and find the correction $\Delta E^{(p)}(R)$. Note that, due to relations (2.3), one may consider for the case $R \to \infty$ both the values of P and p as being equivalent.

As was shown in Sections 4.3 and 4.4, $\Delta E \to 0$ and $\chi \to \varphi$ (in L_2 sense) for large R values. Hence one may replace χ with φ in the integral relations. In particular, it follows from Equations (4.17) and (4.18) that for some constant C one may write:

$$\begin{aligned}\Delta E^{(p)}(R) &= \frac{1}{2}\left[\frac{\partial u(R) - pu(R)}{\partial\varphi(R) - p\varphi(R)}\int_0^R \varphi(s)\chi(s)\mathrm{d}s - \int_0^R u(s)\chi(s)\mathrm{d}s\right]^{-1} \\ &\approx \frac{1}{2}\left[\frac{\partial u(R) - pu(R)}{\partial\varphi(R) - p\varphi(R)} - C - \int_a^R u(s)\varphi(s)\mathrm{d}s\right]^{-1}.\end{aligned} \tag{4.19}$$

In particular, for the Dirichlet problem, one can use $u(R)/\varphi(R)$ as the first term in brackets in Equation (4.19). For the problems under our consideration, the relation u/φ increases, at least exponentially, for large R. This means, that one may ignore any contributions that are polynomial in R. For example, the integral (remember, $c = 0$)

$$\int_a^R u(s)\varphi(s)\mathrm{d}s = \int_a^R \int_a^s \varphi(s)^2\varphi(t)^{-2}\mathrm{d}t\mathrm{d}s \tag{4.20}$$

may be omitted, as $\varphi(s)/\varphi(t) < 1$ for $s > t$ and this integral is not greater than $\sim R^2$. Such terms may be ignored in comparison to the exponentially large terms. As a result, for the Dirichlet problem [66] the shift is

$$\Delta E^{(D)}(R) \approx \frac{1}{2}\frac{\varphi(R)}{u(R)} \approx \frac{1}{2}\left(\int_a^R \varphi(s)^{-2}\mathrm{d}s\right)^{-1}, \tag{4.21}$$

and this expression coincides with the approximate relation (4.15). For the general case

$$\Delta E^{(p)}(R) \approx \frac{1}{2}\left[\frac{\partial u(R) - pu(R)}{\partial\varphi(R) - p\varphi(R)}\right]^{-1} = \frac{p - \partial\ln\varphi(R)}{p - \partial\ln u(R)}\Delta E^{(D)}(R). \tag{4.22}$$

One may estimate $\Delta E^{(D)}$ asymptotically as $-\varphi(R)\partial\varphi(R)$. The use of Equation (4.17) for calculation of $\partial\ln[u(R)]$ allows us also to use for hydrogenic functions the approximation $\varphi(r) \sim Cr^k\exp(-\zeta r)$ (of course, only for the case when p differs from ζ) in order to find

$$\Delta E^{(p)}(R) \approx \frac{p+\zeta}{p-\zeta}\Delta E^{(D)}(R). \tag{4.23}$$

For the hydrogen atom this was shown in [29]. This statement is not correct for the harmonic oscillator when more accurate transformations of Equation (4.22) should be used.

According to Equation (4.23), the energy shifts for the Neumann and Dirichlet boundary conditions have opposite signs for large enough R. This condition is evident when we note that the systems with potential $V(r)$ for the region $[0, R]$ with wavefunctions that vanish at zero and fulfill the Dirichlet and Neumann conditions for the point $r = R$ may be discussed simultaneously as different states for the system on $[0, 2R]$ with a "reflected" potential for the region $[R, 2R]$. At the points $R+x$ and $R-x$ the values of the potential are the same. For the one-dimensional system with the symmetrical "double-well" potential, it is clear that all of the bounded states are ordered as pairs of even-odd states, generated from the "single-well" problem with the energy value E and the wavefunction $\varphi(r)$. The ground state is even and corresponds to a zero derivative at the point $r = R$. The odd states correspond to the Dirichlet condition at the symmetry center. The splitting of the states is described in a quasi-classical approximation in Landau and Lifshitz' classic book [38] (Ch.VII, Section 52), where it is demonstrated that the 2×2 eigenvalue problem gives the levels $E \pm \varphi(R)\partial\varphi(R)$ (the upper state corresponds to the Dirichlet problem and the lower state to the Neumann problem). Compare this result with relation (4.23). This explanation was proposed by Mikhailova [67]. An analogous construction was considered for a similar, but much more complex problem in [68].

4.6. Some historical notes

The asymptotic behavior of $\Delta E^{(D)}(R)$ and the rate of convergence to the corresponding "free" limit has been the subject of a large number of publications. The first estimates of the initial terms for the asymptotic wavefunctions of the 1s state of the hydrogen atom confined at the center of a sphere of radius R were made in Michel's paper [11] and a number of polynomial and rational approximations for energy were derived in [12]. In [14] the ideas of these papers were used for construction of asymptotic power series for the 1s, 2s and 2p states and energy calculations in a wide range of R values. The method used in all the papers mentioned was based on the known radial functions for the hydrogen atom, and the power-series solution of the equation for the cutoff function f in relation (3.21). The generalization of this procedure was made by Wilcox [69], who calculated the series corrections to the free atom energy by one step of the Newton's procedure for the roots of the confluent hypergeometric function ${}_1F_1(\alpha, \gamma, x)$. For example, for the 1s state of the hydrogen atom, one may derive the relation

$$\Delta E_{1s}^{(D)}(R) = \left[\sum_{k=1}^{\infty} \frac{(2R)^k}{k(k+1)!} \right]^{-1}. \tag{4.24}$$

It should be noted, that $R_{1s}(r) = 2e^{-r}$ and the replacement $k \to (k+2)$ in the denominator of the expression in brackets in Equation (4.24) gives relation (4.16). The same result was calculated by the search for asymptotics to the roots of the confluent hypergeometric function in [41] (the same method was applied in [29] to derive relation (4.23)).

The idea of using functions of the form $f\psi$ for approximate solutions of the boundary value problems has been used in a large number of papers. For example, this idea was used in [70] for the one-dimensional Schrödinger equation to prove the Hull–Julius relation (4.21) and to study the confined H_2^+ molecule. The direct use of the Kirkwood–Buckingham relation for variational calculations of the hydrogen atom in a half space was employed in [71]. A different way to organize the variational calculation as a system of iterations for f and ΔE approximations in the spirit of relations (4.18) and (4.21) was developed by Gorecki and Byers-Brown in [72]. This method was generalized as a method of solution of the Euler equations for any system (in particular, for a system in an external field) in [73]. In independent work [74] for a cutoff function f that defines the solution of a one-dimensional equation, the integro-differential equation was derived. The iterative solution of this problem results in relations similar to (4.21) for confined systems on R^1 and in the half-space. Here we note the interesting analytical study of the hydrogen problem in a half-space for the case of a finite potential barrier [75,76].

It is interesting that the equation derived by substitution of the function $f\psi$ into the Schrödinger equation has the form

$$\mathrm{div}[\psi^2 \nabla f] = \varepsilon \psi^2 f, \quad f|_{\partial\Omega} = 0. \tag{4.25}$$

Of course, this is the Euler equation for the energy functional of the form $\langle\psi|\nabla f^2|\psi\rangle$ with the additional normalization condition $\langle\psi|f^2|\psi\rangle = 1$. This is the problem that is solved by Equations (4.13)–(4.15) to find the leading term of the asymptotic relations for ΔE in Sections 4.4 and 4.5. It should be remembered that as a tool to exclude the potential function from all equations, this relation was proposed first in Fröhlich's paper [10].

To end this section, we note that the use of asymptotics is possible for large enough R values, but the choice of what is a "large" value is defined by a number of factors. For example, as accurate data on energy shifts are known for the hydrogen atom in a spherical cavity (see e.g. [77]), it is possible to compare different estimates for large R energy shifts. As was noted before, there are two essentially different types of estimates, namely, the ones given in [69] and compact relations like (4.16). The practical use of such relations is possible for relatively large R values ($\sim$10 au) for lowest states. For example, the energy shift is of the order 10^{-6} au for the 1s state and near 10^{-2} for 2s state. The relative error in the shift is less then -0.002% for the method of [69]. The cruder asymptotic formula (4.16) with derivatives gives an error near 11%. For other states at the same R value, the errors of approximations are much greater (-26, $+11\%$ (2s) and -17, $+23\%$ (2p) for the methods mentioned, respectively). In all cases, for large R values, the errors are of opposite signs. It is interesting to note that if one uses φ/n instead of $-\partial\varphi$ in these relations, the error is much greater, at least a factor of two greater for $R \sim 10$ a.u.

4.7. State ordering in spherically symmetric problems

The asymptotic relation (4.21) is interesting for the case of a large spherically symmetric cavity, where some states of different angular momentum are degenerate for a free system, such as the hydrogen atom. For the hydrogen atom, the normalized radial functions for (n, ℓ) states with the same n value have the asymptotic form $C_{n,\ell} g_n(r)$, where $g_n(r)$ does not depend on the angular momentum. It is clear that the energy shift for large R is proportional to the $C^2_{n,\ell}$ value. For the hydrogen (n, ℓ) state (see [38], Sect.V, Section 36) $g_n(r) = r^{n-1} \exp(-r/n)$ and

$$C^2_{n,\ell} = \frac{2^{2n}}{n^{2(n+1)}(n+\ell)!(n-\ell-1)!}. \tag{4.26}$$

As the ratio of $C^2_{n,\ell}$ to $C^2_{n,\ell-1}$ is less than 1, one has a simple rule: increasing angular momentum results in decreasing of the absolute value of the energy shift $\Delta E^{(D)}_{n\ell}(R)$. Hence, for large R values the state ordering is

$$1\text{s}, 2\text{p}, 2\text{s}, 3\text{d}, 3\text{p}, 3\text{s}, 4\text{f}, 4\text{d}, 4\text{p}, 4\text{s} \ldots. \tag{4.27}$$

It is useful to compare this ordering with the one for the limit $R \to 0$ (see Section 4.2). This idea was used in [41] to analyze the degeneracy of the states for some R values on the basis of properties of confluent hypergeometric functions.

There is a general tendency for degenerate states of the free problem: confinement results in a greater change in states with smaller ℓ than in states with large ℓ. For example, one can demonstrate a similar shift pattern for the isotropic harmonic oscillator which will be taken up in Section 7.

It should also be noted that for general type boundary conditions $\partial\varphi = p\varphi$, the state order is similar to that defined by Dirichlet boundaries. One may note [29] that the functions $E^{(p)}_k(R)$ for a given angular momentum have no common points for $p \neq p'$, and for any finite R value

$$E^{(p)}_k(R) \neq E^{(p')}_{k'}(R), \tag{4.28}$$

as there is only one normalized, bounded radial function, such that $\varphi(0) = 0$ for a given energy and momentum. On the other hand, it easy to find (for example, it is evident from the Hellman–Feynman theorem for the functional of the form of Equation (2.1), see the proof in [29]), that

$$\partial_p E^{(p)}_k(R) = -\frac{1}{2}\varphi(R)^2. \tag{4.29}$$

Hence, the energy values $E^{(p)}_k(R)$ decrease with increasing p. One may also analyze the relation $\partial\varphi(r)/\varphi(r)$ near the point $r = R$ for the Dirichlet problem solution with energy $E^{(D)}_k(R)$. As k enumerates the number of

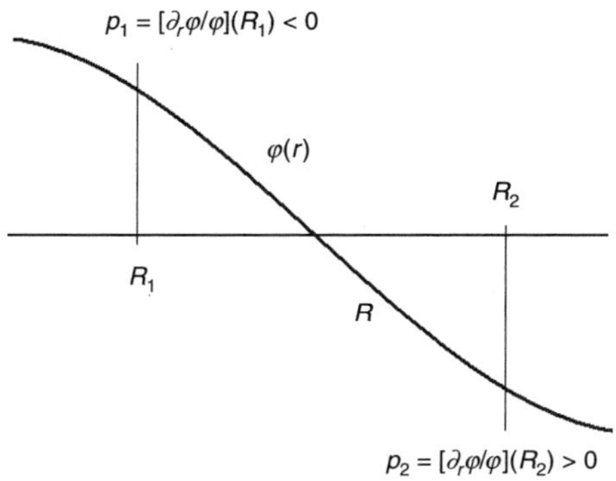

Figure 1 Explanation of inequalities Equation (4.30). For solution of the Schrödinger equation $\varphi(r)$ with the energy $E_k^{(D)}(R)$ for $R_1 < R < R_2$ near the point R, it follows from Equation (2.4) that $E_k^{(D)}(R_1) > E_k^{(D)}(R) = E_k^{(p_1)}(R_1)$. For the interval $[0, R_2]$ φ has an additional root at the point R and $E_k^{(D)}(R) = E_{k+1}^{(p_2)}(R_2) > E_k^{(D)}(R_2)$. Equation (4.30) follows from (4.28) and (4.29).

nontrivial roots of the wavefunction for radial problems, one may use the monotone behavior of the Dirichlet energies and relations (4.28) to prove the inequalities:

$$\begin{aligned} E_k^{(D)}(R) &> E_k^{(p)}(R) \quad \text{for } p < 0, \\ E_k^{(D)}(R) &< E_{k+1}^{(p)}(R) \quad \text{for } p > 0. \end{aligned} \tag{4.30}$$

These inequalities have been graphically displayed in Figure 1. It is clear that the Dirichlet problem separates the solutions for any p value. When the monotonic energy function is bounded (see Equation (4.30)), there exists a corresponding limit for the monotone energy function for $p \to \pm\infty$, and this limit corresponds to the Dirichlet problem. The general picture of state ordering is defined by the relations described and the limits $R \to \infty$. Note that some problems exist with the limit $R \to \infty$ (see below, Section 5.3) [29]. The same methods may be used to analyze the boundary conditions (1.1) at the origin for the harmonic oscillator [78].

5. THE SYSTEM IN AN EXTERNAL POTENTIAL

It is natural to use Dirichlet boundary conditions for situations where the potential function is large in comparison with the energy values and the probability of tunneling into the classically forbidden region is small. It is interesting to analyze this situation in detail. We consider the problem by simple, physically evident methods. One may use a wide variety of external potentials for physical problems. This is why a number of essentially different problems are considered here. Beyond that, we specifically analyze some problems to simplify the discussion, though different effects are simultaneously realized in practice.

5.1. The case of a large external potential

Let us consider the complement $\Omega^c = R^3 \setminus \Omega$ of some region Ω and its characteristic function $W(\mathbf{r})$, such that $W(\mathbf{r}) = 0$ for $\mathbf{r} \in \Omega$ and $W(\mathbf{r}) = 1$ for $\mathbf{r} \in \Omega^c$. One can analyze the spectrum and eigenfunctions for a family of Hamiltonians $\hat{H}_\beta = \hat{H} + \beta \hat{W}$. We use here the simple method of [54]. The detailed mathematical analysis is given in [79]. For the ground state energies E_β and normalized wavefunctions ψ_β we use the Hellmann–Feynman theorem

$$\partial_\beta E_\beta = \langle \psi_\beta | W | \psi_\beta \rangle = \langle \psi_\beta | \psi_\beta \rangle_{\Omega^c}. \tag{5.1}$$

The energy derivative is nonnegative and energy increases monotonically with respect to β. Note that continuation of the Dirichlet boundary value ground state wave function by zero value on R^3 gives rise to estimating the energy functional for any β from the Dirichlet ground state energy $E(\Omega)$ and, in accordance with the variational principle, $E_\beta \leq E(\Omega)$. Hence, when $\beta \to \infty$ there exists a limiting point E_∞ for E_β and $E_\infty \leq E(\Omega)$. As the derivatives $\partial_\beta E_\beta$ converge to zero,

$$\langle \psi_\beta | W | \psi_\beta \rangle = \langle \psi_\beta | \psi_\beta \rangle_{\Omega^c} \to 0 \quad \text{as } \beta \to \infty, \tag{5.2}$$

that is, the large potential "squeezes out" the electron density from the classically forbidden region.

Let Ω_δ be a δ-neighborhood in the region Ω and $f(\mathbf{r})$ be a smooth function, such that $0 \leq f(\mathbf{r}) \leq 1$, $f(\mathbf{r}) = 1$ for $\mathbf{r} \in \Omega$ and $f(\mathbf{r}) = 0$ outside Ω_δ. The relation (3.18) for wavefunction ψ_β means for some constant C that does not depend on β, the inequality

$$E(\Omega_\delta) \leq E_\beta + \frac{C \langle \psi_\beta | W | \psi_\beta \rangle}{1 - \langle \psi_\beta | W | \psi_\beta \rangle}, \tag{5.3}$$

holds. For $\beta \to \infty$ this means $E(\Omega_\delta) \leq E_\infty$ due to Equation (5.2). For δ small enough, the energy values $E(\Omega)$ and $E(\Omega_\delta)$ are close within any desired accuracy. As these energies are upper and lower bounds for E_∞, one may state $E_\infty = E(\Omega)$ in the limit $\delta \to 0$. Hence, in accordance with our physical intuition

$$E(\Omega) = \lim_{\beta \to \infty} E_\beta. \tag{5.4}$$

Katriel's trick [64] gives rise to the generalization of the relation (5.4) for any excited state, while Eckart's estimate [80] means that analogous relations hold for the wavefunctions.

Of course, the same statement is correct for any continuous function $W(r)$, being positive for Ω^c and equal *to zero at* Ω. The detailed analysis of the problem with the rate estimates of the difference $E(\Omega) - E_\beta$ is given in [79].

5.2. The case of separated wells

There is another physically interesting situation that should be considered. Let us consider the Hamiltonian $\hat{H}$ with eigenvalues E_j and eigenfunctions ψ_j, localized in a compact region, say Ω_H, at least for the states with energy values being less then some negative constant. One may suppose the ψ_j to be exponentially small functions outside Ω_H.

Let the potential $W(\mathbf{r})$ be a non-positive function, localized mainly in a compact region Ω^* and disappearing at infinity. For simplicity, one may suppose W to be a function with a compact support Ω^*, that is, $W(\mathbf{r}) = 0$ for $\mathbf{r} \notin \Omega^*$. When the volume of a well is large enough, there are some bounded states for the Hamiltonian operator $\hat{h}_W$ of the form $\hat{h}_W = -\frac{1}{2}\Delta + W(\mathbf{r})$.

One may consider a family of Hamiltonians $\{\hat{H}_\alpha\}$, where $\hat{H}_\alpha = \hat{H} + W(\mathbf{r} + \alpha\mathbf{m})$ for some fixed vector $\mathbf{m}$. Let us consider the case where $\alpha \to \infty$. It is intuitively clear, that for large α the discrete spectrum of the problem consists of the points of the discrete spectra of the operators $\hat{H}$ and $\hat{h}_W$ only. This is easy to prove by calculation of the norms $\|(\hat{H}_\alpha - E_j)f\psi_j\|$ for a function $f(\mathbf{r})$, equal to 1 for some large region $\Omega \supset \Omega_H$ and vanishing outside a δ-neighborhood of Ω, such that this neighborhood does not intersect with the points of the translated well defined by $W(\mathbf{r}+\alpha\mathbf{m})$. We repeat the discussions of Section 4.3 (see the text near Equation (4.11)) to note that for large enough α, one may choose f such that $\|(\hat{H}_\alpha - E_j)f\psi_j\|$ is less then any desired small number. This is why E_j lies near some points of the spectrum of $\hat{H}_\alpha$. The same is true for the spectrum of $\hat{h}_W$.

On the other hand, one may use a similar argument to prove an inverse statement. For any sequence $\{E_\alpha\}$ of eigenvalues of $\{\hat{H}_\alpha\}$ and corresponding wavefunctions $\{\psi_\alpha\}$, one may suppose $E_\alpha \to E_\infty$ for $\alpha \to \infty$. One may divide the space into two half-spaces associated with the region Ω_H and the moved region Ω^*. Any function ψ_α is localized mainly (with the norm being not less then $1/2$) in one of these half-spaces. It is then easy to find, for example, for a sub-sequence of functions $\{\psi_\alpha\}$, localized mainly near Ω_H, the cutoff functions f_α such that

$$\|(\hat{H}_\alpha - E_\alpha)f_\alpha\psi_\alpha\| = \|(\hat{H} - E_\alpha)f_\alpha\psi_\alpha\| \to 0 \quad \text{for } \alpha \to \infty, \tag{5.5}$$

where, for example, $\|f_\alpha\psi_\alpha\| > 1/4$. This statement immediately means that the normalized functions $f_\alpha\psi_\alpha$ converge to an eigenfunction of $\hat{H}$ with the eigenvalue E_∞. The alternative case corresponds to the points of the $\hat{h}_W$-spectrum. (For a more modern treatment and interesting problems on systems of separated wells see [55], Sect. 3).

5.3. Some physical realizations

The outline presented in Section 5.2 explains why one can be sure that the spectrum of $\hat{H}_\alpha$, for large enough α, consists (in the negative region of energies) of the points near the spectrum $\hat{h}_W$ or near the spectrum of $\hat{H}$. In practice, the movement of the wall, described by a potential $W(\mathbf{r})$, is also associated with additional modifications to the potential, but mainly in a more or less localized region. As an illustration, the example described may be realized as the physically interesting problem of an atomic system in a cavity with polarizable walls. That is, in an empty region of a polarizable medium. The method of potential evaluation for a charge in a polarized spherical cavity is well known (see e.g. the review in [81,82] and refs. therein) and it was known to Kirkwood [83]. For example, for a hydrogen-like atom in a sphere of radius R, one can find the total potential of the electron-sphere interaction in the form

$$W(r,R) = -\sum_{\ell=0}^{\infty} \frac{(\varepsilon-1)(\ell+1)}{[\varepsilon(\ell+1)+\ell]} \left[\frac{r^{2\ell} + Z^2 a^{2\ell} - 2Za^\ell r^\ell P_\ell(\cos\theta)}{2R^{2\ell+1}} \right], \tag{5.6}$$

where a is the distance from the nucleus of charge Z to the center, ε is the dielectric constant of the medium *around* the sphere, P_k is the k-th Legendre polynomial, and θ is the angle between the electron and nuclear vectors. In particular, when the nucleus is placed at the center, the essential part of this potential corresponds to the electron-metallic sphere interaction (the metallic limit corresponds formally to $\varepsilon = \infty$):

$$W_m(r,R) = -\frac{R}{2(R^2 - r^2)}. \tag{5.7}$$

It is clear that the potential W_m for $r = R - x$ and small x behaves like $-1/(4x)$. This means that when the problem "atom in a polarized sphere" is solved, one has to use the Dirichlet boundary condition for the atomic wavefunctions at the sphere. This feature does not disappear for any finite ε, and this is one of the reasons why the continuum model is not widely used for molecular systems [81].

As was demonstrated, when the radius R of the sphere is enlarged, one may consider the "free" hydrogen atom problem and the one with the Hamiltonian

$$\hat{h}_W = -\frac{1}{2}\Delta + W_m(\mathbf{r},R) \approx -\frac{1}{2}\Delta - \frac{1}{4(R-r)}. \tag{5.8}$$

The localization region for large R values is far from the origin and one may use the same arguments, as it was done before for the radial problem, to study the system.

Note that the problem with the Hamiltonian (5.8) for large R values can be solved in spherical coordinates. If one uses the radial functions, normalized with weight 1, one can change variables $r \to x = R - r$. Then the radial potential for a state with angular momentum ℓ is

$$\frac{\ell(\ell+1)}{2(R-x)^2} - \frac{1}{4x}. \tag{5.9}$$

In the limit $R \to \infty$ the states with any angular momentum have approximately the same energy as for s-states and the levels are $-1/32n^2$ ($n = 1, 2, \ldots$). Crudely speaking, for a large sphere the inertial moment is too large and when the speed of a particle is not too large, the rotational component of energy vanishes. The other interpretation is that for large R, the problem is practically the same as for an electron near a metallic plane.

The electron in a metallic sphere (and the problem of the electron outside a metallic sphere) requires Dirichlet boundary conditions on the sphere. These problems were analyzed, for example in [84]. In particular, it was noted, that the Rydberg structure of the electron spectrum near a metallic drop can be observed experimentally.

For our discussion, it is important, that the spectrum of an atom in a large metallic sphere consists of two almost independent Rydberg-type systems.

Note also, that when the δ-type potential is used for the radial problem at the point $r = R$, namely, $\gamma\delta\,(R - r)$, there are practically no interesting details for the case of the repulsive potential $\gamma > 0$. In this case, no new bound states appear in the presence of the additional potential. But when $\gamma < 0$, there is only one point in the $\hat{h}_W$ discrete spectrum, namely, $-\gamma^2/2$, due to the delta-function potential. Hence, in the limit of large R values, there exists an additional limiting energy value equal to $-\gamma^2/2$. This limiting point is the same for all the possible moments. This was found when the hydrogen atom in a sphere was studied with a boundary condition of the general type $\partial_n\psi - p\psi|_{\partial\Omega} = 0$; in this case $\gamma = -p$ (see [29] for details). Note, that additional limits for energy levels at $R \to \infty$ exist for boundary value problems of the type (1.1c) for the case $P > 0$ only, that is, when the boundary condition describe the attractive walls.

5.4. Expanded regions

One more type of external potential, closely connected with the ones described in Section 5.3, is the stepwise potential. For example, one may consider an additional potential for the well localized in a sphere of radius α and analyze the case as $\alpha \to \infty$. The same construction may be analyzed as the problem of an additional spherical barrier outside the sphere of radius α.

The two potentials mentioned differ only by a constant, equal to the potential well depth near the barrier.

One may consider the function $w(x)$, that equals 0 for $x < 0$, $w(x) = U > 0$ for $x > d$, and $w(x)$ increases in the interval $(0, d)$. Then the external potential $W(\mathbf{r}, \alpha) = w(|\mathbf{r}| - \alpha)$ describes the extended free region. It is clear that the energy decreases with increasing α and for $\alpha \to \infty$ the energy levels converge to the free energy values. For small enough d, one can also consider the system for $\alpha \to 0$ as a slightly perturbed free system with the levels increased by U. Note that in this case the mean value of any operator is the same in both the limits $\alpha \to \infty$ and $\alpha \to 0$.

It is interesting that one of the possible realizations of the problem described corresponds to the situation when the potential of a system is defined as $V(\mathbf{r})$ for $|\mathbf{r}| < \alpha$ and a constant U for $|\mathbf{r}| \geq \alpha$. This situation (for the case, when $U > V(\mathbf{r})$ for any $\mathbf{r}$) is similar to the one described above, but for small α the potential well may be too small to have any bound states. This condition defines some critical value α_{crit} for the existence of the bounded states: there are no bound states for $\alpha < \alpha_{\text{crit}}$; see e.g. [85] for numerical examples.

5.5. The deep well

There is one more physically interesting situation. Let us consider the family $\left\{\hat{H}_\beta\right\}$ of operators $\hat{H}_\beta = \hat{H} + \beta w$ with a function $w(\mathbf{r})$ equal to 0 outside some compact region Ω_* and $w(\mathbf{r}) = -1$ for points inside Ω^*. We must consider the negative part of the Hamiltonian $\hat{H}$ spectrum. We suppose now that Ω^* and the region Ω_H of the localization for the discrete functions of $\hat{H}$ are well separated in the space.

For any $\beta > 0$ and discrete spectrum eigenvalues $E_k(\beta)$ of $\hat{H}_\beta$, one can see that the classically allowed region consists of two distinct sub-regions that belong either to Ω_H, or to Ω^*, and one may suppose the corresponding wavefunction $\psi_k(\beta)$ to be localized within one of these sub-regions. According to Equation (5.1), the energy derivative with respect to β equals

$$\partial_\beta E_k(\beta) = -(\psi_k(\beta), \psi_k(\beta))_{\Omega^*}. \tag{5.10}$$

It is clear, that there are two types of energy behavior with respect to β. For the states with wavefunctions similar to the eigenstates of $\hat{h}_\beta = -\frac{1}{2}\Delta + \beta w(\mathbf{r})$, one may suppose $\partial_\beta E_k(\beta) \approx -1$. But for the states, similar to the eigenstates of $\hat{H}$, that is localized mainly within Ω_H, one may suppose $\partial_\beta E_k(\beta) \approx 0$.

Hence, there is a set of almost constant energy levels (for eigenstates of $\hat{H}$) and a set of energy levels that monotonically decrease at a more or less constant rate (for eigenstates of $\hat{h}_\beta$). Sometimes the levels of both subsystems "intersect" and, as usual, a regular system of avoided crossings for the pairs of states of the same symmetry appears.

For increasing β values, one can find that the m-th eigenstate of $\hat{h}_\beta$ has a decreasing energy, but for some critical value, the localization region changes and is practically the j-th state of $\hat{H}$. This energy value practically does not change, but at a definite point (that is for some β) this energy coincides with the decreasing energy of $m+1$-th state for $\hat{h}_\beta$ and due to avoided crossings, the energy decreases for greater β values, and so on. Hence, the energy function $E_k(\beta)$ decreases in a ladder manner. Of course, similar effects can be found for localized functions $w(\mathbf{r})$ that change their sign.

The specific structure of the states for $\hat{H}_\beta$ was described in detail in [79], where it is mentioned as a well-known physical effect. For example, it was noted in the theory of disordered semiconductors that a similar "ladder" structure of states is realized for the system where the Coulomb potential is modified within a sphere as a constant potential (see [86,87] for a qualitative discussion and analytical solution of the problem). For quantum chemistry, the situation is interesting, as was shown in a series of publications of Connerade, Dolmatov and others (see e.g. [19,88–91]; note that the series of publications on confined many-electron systems by these authors is much wider). The picture described is realized to some extent for the effective potential of inner electrons in multi-electron atoms, as it is defined by orbital densities with a number of maximal points. The existence of a number of extrema generates a system of the type described above [89]. This situation was modeled and described for the one-electron atom in [88]; it is similar to that one described in Sections 5.2 and 5.3.

One must stress that in the discussion above in Section 5.1 and here, we have used the "rectangular" functions $W(\mathbf{r})$ and $w(\mathbf{r})$ of opposite sign. When these functions are constructed for the same region, one can see that $W = 1 + w$. It means that the difference in qualitative pictures described for different cases is associated with regions of localization of the external potentials, but not with their sign only, as it was proposed in [79].

At the end of this section, one should note, that the quantum mechanical system in a potential cavity of large size may also be described as a system of discrete spectrum states with energies being almost constant and a system of decreasing stationary states for an enlarged "potential box". It is well known that the system of typical "ladder" structures in the spectrum gives rise to resonant states of molecular systems (see e.g. [92,93] and references therein), but the resonant states (the poles of continuation in a complex plane for the resolvent matrix elements) are a special field of Quantum Science and we shall not consider them here (see [5], Sect. XII.6 for discussions, or [55], Sect. 3).

6. ON MEAN VALUES AND OTHER PROPERTIES OF CONFINED SYSTEMS

The study of mean values seems to be an almost trivial problem when one has a regular method to estimate the wavefunctions, but it is not as clear as it is usually supposed to be. For example, when the system of approximate wavefunctions converges to some limit, it is not easy to prove that the mean values converge to the mean value for the limiting function, as was demonstrated by Löwdin [94]. The problem is not evident even for energy estimates within the linear variation method (see theory and examples in [95,96]). For mean values in approximate procedures see [97]. Similar problems for the Dirichlet boundaries are studied in [60–63]. Some analytical statements on the mean value dependence on the size of the region are presented in [98], where the spherically symmetric Dirichlet problems are considered. We use some ideas from this work for more general situations and here consider changes in the density at the center and the mean values for some monotonic function $f(r)$. The statements given below may be proved on the basis of comparison theorems (see [33,34]), but this method seems to be too complex for practical use as it requires a large number of illustrations.

6.1. Density at the origin

There are a number of similar situations where the methods described here may be applied. To be more specific, let us consider some parameter η and the family of spherically symmetric problems for potentials $V(r, \eta)$, Hamiltonians $\hat{H}_\eta$, energies $E(\eta)$ and corresponding normalized radial wavefunctions $\varphi(r, \eta)$ with the boundary condition $\varphi(0, \eta) = 0$ at the center of a sphere of radius R. It follows from the Schrödinger equation that for the derivative $\partial_\eta \varphi$ of the wavefunction $\varphi(r, \eta)$ one may write the equation

$$[\hat{H}_\eta - E(\eta)]\partial_\eta \varphi + [\partial_\eta V - \partial_\eta E]\varphi = 0. \tag{6.1}$$

As $\varphi(0, \eta) = 0$ for any η, one has $\partial_\eta \varphi(0, \eta) = 0$ and integration "by parts" on the interval $[0, r]$ for the relation (6.1) multiplied by $\varphi(r, \eta)$ immediately gives the identity

$$\begin{aligned} &\partial_r \varphi(r, \eta)\partial_\eta \varphi(r, \eta) - \varphi(r, \eta)\partial_r \partial_\eta \varphi(r, \eta) \\ &+2\int_0^r [\partial_\eta V(s, \eta) - \partial_\eta E(\eta)]\varphi^2(s, \eta)\mathrm{d}s = 0, \end{aligned} \tag{6.2}$$

that was derived in [98] in a simplified form. One may check this relation by differentiation. Of course, for $r = R$ this relation reduces to the Hellmann–Feynman theorem, as the terms with derivatives of φ disappear for the boundary conditions (1.1). Later we use the symbol $G(r, \eta)$ for the integral term in this relation:

$$G(r, \eta) = \int_0^r [\partial_\eta V(s, \eta) - \partial_\eta E(\eta)]\varphi^2(s, \eta)\mathrm{d}s. \tag{6.3}$$

According to the normalization condition, one may write $\langle \varphi | \partial_\eta \varphi \rangle = 0$. It is important to note, that when $\eta = R$, this is true for the Dirichlet problem only, but this statement also holds for problems where R does not depend on η.

As $\langle \varphi | \partial_\eta \varphi \rangle = 0$, one may locate some point $0 < r^* < R$ where $\varphi \partial_\eta \varphi(r^*, \eta) = 0$. This is not a result of Equation (6.2), as it was supposed in [98], but it follows from the normalization condition only. Equation (6.2) means at the point $r = r^*$, if φ differs from zero at this point,

$$\varphi(r^*, \eta)\partial_r \partial_\eta \varphi(r^*, \eta) = 2G(r^*, \eta). \tag{6.4}$$

One may suppose the *ground state* function φ to be positive at the point $r^* > 0$, that is, the sign of $\partial_r \partial_\eta \varphi(r^*, \eta)$ equals that of $G(r^*, \eta)$.

All discussions below use the following idea. Let the function $G(r, \eta)$ *conserve* its sign for any internal point of the interval $[0, R]$, at least, for the ground state. It then follows from Equation (6.4) that there is exactly one point r^*, where the function $\varphi \partial_\eta \varphi$ changes sign. Hence, one may calculate the sign of $\varphi \partial_\eta \varphi$ near the origin. When $G(r, \eta)$ is positive, it follows from Equation (6.4) that $\partial_r \partial_\eta \varphi(r^*, \eta)$ is also positive and that near the origin $\varphi \partial_\eta \varphi$ is negative. If $G(r, \eta) < 0$, then $\varphi \partial_\eta \varphi > 0$ near the origin. Hence, the *constant* sign of G for any r values defines the opposite sign of $\partial_\eta [\varphi^2]$ near the origin for the ground state. This sign is interesting for us, as the electron density at the center for the Dirichlet problem differs from zero for radial s-functions and its value $\rho(0, \eta)$ may be estimated as $[\partial \varphi(0, \eta)]^2$, or as the limit of $[\varphi(r, \eta)/r]^2$ as $r \to 0$ (for simplicity we consider here the case of the three-dimensional problems only). It is clear that the sign of $\partial_\eta \rho(0, \eta)$ is that of $\partial_\eta [\varphi(r, \eta)^2]$ for small r. As a result,

$$G > 0 \text{ means } \partial_\eta \rho(0, \eta) < 0; \qquad G < 0 \text{ means } \partial_\eta \rho(0, \eta) > 0. \tag{6.5}$$

Of course, the same statement holds, for the lowest state with a given angular momentum, for the corresponding modifications of density (see Equation (6.13) below).

To illustrate the general relation (6.5), note that for the Dirichlet problem with $\eta = R$, the potential function does not depend on R. One may write (see e.g. Equation (2.4)) $\partial_R E(R) < 0$ and in agreement with Equation (6.3), the function G is nonnegative. Relation (6.5) means that for the smaller cavity the wavefunction is more localized than for the larger one. This physically evident statement was used in [99]. In [98], it was stated for monotonic potentials.

In general, one cannot insist on monotone behavior of the density at the origin for the Dirichlet problem. But as $R \to 0$ (see Section 4.2), the radial s-functions are practically those for a $3D$ box $(ncR^{-1/2}\sin(nr/R)/r)$. Hence, the densities at zero for the ns-state and 1s-state differ by the factor n^2. This gives an insight to suppose that for any s-state, the density at the nucleus decreases with increasing R at small R values. One may also observe a similar effect for the low-lying s-states of the hydrogen atom in a sphere of large radius.

6.2. Mean value of a monotonic function

For the situation described in Section 6.1, one may also insist that the integral

$$\langle f \rangle = \int_0^R f(s)\varphi(s,\eta)^2 \mathrm{d}s, \tag{6.6}$$

is a monotone function of the parameter η. It follows from the statement that, if a function vanishes at the ends of an interval and its derivative vanishes at not more then one internal point, then the function conserves its sign for this interval (otherwise there exists an internal zero for the function and its derivative vanishes at least twice). This statement may be used for the function

$$g(x) = \int_0^x 2\varphi(s,\eta)\partial_\eta\varphi(s,\eta)\mathrm{d}s, \tag{6.7}$$

that vanishes at $x = 0$ and $x = R$, as $g(R) = 2\langle\varphi|\partial_\eta\varphi\rangle = 0$. As was proved in Section 6.1, the function $\varphi\partial_\eta\varphi$ vanishes at some point x^* and for the ground state there exists only one such point when the sign of $G(r,\eta)$ does not alternate. Hence, the sign of $g(x)$ is constant and one may define it for small x as the sign of $\varphi\partial_\eta\varphi$. Note, that

$$\partial_\eta\langle f \rangle = 2\int_0^R f(s)\varphi(s,\eta)\partial_\eta\varphi(s,\eta)\mathrm{d}s = \int_0^R f(s)\mathrm{d}g(s) \tag{6.8}$$

and the use of "the second mean-value theorem" (see e.g. Theorem 6.32. from [100]) allows us to write for some internal point ξ in the interval $[0, R]$

$$\begin{aligned}\partial_\eta\langle f \rangle &= \int_0^R f(s)\mathrm{d}g(s) \\ &= f(0)[g(\xi) - g(0)] + f(R)[g(R) - g(\xi)] \\ &= [f(0) - f(R)]g(\xi)\end{aligned} \tag{6.9}$$

(we do not consider here trivial modifications needed for the case $R = \infty$). The sign of $f(0) - f(R)$ is defined by the monotone behavior of $f(x)$ and hence the sign of the derivative is expressed through the sign of $g(\xi)$, being the sign of $\varphi\partial_\eta\varphi$. According to Equation (6.5), for any monotonically *decreasing* $f(r)$

$$G > 0 \text{ means } \partial_\eta\langle f\rangle < 0; \qquad G < 0 \text{ means } \partial_\eta\langle f\rangle > 0; \tag{6.10}$$

when $f(x)$ increases, the mean value changes have opposite signs.

As was shown at the end of Section 6.1, for the Dirichlet problem, $G > 0$. Hence for any monotonically decreasing $f(r)$, the mean value $\langle f\rangle$ decreases with increasing R. This statement does not depend on the sign of the monotonic function, as was noted in [98] for the 1s-Dirichlet problem. This statement holds for the lowest state with any given angular momentum in a spherically symmetric problem. For example, the monotonically decreasing $\langle r^{-3}\rangle$ data for the Dirichlet problem are presented in [101] for 2p states in the sphere of enlarged radius R.

6.3. Some examples

One must stress, that the statements proved in Sections 6.1 and 6.2 are in general not correct for excited states. For example, let us consider the double-well, spherically symmetric potential. For two approximate functions u_1 and u_2 with approximately the same energy, localized in corresponding regions near the minima of the potential, one may write the functions for two lowest states as

$$\varphi_1 = \cos\alpha u_1 + \sin\alpha u_2, \qquad \varphi_2 = -\sin\alpha u_1 + \cos\alpha u_2 \tag{6.11}$$

for some parameter α. The corresponding values for density at the origin are determined mainly by one of the functions, say u_1, and may be estimated as $\rho_1 \sim \cos^2\alpha$, $\rho_2 \sim \sin^2\alpha$. Hence decreasing the density at the origin for the ground state results in an increase in the density for excited states and vice versa. The same holds for mean values of some decreasing function localized near the origin. The situation described here is realized for the Dirichlet problem, for example, for the lowest states for a simple rectangular potential. In Figure 2, we have shown the behavior of the density at the origin as a function of the confinement radius.

A wide range of situations, where statements (6.5) and (6.10) may be used can be found for the free problems (i.e. $R = \infty$) with some types of external potentials $V(r, \eta)$. The same methods may be used for external potentials in Dirichlet problems and other boundary value problems of the type Equation (1.1), when R does not depend on η. For example, let us suppose that the integral $G(r, \eta)$ (see Equation (6.3)) exists for any $r \in [0, R]$, G is a continuous

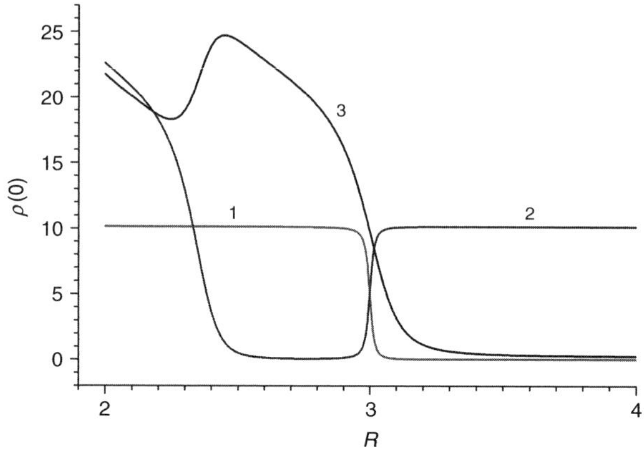

Figure 2 Density at origin $\rho(0)$ for low lying states of the Dirichlet problem vs. sphere radius R. Potential differs from zero for $r \in [1.0, 2.0]$ only, where $V(r) = 10.0$. Note the monotony of density for the ground state (1). The region $R \approx 2.2$ corresponds to avoided crossing of the states 2 and 3; the region near $R \approx 3.0$ is described by Equation (6.11).

function of r, differentiable for $r > 0$ and there exists not more then one point $r^{\#}$, where

$$\partial_\eta E(\eta) = \partial_\eta V(r^{\#}, \eta). \tag{6.12}$$

As was noted (see the text near Equation (6.2)) G vanishes at $r = R$, due to the Hellmann-Feynman theorem and $G(0, \eta) = 0$. The derivative of G vanishes for exactly one point $r^{\#}$ and hence, one may find $G(r, \eta)$ to conserve the sign for $r \in (0, R)$. This sign is defined by the neighborhood of the points $r = 0$ or $r = R$.

Examples of potentials with the properties mentioned may be found for $V(r, \eta) = W_0(r) + \eta W(r)$ with any monotonically increasing, continuous function W that vanishes near the origin and any fixed potential W_0. It is clear that the class described may be extended by some limits to similar potentials and one may use the potentials from Section 5.1 with the Heaviside step function. As E_η is a monotonically increasing function of η and $W(0) = 0$, near the origin $W(r) - \partial_\eta E_\eta \approx -\partial_\eta E_\eta < 0$. Hence, $G < 0$ and increasing the barrier (i.e. increasing η) results in an increase of density at the center and mean values for any monotonically decreasing functions $f(r)$.

This situation is depicted in Figure 3 for the 1s state of the hydrogen atom. Note that one may not insist on monotone density or mean values when the external potential has the form $W(r - R_0)$ and $\eta = R_0$. This may be formally explained as follows: when W is a smooth function (or an approximation to it) for the derivative on R_0 there are at least two solutions for Equation (6.12) and conservation of the sign for the function G is not guaranteed (see

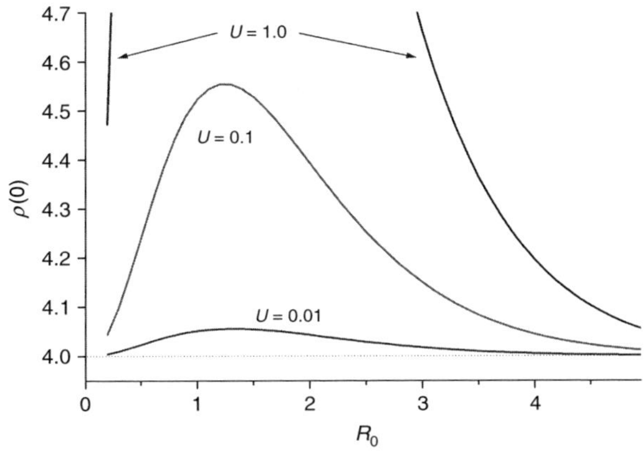

Figure 3 Density at origin for 1s-state of the hydrogen atom in R^3 with an external potential. The external potential has the form $Uw(r - R_0)$ with the Heaviside function w. Note the monotony of density with respect to U. According to Section 5.4 for $R_0 = 0$ and $R_0 = \infty$ wave functions are the same; $\rho(0) = 4.0$ for any U. The region of small R_0 values is not presented (the computational procedure is unstable in this region).

Figure 3). The situation where the bound states move into the continuum for small R_0 values and the density at the origin converges to zero, while it may be positive and decrease monotonically for large parameter values, is described in [85] (see also Section 5.4).

One further example. The problems in R^D for increasing dimension D (or angular momentum ℓ) for the lowest state of a given symmetry may be described by an additional external potential $q(q+1)r^{-2}$, where $q = \ell + (D-3)/2$. For small r values, one concludes that $G > 0$, and hence, for larger D (or larger ℓ) the wavefunctions near the origin decrease and so does any mean value of the monotonically decreasing function $f(r)$.

Nevertheless, analysis of the density at the origin is a much more interesting problem in this case. Density is now defined by the values $[r^{-(q+1)}\varphi(r)]^2$ at the point $r = 0$. But when q is varied, the changes in density are defined not only by the q-dependence of the wavefunction φ. As a result, there are situations where the density at the origin may be an increasing function and systems where it decreases with increasing D (or ℓ).

Very recently, the electronic density and its derivatives at the origin have been used [102] to obtain very accurate energy values of the spherically confined model systems. For the ground or excited state defined by the angular quantum number ℓ we define

$$\eta_\ell(r) = \frac{\rho(r)}{r^{2\ell}}, \tag{6.13}$$

where $\rho(r)$ denotes the spherically averaged density. We shall first consider the hydrogen-like atoms with any additional potential $U(r)$, bounded at the origin. The Kato cusp condition, involving the first derivative of density can be written as

$$\eta'_\ell(0) = -2\frac{Z}{\ell+1}\eta_\ell(0) \tag{6.14}$$

and does not depend on U (see [26]). The similar expression for the second derivative of density at the nucleus includes E as

$$\frac{\eta''(0)}{\eta(0)} = \frac{2}{2\ell+3}\left[\frac{Z^2}{(\ell+1)^2}(4\ell+5) + 2(U(0)-E)\right]. \tag{6.15}$$

This ratio for the confined hydrogen-like atoms can be used to define the critical radius, R_c, at which $E = 0$. Accordingly, at R_c, the *ratio* is exactly given by

$$\frac{\eta''(0)}{\eta(0)} = \frac{2}{2\ell+3}\left[\frac{Z^2}{(\ell+1)^2}(4\ell+5)\right]. \tag{6.16}$$

Equations (6.14)–(6.16) are valid for spherically confined hydrogenic systems inside infinite as well as finite boundary walls, for an arbitrary (n, ℓ) state. We have presented the results of the numerical tests of Equation (6.16) for the 3d state of the spherically confined hydrogen atom at a few representative cases of the confining stepwise potential, that differs from zero and equals V_0 for $r > R_0$ only. The critical values of $R_c = R_0$ are presented [102] in Table 1. The numerically evaluated derivatives agree at least up to 25 decimals with the estimates based on the R.H.S. of Equation (6.14) and Equation (6.16). It follows from Equation (6.15) that specific conditions on E corresponding to the incidental and simultaneous degeneracies can also be associated with the corresponding characteristic values of the ratio $\left[\frac{\eta''_\ell(0)}{\eta_\ell(0)}\right]$. Finally, it is clear from Equation (6.15), that the property of convexity of the scaled density $\eta_\ell(r)$ i.e. $\eta''_\ell(r) \geq 0$ in the neighborhood of the nuclear position, i.e. as $r \to 0$, is determined by the choice of V_0, R_0 for a given (n, ℓ) state of the confined hydrogen atom. In Figure 4, we have plotted the ratio $R2(V_0) = \left[\frac{\eta''_\ell(0)}{\eta_\ell(0)}\right]$ and $\eta_\ell(0)$ for the 2p state of the confined hydrogen atom as a function of R_0 under confining potentials given by V_0 as 10, 50, 100, and ∞. The upper set displays the variation of $\eta_\ell(0)$. The lower set of curves represent the ratio $R2(V_0)$. Thus, the variation of the property of convexity of the confined electron density for the hydrogen atom is clearly displayed [102] in Figure 4, as $R2(V_0)$ changes sign as the radius of confinement varies.

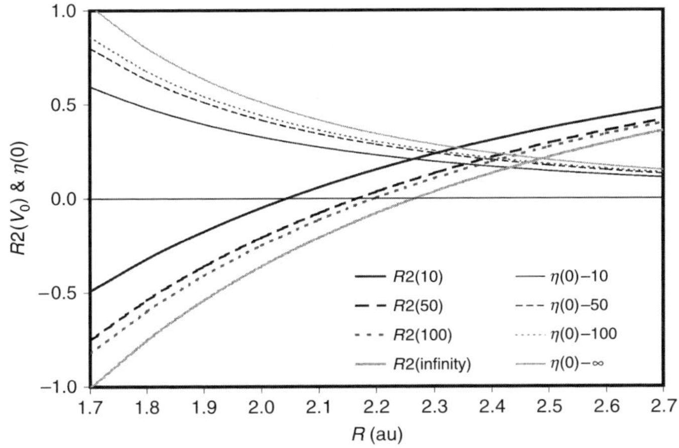

Figure 4 A plot of the ratio $R2(V_0)$, given by $\left[\frac{\eta_\ell''(0)}{\eta_\ell(0)}\right]$ and $\eta_\ell\,(0)$ as a function of the radius of confinement R at the values of V_0 as 10, 50, 100 and ∞ corresponding to the 2p state of the spherically confined hydrogen atom. The lower set of curves at beginning of the graph near the origin correspond to $R2(V_0)$, while the upper curves represent $\eta_\ell\,(0)$.

Relationships analogous to Equation (6.16) can also be derived for spherical screened Coulomb potentials like Yukawa and Hulthén potentials.

6.4. The mean value limits in the Dirichlet problem

As it was noted in the introduction to this Section, convergence of the mean values to some limits is a delicate problem in Quantum Chemistry (see [94–97] and Section XIII.2 in [5]). Nevertheless, the extension of a region for the Dirichlet problem is followed by convergence of some mean-values $\langle r^k \rangle$ to the ones for the free system. Here we consider this statement for the spherically symmetric Dirichlet problem in a sphere of radius R for R^3 and the ground state only, though the results may be generalized to excited states, higher dimensions and more complex regions. The spherically symmetric potential is taken to be the sum of a Coulomb-type potential and a bounded function. For the case of an additional polynomial function bounded from below see [60–62] and references therein. For the bounded 1s-state of the free system, the wave function $\psi(r)$ and radial one $\varphi(r)$ is assumed to decrease exponentially for large enough $r = |r|$ values (see e.g. Section XIII.11 in [5]).

As was demonstrated in Section 4, when $R \to \infty$ the Dirichlet problem 1s energies converge to the energy value of the unbounded system and it follows from the variational principle (see e.g. [95,96]) that the Dirichlet problem wavefunctions $\psi_R(\mathbf{r})$ converge to $\psi(\mathbf{r})$ in $L_2(R^3)$ and the same is true for gradients. Similar relations hold for radial functions $\varphi_R(r) = r\psi_R(|\mathbf{r}|)$, that is, $\varphi_R \to \varphi$ in $L_2(R^1)$ (all the functions are normalized to 1 and ψ_R, φ_R are continued with zero value for $r > R$).

Table 1 A comparison of the estimated values of first and second derivatives of scaled density at the nucleus using Equation (6.14) and Equation (6.16), respectively, with the direct numerical calculations. Four sets of results are shown corresponding to different values of confining potential V_0. At the critical radius R_c in each case the energy value is 0. The $\eta(0)$, $\eta'(0)$ and $\eta''(0)$ values are given under *Density* in columns 1 and 4. All values are in au

$V_0 = 10$			$V_0 = 100$		
R_c	9.39466298624514		R_c	9.54668381169248	
E	0.00000000000000		E	0.00000000000000	
Density	Equations (6.14)–(6.16)	Numerical	*Density*	Equations (6.14)–(6.16)	Numerical
$\eta(0)$		0.0007367880443	$\eta(0)$		0.0007368406561
$\eta'(0)$	−0.0004911920295	−0.0004911920295	$\eta'(0)$	−0.0004912271040	−0.0004912271040
$\eta''(0)$	0.0003040712564	0.0003040712564	$\eta''(0)$	0.0003040929692	0.0003040929692
$V_0 = 1000$			$V_0 = \infty$		
R_c	9.59500625911483		R_c	9.61736604170592	
E	0.00000000000000		E	0.00000000000000	
Density	Equations (6.14)–(6.16)	Numerical	*Density*	Equations (6.14)–(6.16)	Numerical
$\eta(0)$		0.0007368423341	$\eta(0)$		0.0007368423889
$\eta'(0)$	−0.0004912282227	−0.0004912282227	$\eta'(0)$	−0.0004912282593	−0.0004912282593
$\eta''(0)$	0.0003040936617	0.0003040936617	$\eta''(0)$	0.0003040936843	0.0003040936843

One may use the cutoff function technique (see Section 4.3) and change $\psi_R(\mathbf{r})$ near the points $r = R$ to regularize it. The modified function belongs to the domain of definition of the Laplacian operator (an alternative method is mollification, see Sect. 2.17, 2.18 in [22]) and has approximately the same mean energy. Then simple estimates (see e.g. Section V.5.3 in [25]) and relations of Equation (2.4) demonstrate that all of the values $|\psi_R(\mathbf{r})|$, $|\psi(\mathbf{r})|$ are uniformly bounded, for some constant C_0 and large enough R, and the following simple estimates hold for any r:

$$|\varphi_R(r)| \leq C_0 r, \qquad |\varphi(r)| \leq C_0 r. \tag{6.17}$$

According to Equations (4.14)–(4.18) and (4.21), for some constant C_1 and a large enough value of a, one may rewrite relation (4.18) as an inequality that holds for any $R, r > a$:

$$|\varphi_R(r)| \leq C_1 |\varphi(r)|. \tag{6.18}$$

That is, for large r and R, the functions $\varphi_R(r)$ can be estimated as an exponentially decreasing function.

Relations (6.17), (6.18) give rise to estimates of the mean values of r^k for any $k > -3$ with any desired accuracy, uniformly with respect to large enough R values, when small fixed neighborhoods of the points $r = 0$ and $r = \infty$ are excluded from the region of integration. As for $0 < a < b < \infty$ and $r \in [a, b]$, the values for r^k are bounded by some constant C_2 and one may use the notes of paper [97] to find the following relation due to the Cauchy inequality:

$$\begin{aligned} |\langle\varphi_R|r^k|\varphi_R\rangle_{[a,b]} - \langle\varphi|r^k|\varphi_{[a,b]}| &\leq |\langle\varphi_R - \varphi|r^k|\varphi_R\rangle_{[a,b]}| + |\langle\varphi|r^k|\varphi_R - \varphi\rangle_{[a,b]}| \\ &\leq 2C_2 \times \|\varphi_R - \varphi\| \to 0 \end{aligned} \tag{6.19}$$

as $\|\varphi_R - \varphi\| \to 0$ with $R \to \infty$. This means that for large enough R, $\langle\varphi_R|r^k|\varphi_R\rangle$ approximates $\langle\varphi|r^k|\varphi\rangle$ with any accuracy desired. Hence,

$$\langle\varphi_R|r^k|\varphi_R\rangle \to \langle\varphi|r^k|\varphi\rangle \quad \text{for } R \to \infty \text{ and any } k > -3. \tag{6.20}$$

A similar statement holds for any function $f(r)$ that is not too singular at $r = 0$ and not too large at infinity, if it is bounded out of neighborhoods of the points $r = 0$ and $r = \infty$.

According to Section 6.2, the ground state convergence in relation (6.20) is monotonic (increasing for $k > 0$ and decreasing for $k < 0$). Hence, one may use this note to control the accuracy of calculations. For the hydrogen atom, one may find numerical examples of relation (6.20) in [103].

7. SPHERICALLY CONFINED ISOTROPIC HARMONIC OSCILLATOR

In this section we shall state some exact results for the spherically confined isotropic harmonic oscillator inside impenetrable walls. The eigenspectral regularities and the characterization of energy states in terms of the electron density and its derivatives at the equilibrium position will be considered.

7.1. Degeneracy of confined states

The so-called accidental degeneracy of the *unconfined* hydrogen atom and the isotropic harmonic oscillator, is generally understood in terms of the corresponding symmetry groups SO(4) and SU(3), respectively. The level degeneracy patterns in these *unconfined* systems show characteristic differences. For example, while the pair of energy levels $[\{n,\ell\},\{(n+1),(\ell+2)\}]$ for the hydrogen atom are *nondegenerate*, the corresponding isotropic harmonic oscillator levels are found to be *degenerate*. Under spherically confined conditions, choosing the radius of confinement at $R_c = R = (\ell+1)(\ell+2)$, i.e. exactly at the *only* radial node corresponding to the free hydrogen wave functions of the first excited state of a given ℓ, gives rise to degeneracy between the *two confined* states. Thus, the confined state energies $[E_{2\mathrm{s}}(R_c=2), E_{3\mathrm{d}}(R_c=2)]$, $[E_{3\mathrm{s}}(R_c=2), E_{4\mathrm{d}}(R_c=2)]$, etc. are degenerate. Similarly, the entire set of two confined states satisfies the degeneracy condition $[E_{3\mathrm{p}}(R_c=6), E_{4\mathrm{f}}(R_c=6)]$, $[E_{4\mathrm{p}}(R_c=6), E_{5\mathrm{f}}(R_c=6)]\ldots$ etc. [39–41]. As noted above, for the free isotropic harmonic oscillator in three dimensions, the states defined by the quantum numbers $[(n+1,\ell);(n,\ell+2)]$ are degenerate. The only radial node of the first excited state of a given ℓ in this free system occurs at $[(2\ell+3)/2]^{1/2}$. It was found *numerically* [104] that the full set of pairs of confined states defined by the quantum numbers $[(n+1,\ell),(n,\ell+2)]$, $n=0,1,\ldots,$ with $R_c=[(2\ell+3)/2]^{1/2}$ au, displays a constant energy level separation exactly given by twice the harmonic oscillator energy unit. The radial wavefunction for a D-dimensional isotropic harmonic oscillator $\left(V(r)=\frac{\omega^2r^2}{2}\right)$ confined in a hypersphere of radius R is given by

$$\psi_{n\ell}(r) = \left(r\sqrt{\omega}\right)^{\ell} \mathrm{e}^{-\frac{\omega r^2}{2}}\,{}_1F_1\left(\frac{\ell}{2}+\frac{D}{4}-\frac{E}{2\omega},\ell+\frac{D}{2},\omega r^2\right). \tag{7.1}$$

This has been analytically proven [105] using the properties of the confluent hypergeometric functions ${}_1F_1$ in Equation (7.1). For an isotropic harmonic oscillator in D-dimensions, it has been shown that

$$E_{n,\ell}^{(D)}\left(R_c=\sqrt{l+\frac{D}{2}}\right) - E_{n,\ell-1}^{(D+2)}\left(R_c=\sqrt{l+\frac{D}{2}}\right) = 2. \tag{7.2}$$

From the definition of ${}_1F_1$ it can be readily seen that the inter-dimensional states are degenerate under the transformation given by $(n, \ell, D) \rightarrow (n, \ell \pm 1, D \mp 2)$. Very recently [106], similar eigenspectral properties of the $2D$ isotropic harmonic oscillator, centrally enclosed in an axially symmetric box with impenetrable walls, have been derived using the annihilation and creation operators and the infinitesimal operators of the SU(2) group. Extension to the three dimensional case using the SU(3) group has been also completed [107].

7.2. Density at the equilibrium point

Similar to the case of a confined hydrogen atom discussed in Section 6.3, we define the scaled density for the confined isotropic harmonic oscillator by relations (6.13) and (7.1).

Now it is easy to check general relations (6.14), (6.15) for the case when $Z = 0$ and $U(0) = 0$. The only difference is the use for the oscillator in R^D of the quantity $q = \ell + \frac{D-3}{2}$, instead of ℓ in R^3. The simple calculation gives

$$\eta_{n\ell}(0) = \omega^{\ell},\ \eta'_{n\ell}(0) = 0,\ \eta''_{n\ell}(0) = -\frac{4E}{2q+3}\omega^{\ell} = -\frac{4E}{2\ell+D}\omega^{\ell}. \tag{7.3}$$

In Table 2, we have presented the numerical test [105] of Equation (7.3) for the D-dimensional confined harmonic oscillator corresponding to $D = 2 - 4$ cases in the ground state at several values of R_c. As before, in the case of the confined hydrogen atom, it is found that the ratio $\frac{\eta''_{n\ell}(0)}{\eta_{n\ell}(0)}$ can be used to obtain very accurate estimates of E. It is clear from Equation (7.3) that, in general, $\eta''_{n\ell}(0) \leq 0$ for the confined isotropic harmonic oscillator system.

8. INFORMATION THEORETICAL UNCERTAINTY-LIKE RELATIONSHIPS

In this section, we present some general results on the information theoretical uncertainty-like measures applicable to the *standard model* systems of hydrogen-like atoms and the isotropic harmonic oscillator. The characteristic features of the spherically confined systems will be highlighted.

The one-electron Shannon information entropy [108], S_r, of the electron density, $\rho(\mathbf{r})$, in coordinate space is defined as

$$S_r = -\int \rho(\mathbf{r}) \ln \rho(\mathbf{r}) d\mathbf{r}, \tag{8.1}$$

and the corresponding momentum space entropy, S_p, is given by

$$S_p = -\int \gamma(\mathbf{p}) \ln \gamma(\mathbf{p}) d\mathbf{p}, \tag{8.2}$$

Table 2 D-dimensional confined isotropic harmonic oscillator state ($n = 0, \ell = 0$) confined inside an impenetrable well. The cases $D = 2, 3, 4$ at several radii of confinement are given. All values are in au

R_C	E	$\eta(0)''/\eta(0)$	$\eta(0)''/(\eta(0) \times E)$
$D = 2$			
0.5	11.5936192506866	−23.1872385013733	−2.0000000000000
1	3.0000000000000	−6.0000000000000	−2.0000000000000
2	1.1222085296789	−2.2444170593578	−2.0000000000000
3	1.0019367879643	−2.0038735759287	−2.0000000000000
4	1.0000033582152	−2.0000067164304	−2.0000000000000
5	1.0000000006653	−2.0000000013307	−2.0000000000000
10	1.0000000000000	−2.0000000000000	−2.0000000000000
$D = 3$			
0.5	19.7745341792083	−26.3660455722777	−1.3333333333333
1	5.0755820152268	−6.7674426869690	−1.3333333333333
2	1.7648164387806	−2.3530885850408	−1.3333333333333
3	1.5060815272528	−2.0081087030037	−1.3333333333333
4	1.5000146030071	−2.0000194706762	−1.3333333333333
5	1.5000000036716	−2.0000000048954	−1.3333333333333
10	1.5000000000000	−2.0000000000000	−1.3333333333333
$D = 4$			
0.5	29.4056004466975	−29.4056004466975	−1.0000000000000
1	7.5071721804519	−7.5071721804519	−1.0000000000000
2	2.4717752113502	−2.4717752113502	−1.0000000000000
3	2.0149671135031	−2.0149671135031	−1.0000000000000
4	2.0000497838287	−2.0000497838287	−1.0000000000000
5	2.0000000159039	−2.0000000159039	−1.0000000000000
10	2.0000000000000	−2.0000000000000	−1.0000000000000

where $\gamma(\mathbf{p})$ denotes the momentum density. The densities, $\rho(\mathbf{r})$ and $\gamma(\mathbf{p})$ are both normalized to unity and all quantities are given in atomic units. The Shannon information entropies (uncertainty) provide a measure of information about the probability distribution in the respective spaces. A more localized distribution corresponds to a smaller value of the information entropy in the position space.

Using the position and momentum space entropies, Bialynicki-Birula and Mycielski [109] derived a stronger version of the Heisenberg uncertainty principle of quantum mechanics. The entropy sum, in D-dimensions, satisfies the inequality [110]

$$S_T = S_r + S_p \geq D\,(1 + \ln \pi)\,. \tag{8.3}$$

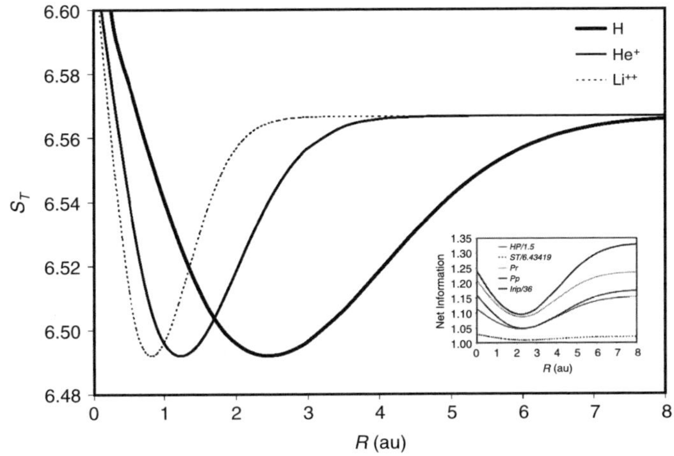

Figure 5 $S_T(R)$ versus R for the confined hydrogen-like atoms. For the confined hydrogen atom, similar behavior is also displayed by the other commonly used uncertainty-like information theoretical measures, as shown in the inset.

The dependence of S_T on D and its bounded region Ω for a given potential, $V(r)$, has been the subject of recent studies [110–112]. The *standard model systems* under unconfined conditions are characterized by *homogeneous potentials*, and in general the potential can be defined as $\lambda V(r)$, where λ denotes the coupling strength. Under these conditions S_T is shown [110, 112] to be independent of λ, for all states. The confined model systems, on the other hand, become dependent on λ, since the truncated potential is no longer homogeneous. Scaling arguments have been used to show [110–112] that for spherically confined hydrogen-like atoms ($Z = \lambda$); (a) an optimum point exists in $S_T\,(R)$, as a function of the confinement radius, $R(= R_{\text{opt}})$, at fixed Z and confinement potential, (b) the optimum value of S_T remains fixed for all Z, and (c) the location of R_{opt} scales as $1/Z$. It is significant to note that while Equation (7.3) is always satisfied, the special appearance of an optimum point in S_T under confined condition in hydrogen-like atoms leads to a set of atomic species for which S_T is bounded from both sides [111]. In Figure 5, we have plotted $S_T(R)$ versus R for the confined hydrogen-like atoms [112]. Similar behavior is also displayed by the other commonly used uncertainty-like information theoretical measures.

9. AFTERWORD

In this review, we have described a set of exact results known for the simplest possible confined models. The applications of confined systems, in general, vary from quantum dots [113] to the core of Jupiter and Saturn [114].

Certainly, novel numerical methods [115–120] have to be further developed to treat the confined model systems which include multiple electrons and/or atoms under a wider variety of boundary conditions and symmetry. *While confined systems present many computational problems not found in free systems, systems prepared under controlled confinement will continue to have their own virtues.*

ACKNOWLEDGMENTS

K.D.S. is grateful to the Department of Science and Technology, New Delhi for funding the research on information theoretical measures on confined atoms and molecules under an Indo-Hungarian scientific collaboration. V.I.P. is grateful for support from the Russian Foundation for Fundamental Research (project 07-03-01021). H.E.M. acknowledges his wife Ruth, whose constant encouragement has made this work more pleasurable.

REFERENCES

[1] E. Wigner, F. Seitz, On the constitution of metallic sodium, Phys. Rev. 43 (10) (1933) 804–810.

[2] E. Wigner, F. Seitz, On the constitution of metallic sodium. II, Phys. Rev. 46 (6) (1934) 509–524.

[3] H. Weyl, Das asymptotishche Verteilungsgesetz der Eigenwerte linearer pertieller Differetialgleichungen, Math. Ann. B.71 (1911) S.441–469.

[4] R. Courant, D. Hilbert, Methoden der Mathematischen Physik. B.1, Springer Verlag, Berlin, 1931.

[5] M. Reed, B. Simon, Methods of Modern Mathematical Physics, V. 1–4, Academic Press, 1972–1978.

[6] H.-J. Stöckmann, Quantum Chaos. An Introduction, Cambridge University Press, 2000.

[7] L. Brillouin, Perturbation d'un problème de valeurs propres par déformation de la frontière, Compt. Rend. 204 (1937) 1863–1865.

[8] H. Frohlich, A quantum mechanical discussion of the cohesive forces and thermal expansion coefficients of the alkali metals, Proc. Roy. Soc. (London) Ser. A, Math. Phys. Sciences 158 (893) (1937) 97–110.

[9] J. Bardeen, Compressibilities of the alkali metals, J. Chem. Phys. 6 (7) (1938) 372–378.

[10] H. Froehlich, A solution of the Schrodinger equation by a perturbation of the boundary conditions, Phys. Rev. 54 (11) (1938) 945–947.

[11] A. Michels, J. de Boer, A. Bijl, Remarks concerning molecular interaction and their influence on the polarisability, Physica 4 (10) (1937) 981–994.

[12] A. Sommerfeld, H. Welker, Künstliche Grenzbedingungen beim Keplerproblem, Ann. Phys. 424 (1–2) (1938) 56–65. [B32, 5F].

[13] A. Sommerfeld, H. Hartmann, Künstliche Grenzbedingungen in der Wellenmechanik. Der beschränkte Rotator, Ann. Phys. 429 (5–6) (1940) 333–343. [B37, 5F].

[14] S.R. de Groot, C.A. ten Seldam, On the energy levels of a model of the compressed hydrogen atom, Physica 12 (9–10) (1946) 669–682.

[15] P.W. Fowler, Energy, polarizability and size of confined one-electron systems, Mol. Phys. 53 (4) (1984) 865–889.

[16] P.O. Fröman, S. Yngve, N. Fröman, The energy levels and the corresponding normalized wave functions for a model of a compressed atom, J. Math. Phys. 28 (8) (1987) 1813–1826.

[17] S. Yngve, The energy levels and the corresponding normalized wave functions for a model of a compressed atom, J. Math. Phys. 29 (4) (1988) 931–936.
[18] W. Jaskólski, Confined many-electron systems, Phys. Rep. 271 (1) (1996) 1–66.
[19] V.K. Dolmatov, A.S. Baltenkov, J.-P. Connerade, S.T. Manson, Structure and photoionization of confined atoms, Radiation Phys. and Chem. 70 (1–3) (2004) 417–433.
[20] J.M.H. Lo, M. Klobukowski, G.H.F. Diercksen, Low-lying excited states of the hydrogen molecule in cylindrical harmonic confinement, Adv Quantum. Chem. 48 (2005) 59–89.
[21] S.L. Sobolev, Some Applications of Functional Analysis in Mathematical Physics, 3rd ed., in: Trans. Math. Monogr., vol. 90, AMS, Providence, 1991.
[22] R.A. Adams, Sobolev Spaces, Academic Press, NY, 1975.
[23] O.A. Ladyzhenskaya, The Boundary Value Problems of Mathematical Physics, Springer-Verlag, New York, 1985.
[24] E. Stein, Singular Integrals and Differentiability Properties of Functions, Princeton Univ. Press, 1970.
[25] T. Kato, Perturbation Theory for Linear Operators, Springer-Verlag, NY, Berlin, 1966.
[26] T. Kato, On the eigenfunctions of many-particle systems in quantum mechanics, Commun. Pure Appl. Math. 10 (2) (1957) 151–171.
[27] M. Hoffmann-Ostenhof, T. Hoffmann-Ostenhof, T.Ø. Sørensen, Electron wavefunctions and densities for atoms, Ann. Henri Poincaré 2 (2001) 77–100.
[28] Á Nagy, K.D. Sen, Ground-and excited-state cusp conditions for the electron density, J. Chem. Phys. 115 (14) (2001) 6300–6308.
[29] V.I. Pupyshev, Wall effects on the state of a hydrogen atom in a cavity, Rus. J. Phys. Chem. 74 (1) (2000) 50–54 (Engl. transl.).
[30] E.B. Wilson, Symmetry, nodal surfaces, and energy ordering of molecular orbitals, J. Chem. Phys. 63 (11) (1975) 4870–4879.
[31] G.M. Zislin, On the nodes of eigenfunctions of the Schrödinger operator, Uspehi Mat. Nauk. 16 (1) (1961) 149–152 (Rus. Engl. transl.).
[32] J.D. Levine, Nodal hydrogenic wave functions of donors on semiconductor surfaces, Phys. Rev 140 (2A) (1965) A586–A589.
[33] T. Hoffmann-Ostenhof, A comparison theorem for differential inequalities with applications in quantum mechanics, J. Phys. A: Math. Gen. 13 (2) (1980) 417–424.
[34] T. Hoffmann-Ostenhof, J.D. Morgan, The qualitative behavior of the ground state wave function of H_2^+, J. Chem. Phys. 75 (2) (1981) 843–846.
[35] V.I. Pupyshev, On the repulsive nature of the Dirichlet boundaries in the confined atom problem, J. Phys. B.: At. Mol. Opt. Phys. 33 (5) (2000) 961–970.
[36] P.V. Yurenev, A.V. Scherbinin, V.I. Pupyshev, Energy levels of the hydrogen atom in a cylindrical cavity, Int. J. Quant. Chem. 106 (10) (2006) 2201–2207.
[37] P.V. Yurenev, A.V. Scherbinin, V.I. Pupyshev, Shifts of the hydrogen atom in a cylindrical cavity, Int. J. Quant. Chem. 108 (14) (2008) 2666–2677.
[38] L.D. Landau, E.M. Lifshitz, Quantum Mechanics: Non-Relativistic Theory. (V. 3), Pergamon, Oxford, 1977.
[39] V.I. Pupyshev, A.V. Scherbinin, Hidden symmetry in the confined hydrogen atom problem, Phys. Lett. A 299 (4) (2002) 371–376.
[40] V.I. Pupyshev, A.V. Scherbinin, The Lenz vector in the confined hydrogen atom problem, Chem. Phys. Lett. 295 (1998) 217–222.
[41] A.V. Scherbinin, V.I. Pupyshev, A.Yu. Ermilov, One-electron atom in a spherical cavity as a model for the electronic structure of the internal atoms in clusters, in: V.D. Lahno, G.N. Chuev (Eds.), Physics of Clusters, World Scientific Publishing Co. Pte. Ltd., Singapore, 1998, pp. 273–292.
[42] H.E. Montgomery, N.A. Aquino, K.D. Sen, Degeneracy of confined D-dimensional harmonic oscillator, Int. J. Quant. Chem. 107 (4) (2007) 798–806.
[43] Dong Shi-Hay, M. Lozada–Cassou, Generalized hypervirial and recurrence relations for radial matrix elements in arbitrary dimensions, Modern Phys. Lett. A 20 (20) (2005) 1533–1540.

[44] F.M. Fernández, E.A. Castro, Hypervirial analysis of enclosed quantum mechanical systems. I. Dirichlet boundary conditions, Int. J. Quant. Chem. 19 (4) (1981) 521–532.
[45] F.M. Fernández, E.A. Castro, Hypervirial analysis of enclosed quantum mechanical systems. II. von Neumann boundary conditions and periodic potentials, Int. J. Quant. Chem. 19 (4) (1981) 533–543.
[46] F.M. Fernández, E.A. Castro, Hypervirial theorems and enclosed quantum-mechanical systems, Phys. Rev. A 24 (5) (1981) 2344–2352.
[47] V.A. Fock, New derivation of the virial theorem in quantum mechanics, Zs. Phys. 63 (11–12) (1930) S.855–858.
[48] S.T. Epstein, The Variational Method in Quantum Chemistry, Academic Press, NY, 1974.
[49] F.M. Fernández, E.A. Castro, Virial theorem and boundary conditions for approximate wave functions, Int. J. Quant. Chem. 21 (4) (1982) 741–751.
[50] S.A. FitzGerald, T. Yildirim, L.J. Santodonato, D.A. Neumann, J.R.D. Copley, J.J. Rush, F. Trouw, Quantum dynamics of interstitial H_2 in solid C_{60}, Phys. Rev. B 60 (9) (1999) 6439–6451.
[51] V.V. Bobrikov, V.I. Pupyshev, Isotope effects for molecules in a cavity, Rus. Chem. Bull. 54 (1) (2005) 55–61 (Engl. transl.).
[52] J.G. Kirkwood, Polarizierbarkeiten, Suszeptibilitäten und van der Waals Kräfte der atome mit mehreren electronen, Phys. Z. B.33 (2) (1932) 57–60.
[53] R.A. Buckingham, The quantum theory of atomic polarization. I. Polarization by a uniform field, Proc. Roy. Soc. London, ser. A 160 (A900) (1937) 94–113.
[54] V.I. Pupyshev, A.V. Scherbinin, N.F. Stepanov, The Kirkwood-Buckingham variational method and the boundary value problems for the molecular Schrödinger equation, J. Math. Phys. 38 (11) (1997) 5626–5633.
[55] H.L. Cycon, R.G. Froese, W. Kirsch, B. Simon, Schrödinger Operators with Applications to Quantum Mechanics and Global Geometry, Springer-Verlag, NY, 1987.
[56] D.H. Berman, Boundary effects in quantum mechanics, Amer. J. Phys. 59 (10) (1991) 937–941.
[57] I. Gonda, B.F. Gray, Quantum-mechanical systems restricted by impenetrable barriers studied by a perturbation method, J. Chem. Soc., Faraday Trans. ser. 2: Molec. and Chem. Phys. 71 (1975) 2016–2024.
[58] M.E. Changa, A.V. Scherbinin, V.I. Pupyshev, Perturbation theory for the hydrogen atom in a spherical cavity with off-centre nucleus, J. Phys. B.: At. Mol. Opt. Phys. 33 (3) (2000) 421–432.
[59] K. Jörgens, J. Weidmann, Spectral properties of Hamiltonian Operators, in: Lecture Notes in Math., vol. 313, Springer-Verlag, NY, 1973.
[60] M.A. Núñez, G.B. Izquierdo, New approximation to the bound states of Schrödinger operators with Coulomb interaction, Int. J. Quant. Chem.: Quant. Chem. Symp 28 (1994) 241–250.
[61] M.A. Núñez, Computation of expectation values with Dirichlet one-dimensional wave functions, Int. J. Quant. Chem. 53 (1) (1995) 15–25.
[62] M.A. Núñez, General criteria for assessing the accuracy of approximate wave functions and their densities, Int. J. Quant. Chem. 53 (1) (1995) 27–35.
[63] M.A. Núñez, Rate of convergence of calculations with one-dimensional Dirichlet wave functions, Int. J. Quant. Chem. 62 (5) (1997) 449–460.
[64] J. Katriel, Reduction of the excited state into the ground state of a super-Hamiltonian, Int. J. Quant. Chem. 23 (5) (1983) 1776–1780.
[65] V.I. Pupyshev, A.V. Scherbinin, Molecular energy level shifts in large boxes: Use of the Kirkwood–Buckingham method, J. Phys. B: At. Mol. Opt. Phys. 32 (19) (1999) 4627–4634.
[66] T.E. Hull, R.S. Julius, Enclosed quantum mechanical systems, Can. J. Phys. 34 (9) (1956) 914–919.
[67] T.Yu. Mikhailova, V.I. Pupyshev, Proton tunneling models for TKHS-family and similar materials. Book of Abstracts of the XVIII Symposium on Modern Chemical Physics, Touapse, Russia, 2006, p. 151.

[68] W. Wilcox, Finite volume effects in self-coupled geometries, Annals of Phys. 279 (2000) 65–80.
[69] W. Wilcox, A formula for energy displacements for the confined hydrogen atom, Amer. J. Phys. 57 (6) (1989) 526–528.
[70] K.K. Singh, Theory of boundary perturbation and the compressed hydrogen molecular ion, Physica 30 (1964) 211–222.
[71] Z. Liu, D.L. Lin, Hydrogenic atom in semi-infinite space, Phys. Rev B 28 (8) (1983) 4413–4418.
[72] J. Gorecki, W. Byers-Brown, Iterative boundary perturbation method for enclosed one-dimensional quantum systems, J. Phys. B: At. Mol. Opt. Phys. 20 (22) (1987) 5953–5957.
[73] J. Gorecki, W. Byers-Brown, Variational boundary perturbation theory for enclosed quantum systems, J. Phys. B: At. Mol. Opt. Phys. 22 (17) (1989) 2659–2668.
[74] G. Barton, A.J. Bray, A.J. McKane, The influence of distant boundaries on quantum mechanical energy levels, Amer. J. Phys. 58 (8) (1990) 751–755.
[75] A.F. Kovalenko, E.N. Sovyak, M.F. Golovko, Electronic States of a hydrogen atom near a hard wall, Phys. Status Solidi (b) 155 (2) (1989) 549–558.
[76] A.F. Kovalenko, E.N. Sovyak, M.F. Holovko, On the quantum properties of adsorbed particles within the model of a hydrogen atom near a hard wall, Int. J. Quant. Chem. 42 (2) (1992) 321–337.
[77] N. Aquino, G. Campoy, H.E. Montgomery, Highly accurate solutions for the confined hydrogen atom, Int. J. Quant. Chem. 107 (7) (2007) 1548–1558.
[78] V.C. Aguilera-Navarro, E. Ley-Koo, Quantum harmonic oscillators with wave functions having a fixed logarithmic derivative at the equilibrium position, Revista Brasileira de Fisica 11 (4) (1981) 825–838.
[79] F. Gesztesy, D. Gurarie, H. Holden, M. Klaus, L. Sadun, B. Simon, P. Vogl, Trapping and Cascading of eigenvalues in the large coupling limit, Comm. Math. Phys. 118 (1988) 597–634.
[80] C. Eckart, The theory and calculation of screening constants, Phys. Rev. 36 (5) (1930) 878–892.
[81] J. Tomasi, M. Persico, Molecular interactions in solution: An overview of methods based on continuous distributions of the solvent, Chem. Rev. 94 (1994) 2027–2094.
[82] J.L. Movilla, J. Planelles, Image charges in spherical quantum dots with an off-centered impurity: Algorithm and numerical results, Comp. Phys. Comm. 170 (2) (2005) 144–152.
[83] J.G. Kirkwood, Theory of solutions of molecules containing widely separated charges with special application to zwitterions, J. Chem. Phys. 2 (7) (1934) 351–361.
[84] A.V. Tulub, The Rydberg spectrum for electron near the spherical surface, Opt. i Spectrosc. 73 (1) (1992) 48–54 (in Russian).
[85] E. Ley-Koo, S. Rubinstein, The hydrogen atom within spherical boxes with penetrable walls, J. Chem. Phys. 71 (1) (1979) 351–357.
[86] Ya.B. Zel'dovich, Energy levels in a distorted Coulomb field, Sov. Phys. Solid State 1 (1960) 1497–1501 (Engl. transl.).
[87] A.A. Grinberg, Energy spectrum of an electron placed in the fields of a small-radius potential well and an attractive coulomb potential, Sov. Phys. Semicond. 11 (1977) 1118–1120 (Engl. transl.).
[88] J.P. Connerade, V.K. Dolmatov, P.A. Lakshmi, S.T. Manson, Electron structure of endohedrally confined atoms: atomic hydrogen in an attractive shell, J. Phys. B: At. Mol. Opt. Phys. 32 (10) (1999) L239–244.
[89] J.P. Connerade, V.K. Dolmatov, P.A. Lakshmi, The filling of shells in compressed atoms, J. Phys. B: At. Mol. Opt. Phys. 33 (2) (2000) 251–264.
[90] V.K. Dolmatov, P. Brewer, S.T. Manson, Photoionization of atoms confined in giant single-walled and multiwalled fullerenes, Phys. Rev A 78 (2008) 013415(6p).
[91] J.P. Connerade, V.K. Dolmatov, S.T. Manson, On the nature and origin of confinement resonances, J. Phys. B: At. Mol. Opt. Phys. 33 (12) (2000) 2279–2285.
[92] R. Lefebvre, Box quantization and resonance determination: The multichannel case, J. Phys. Chem. 89 (20) (1985) 4201–4206.

[93] I.M. Savukov, Simple method for obtaining electron scattering phase shifts from energies of an atom in a cavity, Phys. Rev. Lett 96 (7) (2006) 073202(1-4).
[94] P.-O. Löwdin, Quantum theory of electronic structure of molecules, Ann. Rev. Phys. Chem. 11 (1960) 107–132.
[95] B. Klahn, W.A. Bingel, The convergence of the Rayleigh–Ritz method in quantum chemistry. I. The criteria for convergence, Theor. Chim. Acta 44 (1977) 9–26.
[96] B. Klahn, W.A. Bingel, The convergence of the Rayleigh–Ritz method in quantum chemistry. II. Investigation of the convergence for special systems of Slater, Gauss and two-electron functions, Theor. Chim. Acta 44 (1977) 27–43.
[97] B. Klahn, J.D. Morgan, Rates of convergence of variational calculations and of expectation values, J. Chem. Phys. 81 (1) (1984) 410–433.
[98] G.A. Arteca, F.M. Fernándes, E.A. Castro, Approximate calculation of physical properties of enclosed central field quantum system, J. Chem. Phys. 80 (4) (1984) 1569–1575.
[99] A.L. Buchachenko, Compressed atoms, J. Phys. Chem. B 105 (25) (2001) 5839–5846.
[100] W. Rudin, Principles of Mathematical Analysis, Second Edition, McGraw-Hill Book Co., NY, 1964.
[101] J.A. Weil, Hydrogen atom in a spherical box. II. Effect on hyperfine energy of excited state admixture, J. Chem. Phys. 71 (7) (1979) 2803–2805.
[102] H.E. Montgomery Jr., K.D. Sen, Electron density and its derivatives at the nucleus for spherically confined hydrogen atom, Int. J. Quantum Chem. 109 (4) (2009) 688–692.
[103] N.A. Aquino, Accurate energy eigenvalues for enclosed hydrogen atom within spherical impenetrable boxes, Int. J. Quant. Chem. 54 (2) (1995) 107–115.
[104] K.D. Sen, A.K. Roy, Studies on the 3D confined potentials using generalized pseudospectral approach, Phys Lett. A 357 (2006) 112–119.
[105] K.D. Sen, H.E. Montgomery Jr., N.A. Aquino, Degeneracy of confined D-dimensional harmonic oscillator, Int. J. Quantum Chem. 107 (2007) 798–806.
[106] Lj. Stevanovic, K.D. Sen, A study of the confined 2D isotropic harmonic oscillator in terms of the annihilation and creation operators and the infinitesimal operators of the SU(2) group, J. Phys. A 41 (2008) 265203 (14p).
[107] Lj. Stevanović, K.D. Sen, Eigenspectrum properties of the confined 3D harmonic oscillator, J. Phys. B: At. Mol. Opt. Phys. 41 (22) (2008) 225002 (6p).
[108] C.E. Shannon, A mathematical theory of communication, Bell Syst. Tech. 27 (1948) 379–423; A mathematical theory of communication II, Bell Syst. Tech. 27 (1948) 623–656.
[109] I. Bialynicki-Birula, J. Mycielski, Uncertainty relations for information entropy in wave mechanics, Comm. Math. Phys. 44 (1975) 129–132.
[110] K.D. Sen, J. Katriel, Information entropies for eigendensities of homogeneous potentials, J. Chem. Phys. 125 (2006) 0741171(4p).
[111] K.D. Sen, Characteristic features of Shannon information entropy of confined atoms, J. Chem. Phys. 123 (2005) 074110(4p).
[112] S.H. Patil, K.D. Sen, N.A. Watson, H.E. Montgomery Jr, Characteristic features of net information measures for constrained Coulomb potentials, J. Phys. B 40 (2008) 2147–2162.
[113] R. Henson, L.P. Kouvenhoven, J.R. Petta, S. Tarucha, L.M.K. Vandersypen, Spins in few-electron quantum dots, Rev. Mod. Phys. 79 (2007) 1217(49p).
[114] L. Stixrude, R. Jeanloz, Fluid helium at conditions of giant planetary interiors, Proc. Natl. Ac. Sc. 105 (2008) 11071–11075.
[115] M. Neek-Amal, G. Tayebirad, M. Molayem, M.E. Foulaadvand, L. Esmaeili-Sereshki, A. Namiranian, Ground state study of simple atoms within nanoscale box, Solid State Commun. 145 (2008) 594–599.
[116] C. Diáz-Garciá, S.A. Cruz, Many-electron atom confinement by a penetrable spherical box, Intl. J. Quantum Chem. 108 (2008) 1572–1588.
[117] D. Baye, K.D. Sen, Confined hydrogen atom by the Lagrange-mesh method: Energies, mean radii, and dynamic polarizabilities, Phys. Rev. E 78 (2008) 026701(7p).
[118] S.A. Ndengué, O. Motapon, Electric response of endohedrally confined hydrogen atoms, J. Phys. B 41 (2008) 045001(7p).

[119] T. Sako, G.H.F. Diercksen, Understanding the spectra of a few electrons confined in a quasi-one-dimensional nanostructure, J. Phys.: Condens. Matter 20 (2008) 155202 (13p).
[120] H. Ciftci, R.L. Hall, N. Saad, Study of a confined hydrogen-like atom using asymptotic iteration method, Intl. J. Quantum Chem. 109 (5) (2009) 931–937.

CHAPTER 3

The Hydrogen Atom Confined in Semi-infinite Spaces Limited by Conoidal Boundaries

Eugenio Ley-Koo[a]

Contents

[a] Instituto de Física, Universidad Nacional Autónoma de México Apartado Postal 20-364, 01000 México, D. F., México

E-mail address: eleykoo@fisica.unam.mx.

Advances in Quantum Chemistry, Vol. 57
ISSN: 0065-3276, DOI: 10.1016/S0065-3276(09)00607-8

1. INTRODUCTION

This contribution as a chapter in the special volume of ADVANCES IN QUANTUM CHEMISTRY on Confined Quantum Systems is focussed on (i) the hydrogen atom, (ii) confinement by conoidal boundaries, and (iii) semi-infinite spaces; however, some of its discussions may extend their validity to other physical systems and to confinement in closed volumes. The limitations in the title are given as a point of reference, and also take into account that several of the other chapters deal with confinement in finite volumes. A semantic parenthesis is also appropriate and self-explanatory: Compare conical curves (circles, ellipses, parabolas, hyperbolas and their radial asymptotes) with conoidal surfaces (spheres, spheroids, paraboloids, hyperboloids and their radial asymptotic cones).

Since this is a volume dedicated to the advances in our field of special interest, it is also important to establish some points of reference in time. The Report on Progress in Physics review on Confined Many-Electron Systems by Professor Jaskolski [1] is a good reference in both contents and time. Some articles published later on are representative of advances in the intervening twelve-year period. On the other hand, some of the references not included in that review may also have been followed by other advances.

With these preliminaries in mind, the topics of this chapter and the references have been selected by the author, from the perspective of his own work, in order to illustrate some of the advances. In a natural way, and while examining the literature, the author has also identified pathways that may lead to additional advances; as well as misreadings, mistakes and inconsistencies, which upon clarification, may also allow further advances. The chapter contains a sample of comments, a review of the free hydrogen atom, an overview of the confined hydrogen atom, and a preview of ongoing investigations as discussed next.

The sample of articles and comments presented in Section 2 include the hydrogen atom in a spherical enclosure [2], the hydrogen atom in a semi-infinite space limited by a hyperboloidal boundary [1], the two dimensional hydrogen atom limited by conical curves [3], and models for the binding of an electron by a polar molecule confined by a sphere and by a spheroid [4,5]. The comments in [1,2] serve as the motivation and points of reference for the systematic presentation of the free hydrogen-atom wave functions in different coordinate systems in Section 3, and their modifications under confinement by conoidal boundaries defined by a fixed value of the successive coordinates in Sections 4 and 5. The comments on [3] and [4, 5] have led the author to formulate new situations of confinement of the hydrogen atom in a dihedral angle, and of two-electron atoms and molecules on a plane surface, as well as to analyze the importance of using the complete electrostatic harmonic expansions in the study of atoms and molecules in Section 5.

In fact, Section 3 is a review of the separability and integrability of the Schrödinger equation for the hydrogen atom in 3.1 spherical, 3.2 spheroconal, 3.3 parabolic and 3.4 prolate spheroidal coordinate systems. The emphasis is laid upon the geometry and dynamic constants of motion for each case, recognizing the $O(4)$ symmetry behind the accidental degeneracy of the free hydrogen atom energy levels [6–8].

Next, Section 4 develops an overview of the hydrogen atom confined by the successive conoidal boundaries – 4.1 spherical, 4.2 circular-cone, 4.3 paraboloidal, 4.4 prolate spheroidal and 4.5 hyperboloidal – as a symmetry - breaking effect due to the new boundary condition on the wave functions. The eigenvalues of the respective constant of motion associated with the coordinate defining the boundary, and thereby the energy, are no longer expressed by integer quantum numbers. Correspondingly, the degeneracy of the free hydrogen-atom energy levels is removed.

Section 5 is a preview of current and future investigations on confined atoms and molecules. The one described in 5.1 is the counterpart of 4.2 for confinement of the hydrogen atom by elliptical cones, using spheroconal coordinates. The one anticipated in Section 2 and formulated in 5.2 corresponds to the hydrogen atom confined by a dihedral angle, common to spherical, parabolic and prolate spheroidal coordinates. Two-electron atoms and molecules are the systems of interest in 5.3, with the focus on the atomic hydrogen anion and the hydrogen molecule confined by nodal planes. The initial exploration of the solutions of the Schrödinger equation for an electron in complete spherical, prolate and oblate spheroidal dipole fields is previewed in 5.4.

The discussion in Section 6 summarizes and follows the connections between the comments in Section 2 and their extensions and adaptations to new situations of confinement in Sections 4 and 5.

2. SAMPLE OF COMMENTS ON ARTICLES ABOUT CONFINED ATOMS AND MOLECULES

The articles included in this section and making up the sample, on which the author has made comments at the proper times, serve as points of reference for extensions of his own works [9–11] and those of others [12, 13]; as well as correcting misreadings [1], errors [3], and inconsistencies [4, 5] in these other works about new situations of confinement. Indeed, the basic idea of [2], applied to the ground state of the hydrogen atom in a spherical enclosure at that time, is also valid for the excited s-states of the same system, as commented on in 2.1, as well as for all states of the hydrogen atom in the successive conoidal confinements, as shown and discussed in Section 4. The interpretation of our work on confinement of the hydrogen atom by a hyperboloidal boundary [9] in [1] misses the point of the difference between confinements by open and closed boundaries, as commented on in 2.2 and further developed in Section 4. Reference [3] pioneered the study of the hydrogen atom in two dimensions confined by conical curves using the variational method. Unfortunately, their results are not reliable because their trial function for confinement in an angle includes incompatible angular and radial behaviors, as briefly mentioned in our exact solution of the problem [10], and discussed in more detail in the comments of 2.3, which identify the same situation for their results about confinement by a hyperbola. During the initial review of the recent papers on the specialized topics of this book, the author became aware of the models proposed in [4,5] for the binding of an electron by a confined polar molecule. A comment on conceptual inconsistencies in both articles related to the proper evaluation of the binding energy in confined systems has recently been submitted for publication [11], as described in 2.4.

2.1. Some comments on the hydrogen atom in a spherical enclosure

The writing of our article [2], giving its name to this subsection, was motivated by previous independent works on the hydrogen atom confined in a sphere using Rayleigh–Schrödinger Perturbation Theory (RSPT) to second order [12] and up to fifth order [13], as well as the construction of the exact solutions [14]. In 1954 E. P. Wigner initiated the application of RSPT to the investigation of the ground state of the confined atom, using the kinetic energy of an electron confined in a sphere as the non-perturbed Hamiltonian and the nucleus-electron attractive Coulomb energy as the perturbation. The results for the lowest orders can be expected to be valid for small values of the radius of the sphere. He also conjectured that the ground-state energy for large spheres might asymptotically approach its free hydrogen-atom limit in an exponential form [12]:

$$E(R) = -\frac{e^2}{2a_0}\left[1 - Ce^{-\frac{R}{a_0}}\right]. \tag{1}$$

The work [13] extended RSPT up to fifth order by 1971 and [14] was published before the end of that decade. At that time, some of the authors of the last two references started a collaboration in investigations on confined quantum systems [15], in which the first results were reported in [2].

Here we borrow some of the equations and arguments of the section Properties of Solutions in [2], in order to illustrate the interpolation of the ground-state energy of the confined hydrogen atom between its familiar free-atom value and its zero value:

$$E(R_0) = 0 \tag{2}$$

for a radius of the sphere R_0 to be determined.

A word of warning about terminology is necessary at this point. While $E = 0$ corresponds to the ionization threshold for the free hydrogen atom, when speaking about the confined version the situation is not so clear. The radius in Equation (2) has been called the ionization radius or critical cage radius [4,5]. Here we avoid such usage. An additional and important comment on this point is formulated towards the end of this subsection.

For the $\ell = 0$ or s-states, including the ground state, the Schrödinger equation depends only on the radial coordinate,

$$\left[-\frac{\hbar^2}{2m_e}\frac{1}{r^2}\frac{d}{dr}r^2\frac{d}{dr} - \frac{e^2}{r}\right]\Psi_\nu(r) = E_\nu\Psi_\nu(r) \tag{3}$$

using the parametrization of the energy

$$E_\nu = -\frac{e^2}{2a_0\nu^2}. \tag{4}$$

The well-known exact solutions of Equation (3) are written as

$$\Psi_\nu(r) = M\left(-(\nu-1), 2, \frac{2r}{\nu a_0}\right)e^{-\frac{r}{\nu a_0}} \tag{5}$$

in terms of the Kummer confluent hypergeometric function of the first kind. Here we give its explicit form [16] 13.1.2, for the sake of completeness and later use:

$$M(a, b, z) = \sum_{s=0}^{\infty} \frac{(a)_s z^s}{(b)_s s!} \tag{6}$$

where $(a)_0 = 1, (a)_1 = a, (a)_s = a(a+1)\cdots(a+s-1)$ is the Pochhammer symbol.

The confinement boundary condition requires the vanishing of the functions in Equation (5) at the radius of the sphere $r = R$, or in turn,

$$M\left(-(\nu - 1), 2, \frac{2R}{\nu a_0}\right) = 0. \tag{7}$$

For each value of the radius, the solutions of Equation (7) in ν determine the energy eigenvalue $E_\nu(R)$ using Equation (4). In practice, it is easier to choose the values of ν, between one and infinity and solve Equation (7) for R, between its corresponding values of infinity and R_0, implementing numerically the interpolation of the energy eigenvalue as a function of the radius anticipated in the paragraph of Equation (2). The extrapolation to positive energies and smaller radii is also readily accomplished.

For the value of $E = 0$, corresponding to the limit of infinite ν in Equation (4), the Schrödinger Equation (3) becomes:

$$\left[\frac{\mathrm{d}^2}{\mathrm{d}x^2} + \left(\frac{2}{x}\right)\frac{\mathrm{d}}{\mathrm{d}x} - \left(\frac{2}{x}\right)\right]\Psi_\infty(x) = 0, \tag{8}$$

where $x = r/a_0$. This equation is also integrable in terms of ordinary Bessel functions,

$$\Psi_\infty(x) = \frac{1}{\sqrt{x}} J_1\left(\sqrt{8x}\right), \tag{9}$$

as the reader can readily establish. An alternative way is to take directly the limit of infinite ν in Equation (5). Then the boundary condition of Equation (7), applied to the function of Equation (9), leads to the radii:

$$R_0 = \frac{j_{1,s}^2 a_0}{8} \tag{10}$$

in terms of the zeros of the Bessel function of order one with $s = 1, 2, 3, \ldots$ for the ground and successive excited states.

While in [2] only the ground state was analyzed, here we see the extension of the results for the excited s-states.

Let us now examine the asymptotic limits of Equations (4) and (7) as the radius of the box becomes infinite. For the ground state, the energy parameter ν must be very close to the one from above, so we write it as

$$\nu = 1 + \nu_r \tag{11}$$

where ν_r is positive and very close to zero. Then we use the asymptotic form [16] 13.5.1 of Equation (6):

$$\frac{M(a, b, z)}{\Gamma(b)} = \frac{e^{i\pi a} z^{-a}}{\Gamma(b - a)} + \frac{e^{z} z^{(a-b)}}{\Gamma(a)} \tag{12}$$

for infinitely large values of z. Application of this form to Equation (7) yields:

$$\frac{\left(\frac{2R}{a_0}\right)^{\nu_r}}{\Gamma(2 + \nu_r)} + \frac{e^{\frac{2R}{a_0}} \left(\frac{2R}{a_0}\right)^{-\nu_r - 2}}{\Gamma(-\nu_r)} = 0. \tag{13}$$

Then by taking the limit when ν_r becomes zero, with $\Gamma(-\nu_r) = \frac{\Gamma(1-\nu_r)}{-\nu_r}$, we finally obtain:

$$\nu_r = \left(\frac{2R}{a_0}\right)^2 e^{-\frac{2R}{a_0}}. \tag{14}$$

Then the energy of Equation (4), keeping only linear terms in ν_r, becomes:

$$E(R \longrightarrow \infty) = -\frac{e^2}{2a_0}(1 - 2\nu_r) = -\frac{e^2}{2a_0}\left[1 - 2\left(\frac{2R}{a_0}\right)^2 e^{-\frac{2R}{a_0}}\right]. \tag{15}$$

This result differs from Wigner's conjecture, Equation (1), by the extra quadratic factor multiplying the exponential, and the extra factor of 2 in the exponent of the latter.

Next, we rewrite the result of [13] as a point of reference for two additional comments. The fifth-order RSPT energy is given by

$$E^5(R) = \left(\frac{e^2}{a_0}\right)\left[\frac{\pi^2 a_0^2}{2R^2} - \frac{2.4377 a_0}{R} - \frac{1.0796}{\pi^2} - \frac{1.2112 R}{\pi^4 a_0} - \frac{1.1928 R^2}{\pi^6 a_0^2} - \frac{0.725 R^3}{\pi^8 a_0^3}\right]. \tag{16}$$

This approximation is good for small values of the radius. Its first term is positive and corresponds to the kinetic energy contribution of the confined free electron in the zeroth-order approximation. The following terms are negative and represent the energy contributions of the Coulomb interaction, as the perturbation, in the successive orders of approximation.

The first comment is that the fast increase of the ground-state energy of the hydrogen atom, when the radius of the sphere becomes smaller than R_0, is reproduced well by Equation (16), as numerically illustrated in [2].

The second comment concerns the interpretation of the last five terms in Equation (16) as the corresponding approximation to the binding energy of the electron in the confined hydrogen atom. This interpretation connects with the warning about the names used for R_0 following Equation (2).

More importantly, it leads one to identify the incorporation of the confining degree of freedom in the usual definition of the binding energy, B_e, using the exact energies for both the hydrogen atom and the free electron confined in the same sphere, which we label as CHA and CFe, respectively:

$$B_e(R) = E^{\mathrm{CFe}}(R) - E^{\mathrm{CHA}}(R). \tag{17}$$

The vanishing of this binding energy signals the crossings of the respective energy curves, necessary to identify the unbinding of the electron or ionization of the atom, inside the confining sphere. The practical side of Equation (17) is that all the works that have used $E^{\mathrm{CFe}}(R) = 0$, as warned after Equation (2), need to be reconsidered.

The analysis in [2] was limited to the ground state of the confined hydrogen atom, but most equations in this subsection are valid for all s-states. Section 4 presents the extension of the ideas and methods, introduced in [2] and illustrated here, to the alternative confinements by conoidal boundaries.

2.2. Confined many-electron systems

The author became interested in the models of confinement of the hydrogen atom inside finite volumes [2,14,17,18] in connection with the measurements of the hyperfine structure of atomic hydrogen trapped in α-quartz [19,20]. Ten years later, he extended his interests to confinement in semi-infinite spaces limited by a paraboloid [21], a hyperboloid [9] and a cone [22] in connection with the exoelectron emission by compressed rocks [23,24]. Jaskolski's report [1] cited several of the above-mentioned works [9,14,17, 18,21], each one of which had formulated and constructed exact solutions for new types of confinement for the hydrogen atom. This subsection is focussed on his citation of our article [9]:

"... on the hydrogen atom confined in a semi-infinite space limited not by a plane, but by a hyperboloidal surface defined by $\eta = \eta_0$. The confinement in a semi-infinite

hyperboloidal space is however, substantially different than the confinement in a plane half-space. When the η_0 surface approaches an atom (placed at one of the two foci), the semi-infinite space closes itself leading to an infinite increase in the energy of the levels."

The following comments on this citation are of three kinds, grammatical, mathematical and physical:

(i) It is preferable to use the adjectives hyperboloidal and plane for the boundaries rather than the space.
(ii) The plane boundary is a special case of a hyperboloid for $\eta_0 = 0$, as used in previous investigations with such a boundary [25–30].
(iii) There is no "infinite increase in the energy levels" in the entire interval [-1,1] of η_0 and hyperboloidal boundaries, as Figs. 1–3 in [9] show.
(iv) The infinite degeneracy of the energy levels at zero energy in the limit when η_0 becomes minus one, as the hyperboloid closes around its own axis and its vertex approaches the nucleus, is an interesting result of this work.

Our contemporary article on the hydrogen atom confined by a conical boundary [22] also exhibits the properties commented in (ii)–(iv). In particular, the cone becomes the equatorial plane when the polar angle is ninety degrees, corresponding to Levine's 1965 pioneering model for an impurity atom on the surface of a solid [31].

While the hydrogen atom confined in spaces limited by closed boundaries does show a monotonic and unlimited increase of energy of its levels, as the boundary approaches the nucleus and the volume of confinement is reduced [2,14,17,18]; its confinement by open conoidal boundaries is characterized by the monotonic increasing of energy of its levels only up to zero energy in the corresponding limit situations, with the consequent infinite degeneracy, as it was shown in [9,21,22] and commented on above.

The review and overview of the following sections allow a better appreciation of the common and different properties of the free versus the confined hydrogen atom, including closed versus open boundary confinements, as well as the specific characteristics associated with each conoidal boundary.

2.3. Ground state energy of the two-dimensional hydrogen atom confined with conical curves through the variational method

The investigation in the article with the above title [3] introduced the novelty of the two-dimensionality of the hydrogen atom and aimed at the systematic treatment of the confinement by the different conical boundaries. The authors chose the variational method.

The comment on their work focuses on the limitations of their treatment of the hydrogen atom confined inside an angle. When we became aware of their incorrect results we decided to study the exact solutions for the confinement by circles, angles and circular sectors [10]. We included a brief mention of their choice of an incorrect trial function behind their results while making a comparison of our respective numerical values of the ground state energy as a function of the confining angle. Here we present the exact solution and use it as a reference to identify the inconsistencies in their variational trial function. The same type of inconsistency is identified in their treatment of the confinement with a hyperbola.

The Schrödinger equation in circular coordinates is separable into its radial and circular parts:

$$-\frac{\hbar^2}{2m_e}\left[\frac{1}{\rho}\frac{\mathrm{d}}{\mathrm{d}\rho}\rho\frac{\mathrm{d}}{\mathrm{d}\rho}-\frac{\mu^2}{\rho^2}-\frac{\mathrm{e}^2}{\rho}\right]R(\rho)=ER(\rho) \tag{18}$$

$$-\frac{\mathrm{d}^2\Phi(\phi)}{\mathrm{d}\phi^2}=\mu^2\Phi(\phi). \tag{19}$$

The boundary condition for confinement inside the angle defined by $0\le\phi\le\phi_0$ must be satisfied by the angular part:

$$\begin{aligned}\Phi(\phi=0)&=0\\ \Phi(\phi_0)&=0.\end{aligned} \tag{20}$$

The respective eigenstates and eigenvalues of Equation (19) are

$$\Phi(\phi)=\sqrt{\frac{2}{\phi_0}}\sin\mu\phi \tag{21}$$

$$\mu=\frac{n_\phi\pi}{\phi_0}\qquad n_\phi=1,2,3,\ldots. \tag{22}$$

The radial eigenfunctions of Equation (18) are also known [10], and here we write only their form without radial excitation $n_\rho=0$:

$$R_{0,\mu}(\rho)=\rho^\mu \mathrm{e}^{-\frac{\rho}{(\mu+(1/2))a_0}}, \tag{23}$$

where the power factor removes the singularity associated with the centrifugal potential, and the exponential factor removes the singularity at infinity and leads to the exact energy of the μ-states:

$$E_{0,\mu}=-\frac{\mathrm{e}^2}{2a_0\left(\mu+\frac{1}{2}\right)^2} \tag{24}$$

including the ground state with $n_\phi = 1$ in Equation (22), and the successive excited states with the next values.

The variational trial function for confinement with an angle used in [3] was not explicitly reported in that reference; however, Dr. Rosas kindly provided its form:

$$\Psi = A\rho e^{-\beta\rho} \sin(\phi) \sin(\phi - \phi_0) \tag{25}$$

for the respective intervals of $\rho[0, \infty]$ and $\phi[0, \phi_0]$. Now we compare it with the reference of Equations (23) and (21). The exponent of the radial power factor leads to the identification of the value $\mu = 1$. The arguments of the angular function in Equation (25), rewritten as a superposition of trigonometric functions of $\mu\phi$, Equation (22), leads to the identification of the values of $\mu = 0$ and 2. The correct μ dependence of the functions in Equations (21) and (23) is absent in both the angular and radial factors in Equation (25). Consequently, the numerical results of the variational calculations based on Equation (25) and reported in [3] for the ground state of the hydrogen atom confined in the angular interval of $\phi_0[0, 2\pi]$ are not reliable. Just for the sake of illustration, let us consider the cases of $\phi_0 = \pi$ and 2π, and $n_\phi = 1$ if we want to describe the ground state; according to Equation (22), the respective values of μ are 1 and 1/2, different from the values of 0 and 2 associated with Equation (25). The first case corresponds to a straight-line boundary with an energy from Equation (24) of $-2e^2/9a_0$, as the two-dimensional counterpart of Levine's plane boundary in three dimensions [31]. Also notice that the second case does not correspond to the free hydrogen atom, since the wave function vanishes at the positive x-axis for $\phi_0 = 0$ and 2π and its energy from Equation (24) is $-e^2/2a_0$, and not $-2e^2/a_0$ for $n_\phi = 0$ and $\mu = 0$ in the same equation, nor $-2e^2/9a_0$ as reported in [3].

For consistency in this subsection, it is also necessary to question the validity of the trial variational function and the results of the calculation for the ground-state energy of the hydrogen atom confined by a hyperbolic boundary in [3]. In fact, the corresponding function in their notation is

$$\Psi(\xi, \eta) = A(\eta - \eta_0)e^{-\beta(\xi+\eta)} \tag{26}$$

and their figure shows the energy for the interval of $\eta_0[-1, 1]$ going from zero for the "one-dimensional" limit to $-2e^2/a_0$ for the free hydrogen atom. The latter is again an indication that the boundary condition of the vanishing of the wave function along the x-axis from the focus, where the nucleus is located, to infinity is not being taken into account; and when the focal distance becomes vanishingly small, the results for confinement in an angle, discussed in the previous paragraph, must be reproduced correctly.

The conclusion of these comments is that the trial functions and the results reported in [3] do not provide a reliable description of the two-dimensional hydrogen atom ground-state energy for confinements by an angle and by a hyperbola. The solutions of the Schrödinger equation for the hydrogen atom confined by a hyperbola can also be constructed transparently and accurately using standard methods.

The experience of the study of confining the two-dimensional hydrogen atom in an angle can be extended to its three-dimensional counterparts of confinement by dihedral angles. This is taken up in Section 5.2.

2.4. Electron structure of a dipole-bound anion confined in a spherical box. Addendum to Electron structure of a dipole-bound anion in a spherical box: The case of a finite dipole

The articles with the above titles [4,5] contain obvious conceptual inconsistencies, about which the author has just written and submitted a Comment [11]. At this point, he simply invites the interested parties to make their own reading of the original papers.

It has been through such readings that he has become aware of the need to formulate a consistent description of the ionization of the confined hydrogen atom, as already proposed in 2.1 and to be developed further in Section 4. Some of the topics in the Sections 5.3 and 5.4 were identified in alternative readings.

3. REVIEW OF THE SUPERINTEGRABILITY OF THE SCHRÖDINGER EQUATION FOR THE FREE HYDROGEN ATOM AND ITS IONIZATION LIMIT

The separability and integrability of the Schrödinger equation for harmonic oscillators and the hydrogen atom, in different dimensions and in different coordinate systems, are well-known [6–8]. Such a superintegrability is also familiar in the classical Laplace and Helmholtz equations [32]. Furthermore, in recent years the author and collaborators have identified and constructed common generating functions for the complete wave functions of the respective systems, including their corresponding expansions in the alternative coordinates [33–37].

The review of the exact solutions of the Schrödinger equation for the hydrogen atom in 3.1 spherical, 3.2 spheroconal, 3.3 parabolic, and 3.4 prolate spheroidal coordinates is carried out in this section as the groundwork for the overview in the following section. The review emphasizes the connection between the geometry of the coordinates and the dynamic quantities behind the separability of the Schrödinger equation, leading to the physical meaning of the respective constants of motion and their integer quantum numbers. The "accidental" degeneracy of the energy levels of the hydrogen atom of order n^2, instead of the ordinary $2\ell+1$ degeneracy for any central potential, is explained as a consequence of the $O(4)$ rotational symmetry of the hydrogen atom wave functions in momentum space [6–8].

The separability or factorization of the wave functions of the hydrogen atom, in the respective coordinates, allows us to understand that their nodes correspond to conoidal surfaces or meridian planes. The latter become natural boundaries of confinement for the hydrogen atom, as natural extensions of Levine's plane [31], anticipating their discussion in Section 4.

The ionization limit of the Schrödinger equation and its eigenfunctions for the free hydrogen atom, at a vanishing energy value, corresponds to Bessel functions in the radial coordinate as known in the literature and illustrated in 2.1. The counterparts for paraboloidal [21], hyperboloidal [9], and polar angle [22] coordinates have also been shown to involve Bessel functions. These limits and their counterparts for the other coordinates are reviewed successively in this section.

The Schrödinger equation for the hydrogen atom has the form

$$\left[-\frac{\hbar^2}{2m_e}\nabla^2 - \frac{e^2}{r}\right]\Psi(\vec{r}) = E\Psi(\vec{r}), \tag{27}$$

where $-e$ and m_e are the electric charge and mass of the electron, the nucleus has a positive charge e and its position is taken as the origin, and the infinite nuclear mass approximation is made.

3.1. Spherical coordinates

For this familiar coordinate system $(0 \leq r \leq \infty, 0 \leq \theta \leq \pi, 0 \leq \phi \leq 2\pi)$, Equation (27) becomes

$$\left\{-\frac{\hbar^2}{2m_e}\left[\frac{1}{r^2}\frac{\partial}{\partial r}r^2\frac{\partial}{\partial r} + \frac{1}{r^2}\left(\frac{1}{\sin\theta}\frac{\partial}{\partial\theta}\sin\theta\frac{\partial}{\partial\theta} + \frac{1}{\sin^2\theta}\frac{\partial^2}{\partial\phi^2}\right)\right] - \frac{e^2}{r}\right\}\Psi(r,\theta,\phi) = E\Psi(r,\theta,\phi). \tag{28}$$

The angular part of the Laplacian corresponds to the square of the orbital angular momentum operator $\hat{\ell}^2$. Its second term includes the square of its z-component $\hat{\ell}_z$. Both operators commute with each other, and each one commutes with the Hamiltonian for any central potential, including the

Coulomb potential:

$$
\begin{aligned}
\left[\hat{\ell}^2, \hat{\ell}_z\right] &= 0 \\
\left[\hat{\ell}^2, \hat{H}\right] &= 0 \\
\left[\hat{\ell}_z, \hat{H}\right] &= 0.
\end{aligned}
\tag{29}
$$

These commutabilities are behind the separability of the Schrödinger equation for any central potential. In fact, Equation (27) admits product solutions,

$$
\Psi(r, \theta, \phi) = R(r)\Theta(\theta)\Phi(\phi), \tag{30}
$$

in which the respective factors satisfy the eigenvalue ordinary differential equations:

$$
-\frac{\mathrm{d}^2\Phi(\phi)}{\mathrm{d}\phi^2} = m^2\Phi(\phi) \tag{31}
$$

$$
-\left[\frac{1}{\sin\theta}\frac{\mathrm{d}}{\mathrm{d}\theta}\sin\theta\frac{\mathrm{d}}{\mathrm{d}\theta} - \frac{m^2}{\sin^2\theta}\right]\Theta(\theta) = \ell(\ell+1)\Theta(\theta) \tag{32}
$$

$$
\left\{-\frac{\hbar^2}{2m_e}\left[\frac{1}{r^2}\frac{\mathrm{d}}{\mathrm{d}r}r^2\frac{\mathrm{d}}{\mathrm{d}r} - \frac{\ell(\ell+1)}{r^2}\right] - \frac{\mathrm{e}^2}{r}\right\}R(r) = ER(r). \tag{33}
$$

Here, the familiar eigenvalues of the orbital angular momentum operators have already been incorporated, reminding the reader that their form and numerical values are determined by the periodicity condition $\Phi(\phi + 2\pi) = \Phi(\phi)$ for the first one, and by guaranteeing good behavior for the second at its regular singularity points, $\theta = 0$ and π.

Next, we simply write the respective eigenfunctions and eigenvalues

$$
\Phi_m(\phi) = \begin{cases}\sin m\phi \\ \cos m\phi\end{cases} \tag{34}
$$

$$
m = 0, 1, 2, \ldots \tag{35}
$$

$$
\Theta_{\ell,m}(\theta) = P_\ell^m(\theta) = N_{\ell,m}\sin^m\theta\, F\left(-(\ell-m), \ell-m+1; m+1; \frac{1-\cos\theta}{2}\right) \tag{36}
$$

$$
\ell = 0, 1, 2, \ldots \tag{37}
$$

$$
R_{n_r\ell}(r) = r^\ell \mathrm{e}^{-\frac{r}{na_0}} M\left(-n_r, 2\ell+2, \frac{2r}{na_0}\right) \tag{38}
$$

$$n_r = 0, 1, 2, \dots. \tag{39}$$

The associated Legendre polynomials in Equation (36) are written in their hypergeometric function form [16] 22.5.49 and 15.5.1,

$$F(a, b; c; z) = \sum_{s=0}^{\infty} \frac{(a)_s (b)_s z^s}{(c)_s s!} \tag{40}$$

with a normalization factor $N_{\ell,m}$, in order to facilitate their transcription in Section 4. The radial eigenfunctions in Equation (38) involve the product of the power and exponential functions – which ensure good behavior at the regular singular points at the origin and infinity in Equation (33) – and the confluent hypergeometric function representation of the familiar associated Laguerre polynomials [16] 22.5.54. The energy eigenvalue in Equation (33) has the form

$$E_n = -\frac{\mathrm{e}^2}{2a_0 n^2} \tag{41}$$

involving the principal quantum number,

$$n = n_r + \ell + 1 = 1, 2, 3, \dots. \tag{42}$$

The number of eigenfunctions $\Psi_{n_r,\ell,m}$ of Equations (30), (34), (36) and (38) with the same value of n in Equation (42) and the same energy in Equation (41), is easily counted using the number of complementary combinations of the individual quantum numbers in Equation (35), (37) and (39), leading to the order of the degeneracy of the corresponding energy level,

$$D_n = \sum_{\ell=0}^{n-1} (2\ell + 1) = n^2. \tag{43}$$

This is equal to the number of 4-dimensional spherical harmonics as a representation of the $O(4)$ symmetry group of the hydrogen atom [6,8].

The analysis of the ionization limit, for vanishing E in Equation (33) or infinite n_r in the eigenfunction of Equation (38), leads to the identification of the corresponding radial eigenfunctions in terms of Bessel functions:

$$R_\infty(r) = \sqrt{\frac{a_0}{r}} J_{2\ell+1} \sqrt{\frac{8r}{a_0}} \tag{44}$$

as the extension of Equation (9) introduced in [2].

3.2. Spheroconal coordinates

The spheroconal coordinates (r, χ_1, χ_2) are defined by their transformation equations to cartesian coordinates:

$$\begin{aligned} x &= r\mathrm{dn}(\chi_1 \mid k_1)\mathrm{sn}(\chi_2 \mid k_2) \\ y &= r\mathrm{cn}(\chi_1 \mid k_1)\mathrm{cn}(\chi_2 \mid k_2) \\ z &= r\mathrm{sn}(\chi_1 \mid k_1)\mathrm{dn}(\chi_2 \mid k_2) \end{aligned} \tag{45}$$

in terms of Jacobian elliptical functions, with arguments χ_1 and χ_2 in their amplitude intervals, *am* χ_i $[0, 2\pi]$, for $i = 1, 2$, and complementary parameters k_1 and k_2 such that $k_1^2 + k_2^2 = 1$ [32] page 659 and [16] 16.1.

The corresponding scale factors are

$$\begin{aligned} h_r &= 1 \\ h_{\chi_1} &= h_{\chi_2} = r\sqrt{1 - k_1^2\mathrm{sn}^2(\chi_1 \mid k_1) - k_2^2\mathrm{sn}^2(\chi_2 \mid k_2)}. \end{aligned} \tag{46}$$

Notice the common radial coordinate and scale factor for both spherical and spheroconal coordinates, Equations (45) and (46). Fixed values of χ_i correspond to elliptical cones with a common vertex at the origin, and an axis along the z-axis for $i = 1$ and along the x-axis for $i = 2$.

The Schrödinger equation for the hydrogen atom in these coordinates becomes:

$$\begin{aligned} &\left\{ -\frac{\hbar^2}{2m_e}\left[\frac{1}{r^2}\frac{\partial}{\partial r}r^2\frac{\partial}{\partial r} + \frac{1}{r^2}\frac{1}{1 - k_1^2\mathrm{sn}^2(\chi_1 \mid k_1) - k_2^2\mathrm{sn}^2(\chi_2 \mid k_2)}\left(\frac{\partial^2}{\partial \chi_1^2} + \frac{\partial^2}{\partial \chi_2^2}\right)\right] \right. \\ &\left. \quad - \frac{e^2}{r} \right\} \Psi(r, \chi_1, \chi_2) = E\Psi(r, \chi_1, \chi_2). \end{aligned} \tag{47}$$

It also admits separable solutions with the product form

$$\Psi(r, \chi_1, \chi_2) = R(r)\Lambda_1(\chi_1)\Lambda_2(\chi_2) \tag{48}$$

with the same radial function appearing as in spherical coordinates, Equations (33), (38) and (39).

In fact, the elliptical-cone coordinate dependent part of the Laplacian is identified as the square of the orbital angular momentum, via a direct comparison of Equations (28) and (47):

$$\hat{\ell}^2 = -\frac{\hbar^2}{1 - k_1^2\mathrm{sn}^2(\chi_1 \mid k_1) - k_2^2\mathrm{sn}^2(\chi_2 \mid k_2)}\left[\frac{\partial^2}{\partial \chi_1^2} + \frac{\partial^2}{\partial \chi_2^2}\right]. \tag{49}$$

The works in [36,37] have recently revisited the evaluation of the rotational spectra of asymmetric molecules. Such an evaluation requires the simultaneous solution of the eigenvalue equations for the square of the orbital angular momentum and for the purely asymmetric part of the rotational Hamiltonian,

$$\hat{H}^{\star} = \frac{1}{2}\left(e_1\hat{\ell}_x^2 + e_2\hat{\ell}_y^2 + e_3\hat{\ell}_z^2\right). \tag{50}$$

The parameters of asymmetry e_i are not independent, having to satisfy the restrictions:

$$\begin{aligned} &e_1 > e_2 > e_3 \\ &e_1 + e_2 + e_3 = 0 \\ &e_1^2 + e_2^2 + e_3^2 = \frac{3}{2} \end{aligned} \tag{51}$$

so that the choice of any one of them determines the other two.

It is obvious that the operators in Equations (49) and (50) commute with each other, and each one also commutes with the Hamiltonian in Equation (44), i. e.,

$$\begin{aligned} &\left[\hat{\ell}^2, \hat{H}^{\star}\right] = 0 \\ &\left[\hat{H}, \hat{\ell}^2\right] = 0 \\ &\left[\hat{H}, \hat{H}^{\star}\right] = 0 \end{aligned} \tag{52}$$

as the counterparts of Equation (29). Notice also that the operator $\hat{\ell}_z$ and $\hat{H}^{\star}$, making the difference between the rotations around the z-axis and for asymmetric molecules, do not commute with each other. This leads to the two alternative bases of spherical harmonics and spheroconal harmonics, as eigenfunctions of the square of the angular momentum and the Hamiltonian of any central potential [8,36].

The article [36] illustrates how to use the basis of spherical harmonics to evaluate the common eigenvalues and eigenfunctions of the operators in Equations (49) and (50). On the other hand, the article [37] uses the complementary choices of

$$\begin{aligned} k_1^2 &= \frac{e_2 - e_3}{e_1 - e_3} \\ k_2^2 &= \frac{e_1 - e_2}{e_1 - e_3} \end{aligned} \tag{53}$$

which allow the simultaneous separation of the eigenvalue equations in spheroconal coordinates, in terms of the ordinary Lamé differential equations [38]:

$$\frac{d^2\Lambda_i}{d\chi_i^2} - \left[\ell(\ell+1)k_i^2 \mathrm{sn}^2(\chi_i, k_i) + h_i\right]\Lambda_i = 0. \tag{54}$$

The separation constants h_1 and h_2 are related to the eigenvalues of interest for the molecular rotations as

$$\begin{aligned} h_1 + h_2 &= -\ell(\ell+1) \\ e_1 h_1 + e_3 h_2 &= -\frac{2E^\star}{\hbar^2}. \end{aligned} \tag{55}$$

Whittaker and Watson [38] Chapter 23 contains the analysis of the integrability of Equation (54), identifying and removing its regular singularities and proposing a series expansion for the remaining factor in the solutions:

$$\Lambda^A = f_A(\chi)\sum_{s=0}^{\infty} a_s^A \mathrm{sn}^{2s}(\chi, k), \tag{56}$$

where f_A is the singularity-removing factor, expressed as the product of the respective elliptical Jacobian functions, whose initial letter appears in the label $A = 1, s, c, d, sc, sd, cd, scd$. This leads to the classification of the solutions into two kinds and eight species according to the even or odd number of elliptical functions in the factor. The substitution of Equation (56) into Equation (54) leads to three-term recurrence relations for the expansion coefficients a_s^A.

The numerical evaluation of the eigenvalues of Equation (54) and the expansion coefficients in Equation (56) had not been implemented efficiently in the following eighty year period. In [37] a matrix representation of the three-term recurrence relation was used, following the 1967 suggestion of Hunter and Pritchard [39]. The corresponding matrices turn out to be of finite dimensions $N^A \times N^A$: $N^1 = (\ell/2) + 1$, and $N^{sc} = N^{sd} = N^{cd} = \ell/2$ for ℓ even; and $N^s = N^c = N^d = (\ell+1)/2$, and $N^{scd} = (\ell-1)/2$ for ℓ odd. In both cases the total number of independent solutions are $2\ell+1$. The corresponding diagonalization leads efficiently to accurate results for eigenvalues and eigenfunctions.

Correspondingly, the spheroconal harmonics to be used in the eigenfunctions of the hydrogen atom, Equation (48), are the products of Lamé functions in the respective variables, with common values of ℓ and

$E^\star$, and complementing each other in kind and species:

$$\mathcal{Y}^{AB}_{n_1,n_2}(\chi_1,\chi_2) = \Lambda^A_{n_1}(\chi_1,k_1)|\Lambda^B_{n_2}(\chi_2,k_2) \tag{57}$$

according to the following correspondences in their labels,

A	1	d	c	s	dc	ds	cs	dcs
B	1	s	c	d	sc	sd	cd	scd
AB	1	x	y	z	xy	xz	yz	xyz

The products AB agree with the transformation Equation (45) and their parity properties under reflection in the respective cartesian coordinate planes.The sub-indexes in Equation (57) count the number of nodes of the respective Lamé functions, and their sum is such that

$$n_1 + n_2 = \ell. \tag{58}$$

The radial function in Equation (48) coincides with the one in Equations (38) and (39), being shared by both coordinate systems. The radial quantum number added to the spheroconal quantum numbers in Equation (58) leads again to the same principal quantum number of Equation (42). Correspondingly, the final part of Section 3.1 extends its validity here.

3.3. Parabolic coordinates

The reader of this subsection is warned that the parabolic coordinates used in this chapter are different from those in [21] and coincide with those in [35], $(0 \leq \xi \leq \infty, 0 \leq \eta \leq \infty, 0 \leq \phi \leq 2\pi)$. They are defined by the transformation equations to cartesian coordinates:

$$\begin{aligned} x &= \xi\eta\cos\phi \\ y &= \xi\eta\sin\phi \\ z &= \frac{\xi^2-\eta^2}{2}. \end{aligned} \tag{59}$$

Fixed values of ξ and η define confocal orthogonal paraboloids with the focus at the origin and their common axis as the z-axis, opening in the directions of increasing and decreasing values of z, respectively. The radial coordinate and the scale factors can be evaluated to be

$$r = \frac{\xi^2+\eta^2}{2} \tag{60}$$

$$h_\xi = \sqrt{\xi^2+\eta^2}$$

$$h_\eta = \sqrt{\xi^2 + \eta^2} \qquad (61)$$
$$h_\phi = \xi\eta.$$

The reader may also notice the common ϕ coordinate and scale factor h_ϕ shared with the spherical coordinates.

Then the Schrödinger equation for the hydrogen atom takes the form

$$\left\{-\frac{\hbar^2}{2m_e}\left[\frac{1}{\xi^2+\eta^2}\left(\frac{1}{\xi}\frac{\partial}{\partial\xi}\xi\frac{\partial}{\partial\xi} + \frac{1}{\eta}\frac{\partial}{\partial\eta}\eta\frac{\partial}{\partial\eta}\right) - \frac{1}{\xi^2\eta^2}\frac{\partial^2}{\partial\phi^2}\right] - \frac{2e^2}{\xi^2+\eta^2}\right\}\Psi(\xi,\eta,\phi) = E\Psi(\xi,\eta,\phi). \qquad (62)$$

This equation also admits product solutions of the form

$$\Psi(\xi,\eta,\phi) = \Xi(\xi)H(\eta)\Phi(\phi) \qquad (63)$$

with the angular factor already discussed in Equations (34) and (35). The eigenvalue ordinary differential equations in the paraboloidal degrees of freedom take the forms:

$$\left\{-\frac{\hbar^2}{2m_e}\left[\frac{1}{\xi}\frac{d}{d\xi}\xi\frac{d}{d\xi} - \frac{m^2}{\xi^2}\right] - E\xi^2\right\}\Xi(\xi) = A_\xi\,\Xi(\xi) \qquad (64)$$

$$\left\{-\frac{\hbar^2}{2m_e}\left[\frac{1}{\eta}\frac{d}{d\eta}\eta\frac{d}{d\eta} - \frac{m^2}{\eta^2}\right] - E\eta^2\right\}H(\eta) = A_\eta H(\eta) \qquad (65)$$

where the separation constants are subject to the restriction

$$A_\xi + A_\eta = 2e^2. \qquad (66)$$

Both Equations (64) and (65) have the same form and they can be interpreted as Schrödinger equations in circular-like coordinates for harmonic oscillators [33], as indicated by their respective kinetic energy and quadratic potential energy terms. The identification and interpretation are even more convincing if we parametrize the negative energy of the bound states of the hydrogen atom as

$$E = -\frac{1}{2}m_e\omega^2. \qquad (67)$$

The choice of parabolic coordinates in [35] and Equation (56) is motivated by our interest in exploiting the connection between the superintegrable harmonic-oscillator and atomic-hydrogen systems [33–35]. For instance, the well-known eigenfunctions and energy eigenvalues for the two-dimensional harmonic oscillators can be written immediately by borrowing them from [33]:

$$\Xi_{n_\xi}(\xi) = \xi^m \mathrm{e}^{-\frac{m_e\omega\xi^2}{2\hbar}} M\left(-n_\xi, m+1, \frac{m_e\omega\xi^2}{\hbar}\right) \tag{68}$$

$$A_\xi = \hbar\omega(2n_\xi + m + 1) \tag{69}$$

$$H_{n_\eta}(\eta) = \eta^m \mathrm{e}^{-\frac{m_e\omega\eta^2}{2\hbar}} M\left(-n_\eta, m+1, \frac{m_e\omega\eta^2}{\hbar}\right) \tag{70}$$

$$A_\eta = \hbar\omega(2n_\eta + m + 1). \tag{71}$$

Substitution of the separation constants from Equations (69) and (71) in Equation (66) leads to the identification of the oscillator frequency,

$$\omega = \frac{\mathrm{e}^2}{\hbar(n_\xi + n_\eta + m + 1)}. \tag{72}$$

In turn, substitution of this frequency in Equation (67) reproduces the familiar expression for the energy eigenvalues and the identification of the principal quantum number in terms of the parabolic quantum numbers:

$$n = n_\xi + n_\eta + m + 1. \tag{73}$$

The complete wave functions from Equations (63), (68), (69) and (34) become

$$\Psi_{n_\xi,n_\eta,m}(\xi,\eta,\phi) = (\xi\eta)^m \mathrm{e}^{-\frac{\xi^2+\eta^2}{2na_0}} M\left(-n_\xi, m+1, \frac{\xi^2}{na_0}\right) \cdot M\left(-n_\eta, m+1, \frac{\eta^2}{na_0}\right)\Phi_m(\phi). \tag{74}$$

The reader can be convinced that the number of independent states with the same principal quantum number n is the same as that of Equation (40) by counting the different combinations of the parabolic quantum numbers in Equation (73).

The ionization limit for either paraboloidal degree of freedom follows from the corresponding Equation (64) or (65) with $E = 0$, which become the ordinary Bessel differential equation with an argument that can be read off directly from the same equations. Their solutions can also be obtained as the limits of the respective factors in Equation (74) for infinite n and n_i, with $i = \xi$ or η. The final results are expressed by

$$\Xi_\infty(\xi) = J_m\left(\sqrt{\frac{2m_e A_\xi}{\hbar^2}}\xi\right) \tag{75}$$

and the corresponding one for η, as the counterparts of Equation (44).

3.4. Prolate spheroidal coordinates

The prolate spheroidal coordinates used in this chapter are the same as those in [18,40] represented with different letters ($1 \le u \le \infty, -1 \le v \le 1, 0 \le \phi \le 2\pi$). They are defined by their transformation equations to cartesian coordinates:

$$\begin{aligned} x &= f\sqrt{(u^2-1)(1-v^2)}\cos\phi \\ y &= f\sqrt{(u^2-1)(1-v^2)}\sin\phi \\ z &= fuv. \end{aligned} \tag{76}$$

Fixed values of u define confocal prolate spheroids with the origin as their center, common foci at ($x = 0, y = 0, z = \pm f$) and the z-axis as their common axis; each one has an eccentricity $1/u$. Fixed values of v define confocal two-sheet hyperboloids sharing the same center, foci and axis with the orthogonal spheroids; each one has an eccentricity $1/v$. The angle ϕ is shared with the spherical and parabolic coordinates, including its conjugate momentum and its associated eigenfunctions and eigenvalues.

The corresponding scale factors are

$$\begin{aligned} h_u &= f\sqrt{\frac{u^2-v^2}{u^2-1}} \\ h_v &= f\sqrt{\frac{u^2-v^2}{1-v^2}} \\ h_\phi &= f\sqrt{(u^2-1)(1-v^2)}. \end{aligned} \tag{77}$$

The position of the nucleus is chosen to be at the focus F_1 on the z-axis with $z = -f$. Then the distance from the nucleus becomes

$$r_1 = \sqrt{x^2+y^2+(z+f)^2} = f(u+v). \tag{78}$$

The Schrödinger Equation (24) takes the corresponding form

$$\left\{-\frac{\hbar^2}{2m_e f^2}\left[\frac{1}{u^2-v^2}\left(\frac{\partial}{\partial u}(u^2-1)\frac{\partial}{\partial u}+\frac{\partial}{\partial v}(1-v^2)\frac{\partial}{\partial v}\right)\right.\right.$$
$$\left.\left.+\frac{1}{(u^2-1)(1-v^2)}\frac{\partial^2}{\partial\phi^2}\right]-\frac{e^2}{f(u+v)}\right\}\Psi(u,v,\phi)=E\Psi(u,v,\phi). \quad (79)$$

It is also separable admitting the product solutions

$$\Psi(u,v,\phi)=U(u)V(v)\Phi(\phi) \quad (80)$$

for which the spheroidal and hyperboloidal factors satisfy the eigenvalue ordinary differential equations:

$$\left\{-\left[\frac{d}{du}(u^2-1)\frac{d}{du}-\frac{m^2}{u^2-1}\right]-\frac{2f}{a_0}u+\frac{f^2}{n^2a_0^2}u^2\right\}U(u)=-AU(u) \quad (81)$$
$$\left\{-\left[\frac{d}{dv}(1-v^2)\frac{d}{dv}-\frac{m^2}{1-v^2}\right]+\frac{2f}{a_0}v-\frac{f^2}{n^2a_0^2}v^2\right\}V(v)=AV(v). \quad (82)$$

The linear terms come from the Coulomb interaction, the quadratic terms come from the energy with the parametrization of Equation (41), and A is the separation constant.

The reader may notice that both equations are the same one, each in the domain of its independent variable. Their regular singularities at the end points of these domains, $v=-1$ and 1 and $u=1$ and ∞, are removed by the powers of the factors entering in the distance from the z-axis, taking into account the effect of the centrifugal potential, and by the asymptotic exponentials, determined by the energy as coefficients of the quadratic terms:

$$U(u)=(u^2-1)^{\frac{m}{2}}e^{-\frac{fu}{na_0}}S(u) \quad (83)$$
$$V(v)=(1-v^2)^{\frac{m}{2}}e^{-\frac{fv}{na_0}}S(v) \quad (84)$$

showing the counterparts of the corresponding behaviors in spherical and parabolic coordinates, Equations (36), (38) and (74). Coulson and Robinson [41] investigated this type of solution by including the power series solutions for the remaining factors,

$$S(u)=\sum_{s=0}^{\infty}c_s(u-1)^s \quad (85)$$

$$S(v) = \sum_{s=0}^{\infty} c_s (1-v)^s \tag{86}$$

involving the same expansion coefficients. In fact, the substitutions of the proposed solutions of equations (83) and (85) or (84) and (86) in Equation (81) or (82), for chosen values of f, n and m, lead to the same three-term recurrence relation:

$$2(s+1)(s+m+1)c_{s+1} + \left[s\left(s+2m+1-\frac{4f}{na_0}\right) - A^{nm}\right] c_s \\ + (n-m-s)c_{s-1} = 0, \tag{87}$$

where

$$A^{nm} = A + \frac{f^2}{n^2 a_0^2} - \left(\frac{2f}{na_0}\right)(n-m-1) - m(m+1). \tag{88}$$

The recurrence relation in Equation (87) is equivalent to a tridiagonal matrix [39], of order $(n-m)\times(n-m)$ as determined by the last term on the l.h.s. of Equation (87). Its diagonalization yields the increasing eigenvalues of A_k^{nm} with $k = 1, 2, 3, \ldots, (n-m)$, and the associated expansion coefficients c_s^{nmk}. For each set of values of (n, m, k), the complete wave functions of the type of Equations (80) and (83)–(86), with their quantum numbers counting the respective number of nodes, can be written as

$$\Psi^{nk}_{n_u n_v m}(u, v, \phi) = U^{nmk}_{n_u}(u) V^{nmk}_{n_v}(v)\Phi_m(\phi) \tag{89}$$

with the restriction

$$n = n_u + n_v + m + 1 = k_{\max} + m \tag{90}$$

as the counterpart of Equations (42) and (73) in the previous coordinates. The labeling in Equation (89) is redundant, but helps to understand the reasons behind Equation (90).

The ionization limit for vanishing energy and infinite value of n also follows from Equations (81) and (82). The analysis of the cases of n_u becoming infinite for fixed values of n_v and m, and n_v infinite for fixed n_u and m is postponed to Sections 4.4 and 4.5, respectively.

We refer the reader to the original work of Erikson and Hill [42] and its adaptation in [9] in order to illustrate the unity, connections and differences in the solutions reviewed in this section. We simply cite equation (14) from [9] in our notation:

$$\hbar^2 \hat{A} = \hat{\ell}_1^2 + 2\vec{f} \cdot \left[\hat{\vec{p}} \times \hat{\vec{\ell}}_1 - \hat{\vec{\ell}}_1 \times \hat{\vec{p}} - 2m_e e^2 \hat{r}\right] - 2m_e f^2 \hat{H} \tag{91}$$

for the dynamic quantity associated with the separation constant in Equations (81) and (82). The limit of vanishing focal distance allows us to recover the square of the angular momentum as the constant of motion in spherical and spheroconal coordinates. The quantity inside the brackets in the second term is the Runge–Lenz vector operator, which is the constant of the motion in parabolic coordinates and the surviving non-trivial term in the limit of an infinite focal distance. The third term, proportional to the Hamiltonian, is included in order to get the complete quantity $\hat{A}$. The linear combination of operators in Equation (91) interpolates between the conoids with eccentricities going from zero to one in this discussion. The extrapolation beyond, for hyperboloids [9] and circular cones [22], is further explored in Sections 4.5 and 4.2 respectively.

4. OVERVIEW OF THE CONFINEMENT OF THE HYDROGEN ATOM BY A CONOIDAL BOUNDARY AS A SYMMETRY-BREAKING EFFECT

The separability and integrability of the Schrödinger equation for the hydrogen atom confined by impenetrable conoidal boundaries, coinciding with one of the constant conoidal coordinate surfaces, allows the direct use of one of the corresponding eigenfunctions reviewed in Section 3, when the boundary coincides with one of its nodal surfaces. There are many other solutions for the chosen boundary, which in general are different from the ones included in the review. This section contains an overview of the systematic construction of such eigenfunctions, which must satisfy the boundary condition of vanishing at the position of the confining conoidal boundary: sphere [14], circular cone [22], paraboloid [21], prolate spheroid [18], and hyperboloid [9]. The familiar degeneracy of the energy levels of the free hydrogen atom is removed by the new boundary condition for the hydrogen atom in the respective situations of confinement, which may be viewed as a symmetry-breaking effect on the $O(4)$ symmetry [6–8].

Indeed, the selection of the confining boundary, by a fixed value of one of the coordinates of those systems in which the Schrödinger equation is separable and integrable, allows the analysis to be carried out for each type of conoidal boundary successively. The forms of the solutions reviewed in Section 3 are still valid, but the eigenvalues of the constant of the motion associated with the boundary coordinate cease to be integers, in general, thereby removing the original degeneracy of the free hydrogen atom. The other functions entering as factors in the complete wave function for the confined atom are also affected by the change described in the previous sentence, but keep their own intrinsic integer quantum numbers.

The overview of this section is restricted to the geometries previously investigated by the author and his collaborators. It is also being extended in a natural way to the confinement in the complementary geometries of elliptical cones and dihedral angles, already included in Section 3 and previewed in Section 5.

As a guide to the reader, in the following subsections, free use, adaptations and extensions of the equations in Section 3, for the successive coordinates are made. Following the order of the comments in 2.1, the exact solutions are written out, taking into account the confinement boundary condition and incorporating the non-integer eigenvalues; the breaking of the symmetry or removal of the degeneracy of the free hydrogen energy levels is analyzed at the end points of the domain of the confining coordinate; the zero energy eigenstates and the positions of the confining boundary are identified; the energies of the confined free electron for any position of the confining boundary are also identified, providing the physical reference for the analysis of the ionization of the confined atom.

4.1. Confinement by a sphere

The analysis is for concentric confinement of the atom in a sphere of radius $r = r_0$, at which the wave function must vanish. The complete wave functions are products of the radial function of Equation (38) and spherical harmonics of Equations (36) and (34) or spheroconal harmonics of Equation (57). The boundary condition depends only on the radial function, becoming

$$M\left(-\nu_r, 2\ell+1, \frac{2r_0}{\nu a_0}\right) = 0. \tag{92}$$

Notice the replacement of n_r for the free atom by ν_r for the confined atom, which in general is non-integer and leads to the corresponding changes in the energy via Equations (41) and (42),

$$E_\nu = -\frac{\mathrm{e}^2}{2a_0\nu^2} \tag{93}$$

$$\nu = \nu_r + \ell + 1. \tag{94}$$

While the radial function in Equation (38) with integer n_r involves the Laguerre polynomials, the corresponding function in Equation (92) with non-integer ν_r becomes an infinite series. This poses no difficulty in the accurate evaluation of its zeros as already discussed in Section 2.1, in connection with the s-states. The reader may also take notice of the ℓ-dependence of the function and the eigenenergies in Equations (92) and

(93) via Equation (94), which translates into the removal of the degeneracy of the states of the free hydrogen atom upon confinement in a sphere. Also notice that the quantum numbers associated with the angular degrees of freedom keep their integer values.

We go on to the analysis of the asymptotic limit of the solutions of Equation (92) for infinitely large spheres and states with a fixed value of ℓ and $\nu_r = n_r + \beta$ with vanishing values of β. The difference with respect to the analysis for s-states in Section 2.1 resides in the differences in their values of n_r and ℓ, with the more general result:

$$\beta = \left(\frac{2r_0}{na_0}\right)^{2n_r+2\ell+2} \frac{e^{-\frac{2r_0}{na_0}}}{n_r!\Gamma(n_r + 2\ell + 2)}. \tag{95}$$

Then the energy from Equation (93), keeping only the linear term in β, takes the form,

$$E_{\nu_r\ell}(r_0) = -\frac{e^2}{2a_0n^2}\left[1 - \frac{2}{n}\left(\frac{2r_0}{na_0}\right)^{2n} \frac{e^{-\frac{2r_0}{na_0}}}{n_r!\Gamma(n + \ell + 1)}\right], \tag{96}$$

as the extension of Equation (15). The originally degenerate states of the free hydrogen atom with a chosen value of n have their energies ordered according to the decreasing values of $\ell = n-1, n-2, \ldots, 0$ and increasing values of $n_r = 0, 1, \ldots, n-1$, determined by the argument of the Gamma function.

Next, Equation (44) is used for the analysis of the states with zero energy, for which the boundary condition becomes

$$J_{2\ell+1}(\sqrt{8r_0/a_0}) = 0. \tag{97}$$

Then the radii of the spheres leading to zero energy for the originally $n_r\ell$ states are expressed in terms of the zeros of the respective ordinary Bessel functions as

$$r_0(E=0) = \frac{a_0}{8} j^2_{2\ell+1,n_r+1}. \tag{98}$$

Since n_r counts the radial excitations in the free hydrogen atom, the reader must notice the shift in counting the number of zeros in Equation (98). This insures that the states involved in the interpolation of the energy levels, between their values from Equations (96) and (98), have the same number of radial nodes in the entire range of the confining sphere radius.

The free electron confined in spherical boxes is included here for two reasons. Their eigenfunctions provide complete orthonormal bases for accurate evaluations of confined potentials, and their energy eigenvalues are also valid references to analyze binding and unbinding processes as already recognized in Section 2.1.

For spherical confinement we only need to cite the radial function in terms of spherical Bessel functions and their roots:

$$R^{\mathrm{CFe}}(r) = j_\ell\left(\frac{x_{\ell,s} r}{r_0}\right) \tag{99}$$

and its energy eigenvalues in terms of its quantized wave numbers:

$$E^{CFe}(r_0) = \frac{\hbar^2 x_{\ell,s}^2}{2m_e r_0^2}. \tag{100}$$

For s-states $x_{0,1} = \pi$ is the value used in [12,13] and 2.1. This subsection of the overview contains the natural extension for all the states of the hydrogen atom in spherical enclosures.

4.2. Confinement by a circular cone

The same eigenfunctions in spherical coordinates used in the previous subsection and introduced in 3.1 are the basis for the analysis of the hydrogen atom confined by a circular cone defined by a fixed value of the polar angle $\theta = \theta_0$. The boundary condition requiring the vanishing of the wave function at such an angle must be satisfied by the hypergeometric function in Equation (36),

$$F\left(-(\lambda - m), \lambda - m + 1; m + 1; \frac{1 - \cos\theta_0}{2}\right) = 0. \tag{101}$$

While the spherical harmonics, when all directions are included, involve associated Legendre polynomials with integer values of ℓ and m, Equations (37) and (39), the function in Equation (101) becomes an infinite series for non-integer values of λ, which replaces ℓ. The numerical evaluation of its zeros in θ_0, between 0 and π, for fixed values of m and chosen values of λ, was implemented accurately in [22], by including a large enough number of terms in the series so that convergence was guaranteed.

The eigenvalues of λ have to be incorporated in the radial part of the wave function, Equation (36), maintaining its integer quantum number n_r from Equation (37). Then the energy also maintains its form of Equation (93), with the replacement of Equation (94) by

$$\nu = n_r + \lambda + 1. \tag{102}$$

The polar angle θ in the domain $[0, \pi]$, or its $\cos\theta$ function in the domain $[-1, 1]$, are alternative variables for the description of the confining cones. The end points of the domain correspond to the two directions of the polar axis, where the regular singularities of the associated Legendre functions are located. We assume the nucleus to be at the origin infinitesimally to the north. This situation corresponds to the chosen representation of the function in Equation (101): it is regular in the northerly direction, where its argument is zero, and singular in the southerly direction, where its argument becomes one. These situations correspond to the "one-dimensional" limit of the cone closing around the positive z-axis with the nucleus inside at the vertex, and the "almost-free" hydrogen atom limit in which the cone closes around the negative z-axis with the nucleus outside at the vertex and the electron has the open domain $(-1, 1]$ to move about in. The limit of the cone becoming the equatorial plane for $\theta_0 = \pi/2$ corresponds to Levine's [31].

We invite the reader to follow the changes of the energy levels of the hydrogen atoms confined by circular cones, starting infinitesimally close to the south pole, passing the equator and approaching the north pole, by referring them to [22]. At the end we will get back to the south pole itself to clarify the difference between "almost-free" and free.

In fact, in Figs. 1(a), (b), and (c) of [22] the energy levels for the states $(n_r = 0, \lambda, m)$ are displayed as functions of $\cos\theta_0$. At the left end, the degenerate energy levels of the free hydrogen atom may be identified; in the middle, the plane nodal energy levels and their degeneracies of [31] are also identified; and at the right end, the energy levels approach their common limit of zero energy. This interpolating journey is guided by Equation (101), and next we carefully visit its starting and finishing poles.

The singularity of the function in Equation (101) is of the logarithmic type [16], allowing accurate numerical evaluation of its zeros as $\cos\theta_0$ approaches minus one [22]. The energy spectrum in that limit is practically the same as that of the free hydrogen atom.

However, the eigenfunctions are not. It is the same situation already commented on in Section 2.3 in the two-dimensional case. The boundary condition of Equation (101) is not satisfied by the states with integer ℓ and $m = 0$ since $P_\ell(\cos\pi = -1) = (-1)^\ell$. The exclusion of these states has just been realized by the author and justifies the adjective of "almost-free" used in a previous paragraph.

The visit to the north pole does not include any surprises. It involves the limit of infinitely large values of λ, for which the associated Legendre functions become Bessel functions [16,22] of order m and argument $\lambda\sqrt{2(1 - \cos\theta)}$. Correspondingly, its zeros become

$$\theta_0 = \cos^{-1}\left(1 - \frac{j_{m,s}^2}{2\lambda^2}\right) \to 0 \tag{103}$$

behind the infinitely degenerate energy levels at zero energy, Equations (93) and (102), for the "one-dimensional" confined hydrogen atom.

The free electron confined by a circular cone shares the same harmonic angular functions with the hydrogen atom, and its radial function is the adaptation of Equation (99) with a spherical Bessel function of order λ and wave number determined by its energy:

$$E^{CFe} = \frac{\hbar^2 k^2}{2m_e}. \tag{104}$$

For open boundaries, the energy and wave numbers of the type of this equation are not quantized, in contrast with the situation of the closed boundaries as in Equation (100). Correspondingly, the threshold value of the energy in Equation (104) is zero, validating it as the reference for the analysis of ionization of the hydrogen atom confined by an open boundary.

4.3. Confinement by a paraboloid

We consider the confinement of the hydrogen atom by a paraboloid defined by a fixed value of $\xi = \xi_0$. Its wave functions have the form of Equation (74), and the condition that it vanishes at the position of the confining boundary is expressed as

$$M\left(-\nu_\xi, m+1, \frac{\xi_0^2}{\nu a_0}\right) = 0. \tag{105}$$

Notice the change of the integer quantum number n_ξ in Equation (74) to the non-integer values ν_ξ here. Correspondingly, the principal quantum number in Equation (73) becomes

$$\nu = \nu_\xi + n_\eta + m + 1, \tag{106}$$

while the energy keeps its form of Equation (93).

The evaluation of the zeros of the infinite series in Equation (105) or its equivalents in [21] was implemented in the latter for fixed values of m and n_η, and changing values of ν_ξ, yielding the energy eigenvalues via Equations (93) and (106), and the positions of the confining paraboloids. Figs. 1(a), (b) and (c) in [21] illustrate the variations of the energy levels from their free-hydrogen atom values for large values of ξ_0, with increasing values towards zero as ξ_0 decreases for increasing values of ν_ξ and fixed values of $n_\eta = 0, 1, 2, 3$ and $m = 0, 1, 2, 3$. The states with common values of m show their energies approaching the zero value at common positions of the boundary. Such an overall behavior is quantitatively explained next, on the basis of the asymptotic behaviors of the function in Equation (105) for large values of ξ_0, according to Equation (12), and for large values of ν_ξ, according to Equation (75), respectively.

Indeed, the removal of the degeneracy of the free hydrogen atom energy levels due to the presence of a paraboloidal boundary far away from the nucleus follows from the application of Equation (12) in Equation (105), and here we limit ourselves to writing the final result:

$$E_{\nu_\xi m}(\xi_0) = -\frac{e^2}{2a_0 n^2}\left[1 - \frac{2}{n}\left(\frac{\xi_0^2}{na_0}\right)^{2n_\xi+m+1} \frac{e^{-\frac{\xi_0^2}{na_0}}}{(n_\xi)!\Gamma(n_\xi+m+1)}\right], \tag{107}$$

where $\nu_\xi = n_\xi + \beta$ and the limit of vanishing β has been taken just as in 2.1 and 4.1. The exponent in the power factor in the second term inside the brackets favors the ordering in the splitting of the energy levels according to increasing values of n_ξ.

The positions of the paraboloidal boundaries where the successive energy levels become degenerate at zero energy follow from Equation (75) with $A_\xi = 2e^2$, with the explicit result,

$$\xi_0(E=0) = \frac{j_{m,n_\xi+1}\sqrt{a_0}}{2}. \tag{108}$$

The remarks made after Equation (98) about the shifting of n_r by one unit also apply here about n_ξ. The reader is also reminded about the difference in the parabolic coordinates used in [21] and here.

The free electron confined by a paraboloidal boundary is described by Weber parabolic functions [16] Chapter 9, which may be obtained from Equations (64) and (65), with the following changes in the separation constants:

$$A_\xi + A_\eta = 0 \tag{109}$$

instead of Equation (66), since there is no longer any Coulomb interaction; and since the energies are now positive,

$$E = \frac{\hbar^2 k^2}{2m_e}, \tag{110}$$

the eigenfunctions are expressed as analytical continuations of harmonic oscillator wave functions, Equations (68)–(71), with an imaginary frequency.

For our purposes the important point to recognize is that the energies in Equation (110) are not quantized, because the boundary is open as in 4.2.

Correspondingly, its lowest value of zero is the reference for the analysis of ionization.

4.4. Confinement by a prolate spheroid

The confining boundary is defined by a fixed value of $u = u_0$ and a chosen value of f in Equation (76). The wave functions have the form of Equation (89) and must satisfy the boundary condition of vanishing at the position of the confining spheroid. This is equivalent to the vanishing of the factor in Equation (85):

$$S_{\nu_u}^{\nu m}(u = u_0) = 0, \tag{111}$$

which in general will involve non-integer values of ν_u, with the corresponding changes in Equation (90),

$$\nu = \nu_u + n_v + m + 1, \tag{112}$$

as well as in Equations (87) and (88), where $\nu - m$ becomes $\nu_u + n_v + 1$; no longer an integer. While in Section 3.4 the three-term recurrence relation for the expansion coefficients could be represented by finite size matrices, now the matrix size becomes infinite. In [18] an alternative representation of the spheroidal wave function was used, also involving infinite series and matrices. In practical terms, taking a finite number of terms in the series and diagonalizing the corresponding matrices leads to accurate and convergent results for the eigenvalues, the eigenvectors and the zeros in Equations (88), (87) and (111), respectively.

For a fixed value of the semi-focal distance f, the energy of each state of the hydrogen atom increases monotonically, as u_0 takes decreasing values starting from infinity, from its free atom value (slowly) to zero (increasingly fast) and on with positive values (faster). The removal of the initial degeneracy of the free hydrogen energy levels is governed by their ordering of increasing values according to the quantum number associated with the confining degree of freedom $n_u = 0, 1, 2, \ldots$. When the latter approaches infinity, and the energy in Equation (93) goes to zero, the situation is analyzed next by going back to Equation (81).

Indeed, the quadratic term in f representing the energy vanishes, the separation constant A becomes infinitely large, the confining spheroids become very eccentric, so that u becomes almost one. This suggests using the transformation complementary to the one introduced in equation (11) of [9] for the hyperboloidal boundary:

$$x = \sqrt{2A(u - 1)}, \tag{113}$$

and $u + 1$ is slightly larger than two. Then, Equation (81) is transformed into the ordinary Bessel differential equation,

$$\left[x^2 \frac{\mathrm{d}^2}{\mathrm{d}x^2} + x \frac{\mathrm{d}}{\mathrm{d}x} + (x^2 - m^2)\right] U = 0. \tag{114}$$

Consequently, the spheroidal boundaries for which the eigenenergies vanish are given in terms of the zeros of the Bessel functions, via Equations (113) and (114),

$$u_0(E = 0) = 1 + \frac{j^2_{m,n_u+1}}{2A}. \tag{115}$$

This is the counterpart of Equations (98), (103) and (108) in 4.1, 4.2 and 4.3 of the confinement by the previous boundaries. The reader is also advised about the difference in signs of the separation constant in Equations (81) and (82) and in equations (5a, b) of [9].

The eigenfunctions of the free electron confined in the same prolate spheroids are expressed as products of regular radial and angular spheroidal wave functions [16] Chapter 21, in the respective coordinates u and v, and the eigenfunctions of Equations (34) and (35). The radial functions are expressed as infinite series of spherical Bessel functions of order $m + s$ and argument kfu. Its eigenvalues are determined by the boundary condition on the radial factor,

$$R^1_{mn}(kf, u = u_0) = 0. \tag{116}$$

The corresponding energy eigenvalues are determined by the quantized values of the wave number, as the zeros in Equation (116),

$$E_{mn_u}(u_0) = \frac{\hbar^2 k^2_{mn_u}}{2m_e}. \tag{117}$$

This is the counterpart of Equation (100), which is recovered in the limit of vanishing f and infinite u, so that fu becomes r, the radial coordinate. Correspondingly, it is the reference for the analysis of the ionization of the hydrogen atom confined by prolate spheroids.

4.5. Confinement by a hyperboloid

This is the situation already commented on in 2.2 apropos of [1] and [9]. The hyperboloid is defined by $v = v_0$. The eigenfunctions of the form in

Equation (89) must vanish at this confining boundary. Now it is the function from Equation (86) that must satisfy such a condition,

$$S_{\nu_v}^{\nu m}(v_0) = 0. \tag{118}$$

Again here, non-integer values of ν_v and ν related by

$$\nu = n_u + \nu_v + m + 1 \tag{119}$$

are obtained, determining the energy eigenvalues via Equation (93).

The evaluation of the expansion coefficients and separation constant in Equations (87) and (88), with $\nu - m = n_u + nu_v + 1$, needs the diagonalization of infinite size matrices. In [9] the diagonalization of finite size matrices provided accurate and convergent results for the domains of the confining boundary $(-1, 1)$ and the energy eigenvalues $(-e^2/2a_0, 0)$. The nucleus is placed in the southern focus, so that the upper end of the boundary domain and the complete energy domain correspond to the "almost-free" hydrogen atom, while the combination of the lower and upper ends of the respective domains corresponds to the "one-dimensional" limit where the hyperboloid closes around its lower axis with the nucleus at its vertex. The intermediate value $v_0 = 0$ corresponds to the equatorial plane confining the distant hydrogen atom [25–30].

The terminology of the previous paragraph is the same as the one introduced in 4.2, and with similar explanations. The free hydrogen-atom states with $m = 0$, and n_v integer are not included among those of the "almost-free" version, because they do not satisfy the boundary condition of Equation (118). The other ones get their degeneracies removed with energy eigenvalues ordered according to increasing values of n_v, for values of v_0 close to one. The approach to the "one-dimensional" hydrogen atom is described by analyzing Equation (82) in the limits of v_0 approaching minus one, the energy becoming zero and the separation constant getting very large, in such a way that

$$x = \sqrt{2A(1+v)}. \tag{120}$$

By recalling that Equations (81) and (82) are the same, but in different domains, it should not be surprising that the change of variable in Equation (120) transforms Equation (82) into Equation (114). Consequently, the boundary condition on the corresponding ordinary Bessel function leads to the positions of the hyperboloids for which the energy is zero:

$$v_0(E = 0) = -1 + \frac{j_{m,n_v+1}^2}{2A}, \tag{121}$$

which become minus one for infinitely large A. The zero energy level becomes infinitely degenerate for the hydrogen atom confined by the hyperboloid closing around the negative z-axis from the position of the nucleus to minus infinity.

The free electron confined by a hyperboloid has the same type of wave function described in the paragraph of Equation (116). The difference consists in the boundary condition applied to the angular spheroidal functions, which are expressed as infinite series of associated Legendre polynomials of order $m+r$, degree m, and argument v:

$$S^1_{mn}(kf, v = v_0) = 0. \tag{122}$$

Here, as in 4.2 the boundary condition is satisfied via the order $n = \nu_v$ for any value of the wave number k or the energy in Equation (103). Correspondingly, the discussion after that equation also holds here, reinforcing the conclusion that for open boundaries the zero energy of the confined free electron is the reference as the ionization threshold.

This overview has been developed following the order of the successive coordinates, which define the respective confining boundaries. Both spherical and prolate spheroidal coordinates include closed and open boundaries, while the parabolic coordinates are all open. An alternative order of reading is guided by the value of the eccentricity, as pointed out in the closing of Section 3, for which the sphere and the spheroid appear together, and the paraboloid, hyperboloid and circular cone follow, covering the values from zero to infinity. The geometrical elements of the confining boundaries are accompanied by the dynamical constants of the motion as exhibited by Equation (91). The common features and the specific ones for each boundary can be appreciated by comparing them in succession. The breaking of the symmetry, or removal of the degeneracy of the free-hydrogen atom energy levels, by the presence of the boundary has been emphasized in this overview. At the same time, the surviving symmetry or degeneracy associated with the sum of the two other quantum numbers must not be forgotten. While the hydrogen atom in spherical confinement remains non-polar, in the eccentric confinements it acquires an electric dipole moment.

5. PREVIEW OF PROBLEMS ON CONFINED ATOMS AND MOLECULES OF CURRENT AND FUTURE INVESTIGATIONS

The problems on new situations of confinement of atoms and molecules previewed in this section are in different stages in their respective investigations. The hydrogen atom confined by an elliptical cone was formulated between the completion of [37] and the invitation to write this Chapter, as described in 5.1. Preliminary numerical results were presented at the LI National Physics Congress of Sociedad Mexicana de Física in Zacatecas, Mexico, November 2008. The problem of the hydrogen atom confined by dihedral angles is a natural geometrical extension of [31], and an extension of [3,10] from two to three dimensions. Its formulation was decided while preparing Section 4 and its presentation in 5.2 is quite preliminary. The problems on confined two-electron atoms and molecules formulated in 5.3 are focused on the atomic hydrogen anion confined by a nodal plane and the molecular hydrogen confined by a dihedral angle.

The first one, on the binding of an electron by a confined polar atom, is a challenging alternative to [4,5]. The second one is an alternative in the choice of boundary to [43] for the molecular hydrogen close to a confining plane. Section 5.4 emphasizes the importance of using the complete harmonic expansions, outside and inside the sources, of the electrostatic potential of atoms and molecules. In particular, the dipole fields in spherical, prolate and oblate spheroidal coordinates define other alternatives to [4,5].

5.1. Hydrogen atom confined by an elliptical cone

This type of confinement is alternative to that by a circular cone in [22] and Section 4.2, using spheroconal coordinates instead of spherical coordinates. The radial coordinate is common to both sets, but the angular coordinates are different. Here we first identify the set of cones by fixed values of the coordinate χ_1 with their common vertex as the origin, their common axis as the z-axis, and elliptical cross sections as determined next. In fact, by using the relationships

$$\begin{aligned} \mathrm{cn}^2(\chi_1 \mid k_1) &= 1 - \mathrm{sn}^2(\chi_1 \mid k_1) \\ \mathrm{dn}^2(\chi_1 \mid k_1) &= 1 - k_1^2 \mathrm{sn}^2(\chi_1 \mid k_1) \end{aligned} \tag{123}$$

and the transformation equations (45), the following equation is obtained:

$$\frac{k_1^2 \mathrm{sn}^2(\chi_1 \mid k_1) x^2}{z^2 \mathrm{dn}^2(\chi_1 \mid k_1)} + \frac{\mathrm{sn}^2(\chi_1 \mid k_1) y^2}{z^2 \mathrm{cn}^2(\chi_1 \mid k_1)} = 1. \tag{124}$$

For fixed values of χ_1 and z, the horizontal elliptical cross sections with major and minor semi-axes of lengths $z\mathrm{dn}(\chi_1 \mid k_1)/[k_1 \mathrm{sn}(\chi_1 \mid k_1)]$ parallel to the x-axis, and $z\mathrm{cn}(\chi_1 \mid k_1)/\mathrm{sn}(\chi_1 \mid k_1)$ parallel to the y-axis, respectively, are directly read off.

Similarly, the fixed values of χ_2 define elliptical cones with vertex at the origin, axis along the x-axis, and elliptical cross sections — in vertical planes with a fixed value of x — and with major axis parallel to the z-axis and minor axis parallel to the y-axis, with the corresponding lengths as in the family of cones in the previous paragraph with the change of the index 1 to 2 in the coordinate χ and parameter k, and the exchange of coordinates x and z.

Both sets of families span the shapes from circular cones, for $k_i = 1$ and eccentricity zero, to dihedral angles, for $k_i = 0$ and eccentricity 1, keeping in mind the complementarity between their parameters, Equation (50). While in [37] these parameters are related to the asymmetry of the molecules, here they determine the shape of the confining elliptical cones. The spheroconal harmonics borrowed from [37] in Section 3.2 are some of the solutions for the hydrogen atom confined by an elliptical cone, when the latter coincides with one of the elliptical-cone nodes in Equation (57). Of course, just as already discussed for the other conoidal boundaries in Section 4, here there are other solutions for the same elliptical-cone boundary, and their evaluation is discussed next.

We formulate the confinement of the hydrogen atom by an elliptical cone defined by $\chi_1 = \chi_{10}$. The boundary condition on the complete wave function of Equation (48), with its factors from Equations (39) and (57), is

$$\Psi_{n_r n_1 n_2}(r, \chi_1 = \chi_{10}, \chi_2) = 0, \tag{125}$$

and can be satisfied by finding the zeros of the corresponding Lamé function:

$$\Lambda^A_{\nu_1}(\chi_1 = \chi_{10}) = 0. \tag{126}$$

In [36,37] the zeros of these functions were determined for specific values of k_1, corresponding to the asymmetry of the molecule, and for integer values of the angular momentum quantum number ℓ and of its components n_1 and n_2 in Equation (58). In general, the zeros of the function in Equation (126), for an elliptical cone with fixed values of k_1 and χ_{10}, become non-integer values represented as ν_1, with the corresponding change in the value of ℓ in Equation (58), just like in [22] and Section 4.2 for the circular cone.

The eigenvalue problem of Equations (54) and (55) with the new boundary condition has been reformulated accordingly. The same matrix method can be used, with the important change that the matrix becomes of infinite size, which does not present any practical difficulty. In fact, by choosing a large enough size of the matrix, convergence and accuracy in the eigenvalues and eigenfunctions can be achieved, as we have learned from the experience of evaluating Mathieu and spheroidal wave functions [44,45] and their variants. The presence of the boundary also breaks the original parity symmetry under $z \to -z$.

5.2. Hydrogen atom confined by a dihedral angle

The confinement of a three-dimensional hydrogen atom by a dihedral angle, defined by its meridian half-planes — $\phi = 0$ and $\phi = \phi_0$ in spherical, parabolic and prolate spheroidal coordinates — is the natural extension of the confinement by an angle of the two-dimensional hydrogen atom discussed in Section 2.3 on the basis of [3,10]. By using the common ϕ-dependent eigenfunctions and eigenvalues in the successive coordinates, the respective complete eigenfunctions become:

$$\Psi_{n_r n_\theta n_\phi}(r, \theta, \phi) = r^\lambda e^{-\frac{r}{\nu a_0}} M\left(-n_r, 2\lambda + 2, \frac{2r}{\nu a_0}\right) \sin^\mu \theta \times F\left(-n_\theta, n_\theta + 1; \mu + 1; \frac{(1 - \cos\theta)}{2}\right) \sin \mu\phi \quad (127)$$

$$\Psi_{n_\xi n_\eta n_\phi}(\xi, \eta, \phi) = (\xi\eta)^\mu e^{-\frac{\xi^2+\eta^2}{2\nu a_0}} M\left(-n_\xi, \mu + 1, \frac{\xi^2}{\nu a_0}\right) \times M\left(-n_\eta, \mu + 1, \frac{\eta^2}{\nu a_0}\right) \sin \mu\phi \quad (128)$$

$$\Psi_{n_u n_v n_\phi}(u, v, \phi) = \left[(u^2 - 1)(1 - v^2)\right]^{\frac{\mu}{2}} e^{-\frac{f(u+v)}{\nu a_0}} S_{n_u}(u) S_{n_v}(v) \sin \mu\phi \quad (129)$$

in terms of the eigenvalue μ and its quantum number n_ϕ from Equations (21) and (22).

In the case of spherical coordinates, we introduce the notation of λ such that $\lambda(\lambda + 1)$ is the eigenvalue of the square of the angular momentum. Good behavior of the polar angle dependent function requires $\lambda - \mu = n_\theta$ to be a non-negative integer. Consequently, the energy eigenvalue and the eigenfunctions are modified by the replacements of the successive constants of motion, including the replacement of the principal quantum number by

$$\nu = n_r + \lambda + 1 = n_r + n_\theta + \mu + 1, \quad (130)$$

with an explicit dependence on μ and the angle ϕ_0, reflecting the breaking of the original rotational symmetry by the presence of the dihedral boundary.

For the parabolic and prolate spheroidal coordinates, the changes in the eigenfunctions and eigenvalues are reduced to the replacement of m by the value of μ. The respective energy parameters become

$$\nu = n_\xi + n_\eta + \mu + 1, \quad (131)$$

$$\nu = n_u + n_v + \mu + 1. \quad (132)$$

The reader may appreciate the common form of the parameter and its values in the three coordinate systems.

The changes of the energy eigenvalues and eigenfunctions can be evaluated systematically as ϕ_0 changes from zero to π and on to 2π, making use of the experiences from [10] and Section 2.3. The case of $\phi_0 = \pi$ corresponds to the confinement by a plane [31]. However, the three sets of solutions discussed above are alternative solutions to Levine's $\Psi_{n_r,\ell,m}(r,\theta,\phi)$ vanishing at the equatorial plane $\theta_0 = \pi/2$, instead of a meridian plane.

The discussion has been focussed on the hydrogen atom, but some of the results are applicable also for the confinement by a dihedral angle of any system with a central potential, or simply with rotational symmetry around the z-axis, including isotropic and anisotropic harmonic oscillators.

Further extensions can be made when additional confinements are introduced. Following the example of [10], in which confinement in circular sectors was investigated, the dihedral confinement may be complemented with additional spherical and/or conical, one or two paraboloidal, and spheroidal and/or hyperboloidal boundaries.

5.3. Confined two-electron atoms and molecules

Our early work on the hydrogen atom confined inside prolate spheroidal boxes also dealt with the molecular hydrogen H_2^+ and molecular HeH^{++} ions [18]. The investigation of the hydrogen molecular ion was extended recently for confinement in boxes with the same shape with penetrable walls [40]. On the other hand, more than ten years ago, we investigated the ground state of the helium atom confined in a semi-infinite space [46] and inside boxes [47] with paraboloidal boundaries.

The title and contents of this Chapter were chosen from the very beginning with the restrictions of the one-electron systems and semi-infinite spaces, knowing that among the other Chapters, several cover two-electron atoms, a few deal with many-electron systems, and most of them are restricted to confinement in finite volumes. This subsection is intended to go beyond the self-imposed restrictions just mentioned, in order to identify some interesting problems involving two-electron atoms or molecules in the diverse situations of confinement. It should not be surprising if some of the authors of the other chapters have already zeroed in on them from their own perspectives.

The helium atom and the helium-like ions confined in semi-infinite spaces are worth investigating for the interest in surface effects [9,21,22,25–31,43, 46]. Here I choose the atomic hydrogen anion H^-, which involves new and interesting physics and quantum chemistry and is the most challenging computationally. In fact, while the free atom, or the atom in concentric spherical confinement, do not have an electric dipole moment, other types of confinement force it to become polar. For instance, the ground state of the hydrogen atom confined by a nodal plane has a dipole moment of $3.5ea_0$ [31]. Does this property facilitate its binding an electron to become an atomic hydrogen anion? The answer requires rigorous quantum chemistry calculations for the chosen confining boundary.

The electronic structure of the neutral hydrogen molecule is well-known in its free form [48], and has been investigated under different situations of confinement [43]. Here the confinement by a nodal plane or a dihedral angle are suggested as relatively simple situations susceptible to analysis à la Levine [31]. The readers may choose other situations of confinement adapted to their interests and applications.

5.4. Complete electrostatic harmonic expansions for electronic structure of atoms and molecules

In Section 3 reference was made to the superintegrability of the Laplace equation in parallel to that of the Schrödinger equation for the hydrogen atom. This subsection goes back to that common feature of the respective harmonic functions and atomic wave functions in their common systems of coordinates. This is particularly important in the investigation of two or more electron systems, for which the respective harmonic expansions of the Coulomb inter-electronic potential becomes the important tool for physically and computationally tractable calculations.

As a specific experience, we can point out that our works on the helium atom confined by paraboloids [46,47] were preceded by variational calculations for the hydrogen atom in the respective confinement situations [49] and by the construction of the paraboloidal harmonic expansion of the Coulomb potential [50]. The last reference also includes the corresponding expansions in prolate and oblate spheroidal coordinates.

The additional word "complete" may be noticed in the title of this subsection, and a few words of explanation are appropriate. Textbooks usually discuss the multipole expansion as an approximation for the far-away fields [51]. However, the expansion is valid and complete, outside and inside the region of the sources [52].

In general terms, we may say that the atomic or molecular calculations in quantum chemistry have the aim of finding self-consistent solutions of the Schrödinger equation and Poisson equation for the distributions of nuclei and electrons making up the system of our interest in the chosen state.

The lowest harmonic components of the electrostatic field may determine the asymptotic behavior of the wave functions for the outer electron(s), but the behavior at smaller distances is also determined by the higher components of the electrostatic field.

The pioneers of the investigation of the binding of an electron by a dipole moment using the interaction energy between them,

$$V(\vec{r}) = -\frac{e\vec{D}\cdot\hat{r}}{r^2}, \tag{133}$$

were perfectly aware that this dominant dipole moment approximation is valid only in the far-away asymptotic region, and that if it is followed for shorter distances it will lead to the falling in of the electron to the center of the dipole [53]. For a realistic distribution of charges, the same dipole approximation gives the energy of the electron as

$$V(\vec{r}) = \frac{e\vec{D} \cdot \vec{r}}{d^3} \tag{134}$$

for the interior of a surface with a pure dipole charge distribution [52] with an effective diameter d. In such a case, the electron does not fall into the origin, since the potential energy function inside does not have the singularity associated with the potential outside. Solutions of the Schrödinger equation for this complete dipole approximation in spherical, prolate and oblate spheroidal coordinates may provide a more realistic way of modeling polar molecules with the respective geometries. Work on such solutions is at the exploratory stage.

6. DISCUSSION

The sample of comments in Section 2 have provided the basis and motivation for the unified study of the hydrogen atom in its free form in the review of Section 3, and in its conoidal confinements in the overview of Section 4 and in part of the preview of Section 5.

Indeed, the first comment on the ground state of the hydrogen atom in a spherical enclosure contains the idea and the methodology, which have proven to be valid for all the bound states in the successive situations of confinement, including the removal of the degeneracy of their energy levels by far-away boundaries as well as the quantitative evaluation of the positions of each boundary for which the specific states have zero energy.

The distinctions between confinement by closed versus open boundaries, free versus "almost-free" hydrogen atom, and the criteria for binding and unbinding of an electron in the comments of 2.2, 2.3 and 2.4 have also been addressed, with new and, originally unexpected, results.

Specifically, the confinements of the hydrogen atom by spheres and spheroids do lead to the increase in the energy eigenvalues without any limit as the confining boundary approaches the nucleus, [14,18], 4.1 and 4.4. In contrast, the confinement by paraboloids [21] and 4.3, hyperboloids [9] and 4.5, and circular cones [22] and 4.2, show increases of the energies only up to zero for common positions of the respective boundaries, exhibiting infinite degeneracies; for the last two shapes the limit situations of the "one-dimensional" hydrogen atom are restricted to states with $m = 0$, since the probability of finding the electron at the position of the z-axis becomes zero for the other states with different values of m.

The confinement of the two-dimensional hydrogen atom by angles and hyperbolas [3,10] and 2.3, has its counterparts with circular cones [22] and 4.2, and hyperboloids [9] and 4.5 in the three-dimensional case. The "almost-free" hydrogen atom limit happens at the other end of the domain of the respective coordinates in comparison with that of the previous paragraph. It cannot be the free hydrogen atom limit, because the eigenstates of the latter with zero angular momentum do not satisfy the condition of vanishing at the boundary.

The two types of remaining coordinates included in the review of Section 3, representing elliptical cones and dihedral angles, complete the list of natural boundaries for the confinement of the hydrogen atom. Their investigation has been incorporated in the first half of the preview in Section 5. It may be reiterated that these types of confinement are also natural for any central potential, and the dihedral one also for any potential with rotational symmetry around the z-axis.

The comment in 2.4 concerning the binding of an electron by a confined polar molecule made the author aware of the need to formulate such a process, and its complementary unbinding process, in a consistent way for confined quantum systems in general. The specific case of the hydrogen atom has not been properly treated as recognized in 2.1, and the possibility of using Equation (17) as a reference to analyze its ionization under confinement is proposed. The additional element to implement such an analysis is the energy of the confined free electron incorporated in the overview of Section 4, with Equations (100) and (117) for the closed boundaries and (104) for the open ones. This is also a practical and important difference between the confinements by closed and open boundaries.

In a natural way the problems of 5.3 combine some of the above elements of confinement and the binding of an electron by the hydrogen atom and by the hydrogen molecular ion, respectively. The superintegrability of both the Poisson and Schrödinger equations in common coordinate systems provide the tools for the analysis of alternative models for the binding of an electron in pure and complete dipole fields in 5.4.

ACKNOWLEDGEMENTS

The author was invited to write this Chapter for the special volume of ADVANCES IN QUANTUM CHEMISTRY as he was about to start his sabbatical Spring semester in China. He has the pleasure of thanking the Editors of the volume, Professors John Sabin and Salvador Cruz, for their invitation. Informal and fruitful e-mail discussions throughout the period with co-authors Norberto Aquino, Carlos F. Bunge, Lorea Chaos-Cador, Salvador Cruz, Araceli Góngora, Sergio Mateos-Cortés and Ricardo Méndez-Fragoso were very helpful in the successive stages. Financial support from DGAPA-UNAM and CONACYT-SNI 1796 is gratefully acknowledged. The final typescript is the professional work of Ms. María Luisa Araujo, just as many of the references have been in the past.

He also wishes to express his appreciation for the opportunities to give seminars and lectures and hold discussions with colleagues and students at Central China Normal University, Wuhan, Hubei; Hunan University of Science and Technology, Hunan Normal University, Central South University, and South China University under the sponsorship of the Department of Science and Technology of Hunan Province; and the Zhongshan University of Electronic Science and Technology under the joint sponsorship of Guangdong Zhuhai Office of Science and Technology and Zhongshan Overseas Exchange Association. The warm hospitality and support of the Zheng-Gao family during the writing period in Zhongshan deserves a special mention. Also, during this period the Sichuan earthquake occurred and we have shared the pain and sadness of the Chinese people.

Last, but not least, the constant and continuous attention and support of my wife, Professor Jocele Allegra Wild Wolf, has made the tasks easier and more enjoyable, and also the pain of loss and sadness more bearable.

REFERENCES

[1] W. Jaskolski, Phys. Rep. 271 (1996) 1.
[2] V. Aguilera-Navarro, E. Ley-Koo, A. Zimerman, Rev. Brasil. Fis. 10 (1980) 251.
[3] R.A. Rosas, J.L. Marin, R. Riera, R. Núñez, Phys. Low-Dimen. Struct. 5–6 (1999) 145.
[4] S. Ronen, Phys. Rev. A 68 (2003) art. no. 012106.
[5] S. Ronen, Phys. Rev. A 68 (2003) art. no. 064101.
[6] V. Fock, Z. Physik 98 (1935) 145.
[7] H.V. McIntosh, Am. J. Phys. 27 (1959) 620.
[8] E.G. Kalnins, W. Miller, P. Winternitz, SIAM, J. Appl. Math. 30 (1976) 630.
[9] E. Ley-Koo, S. Mateos-Cortés, Int. J. Quant. Chem. 46 (1993) 609.
[10] L. Chaos-Cador, E. Ley-Koo, Int. J. Quant. Chem. 103 (2005) 369.
[11] E. Ley-Koo, Phys. Rev. A 78 (2008) 036102.
[12] E.P. Wigner, Phys. Rev. 94 (1954) 77.
[13] V.C. Aguilera-Navarro, W.M. Kloet, A.H. Zimerman, Rev. Brasil. Fis. 1 (1971) 55.
[14] E. Ley-Koo, S. Rubinstein, J. Chem. Phys. 71 (1979) 351.
[15] E. Ley-Koo, V.C. Aguilera-Navarro, Topics in Theoretical Physics II. Festschrift for Prof. Abraham Zimerman on his 70th Birthday IFT/UNESP, São Paulo, Brasil, 1998, p. 184.
[16] M. Abramowitz, I.A. Stegun, Handbook of Mathematical Functions, Dover, New York, 1965.
[17] E. Ley-Koo, S. Rubinstein, J. Chem. Phys. 73 (1980) 887.
[18] E. Ley-Koo, S.A. Cruz, J. Chem. Phys. 74 (1981) 4603.
[19] D. Suryanarayana, J.A. Weil, J. Chem. Phys. 64 (1976) 510.
[20] J.A. Weil, J. Chem. Phys. 64 (1976) 510.
[21] E. Ley-Koo, R.M.G. García-Castelán, J. Phys. A 24 (1991) 1481.
[22] E. Ley-Koo, S. Mateos-Cortés, Am. J. Phys. 61 (1993) 246.
[23] Z.Q. Guo, E. Ley-Koo, J.J. You, X.J. Shih, Diqiu Wuli Xuebao 31 (1988) 566.
[24] E. Ley-Koo, Z.Q. Guo, J.H. You, X.J. Shih, Acta Geophys. Sin. 32 (1989) 105.
[25] Z. Liu, D.L. Lin, Phys. Rev. B 28 (1983) 4413.
[26] S. Satpathy, Phys. Rev. B 28 (1983) 4584.
[27] Y. Shan, T.F. Jiang, Y.C. Lee, Phys. Rev. B 31 (1985) 5487.
[28] Y. Shan, J. Phys. B 20 (1987) 4275.
[29] Y. Shan, P.C. Lee, H.C. Tseng, J. Phys. 20 (1987) 4285.
[30] A.F. Kovalenko, E.N. Sovyak, M.F. Golovko, Int. J. Quant. Chem. 42 (1992) 321.
[31] J.D. Levine, Phys. Rev. A 140 (1965) 586.
[32] P.M. Morse, H. Feshbach, Methods of Theoretical Physics, Mc Graw-Hill, New York, 1953.
[33] L. Chaos-Cador, E. Ley-Koo, Int. J. Quant. Chem. 97 (2004) 844.

[34] L. Chaos-Cador, E. Ley-Koo, Int. J. Quant. Chem. 103 (2005) 369.
[35] E. Ley-Koo, A. Góngora-T, Int. J. Quantum Chem. 109 (2009) 790.
[36] E. Ley-Koo, R. Méndez-Fragoso, Rev. Mex. Fís. 54 (2008) 69.
[37] E. Ley-Koo, R. Méndez-Fragoso, Rev. Mex. Fís. 54 (2008) 162.
[38] E.T. Whittaker, G.N. Watson, A Course of Modern Analysis, Cambridge University Press, United Kingdom, 1927.
[39] G. Hunter, H.O. Pritchard, J. Chem. Phys. 46 (1967) 2146.
[40] S. Mateos-Cortés, E. Ley-Koo, S.A. Cruz, Int. J. Quant. Chem. 86 (2002) 376.
[41] C.A. Coulson, P.D. Robinson, Proc. Phys. Soc. 71 (1958) 815.
[42] H.A. Erikson, E.L. Hill, Phys. Rev. 75 (1949) 29.
[43] A.F. Kovalenko, E.N. Soyak, M.F. Holovko, Int. J. Quant. Chem. 42 (1992) 321.
[44] L. Chaos-Cador, E. Ley-Koo, Rev. Mex. Fís. 48 (2002) 67.
[45] N. Aquino, E. Castaño, E. Ley-Koo, Rev. Mex. Fís. 48 (2002) 277.
[46] E. Ley-Koo, K.P. Volke-Sepúlveda, Int. J. Quant. Chem. 65 (1997) 269.
[47] E. Ley-Koo, A. Flores-Flores, Int. J. Quant. Chem. 66 (1998) 122.
[48] W. Kolos, K. Szalewicz, H.J. Monkhorst, J. Chem. Phys. 84 (1986) 3278.
[49] S.A. Cruz, E. Ley-Koo, J.L. Marín, A. Taylor-Armitage, Int. J. Quant. Chem. 54 (1995) 3.
[50] E. Ley-Koo, A. Góngora-T, Rev. Mex. Fís. 39 (1993) 785.
[51] J.D. Jackson, Electrodynamics, 2nd ed., John Wiley and Sons, New York, 1975.
[52] E. Ley-Koo, A. Góngora-T, Rev. Mex. Fís. 34 (1988) 685.
[53] J.E. Turner, Am. J. Phys. 45 (1977) 758.

CHAPTER 4

The Hydrogen and Helium Atoms Confined in Spherical Boxes

N. Aquino [a]

Contents

1. INTRODUCTION

In this work we aim at presenting a review of the most relevant methods utilized in the study of one and two-electron atoms confined in spherical boxes, from the early works published in the 1930s up to the most recent

[a] Departamento de Física, Universidad Autónoma Metropolitana-Iztapalapa, Apartado Postal 55-534, 09340 México D. F

E-mail address: naa@xanum.uam.mx.

Advances in Quantum Chemistry, Vol. 57
ISSN: 0065-3276, DOI: 10.1016/S0065-3276(09)00608-X

investigations. We will also show some new results for the hydrogen atom confined in soft spherical boxes and when the compression regimes are considered within Neumann boundary conditions.

Seventy years ago Michels et al. [1] proposed a model consisting of a hydrogen atom confined at the centre of an impenetrable spherical box, in order to study how the pressure and polarizability evolve as a function of the compression. The model of confinement in boxes of different sizes and geometrical forms has become very popular and it is widely used in a variety of quantum systems.

The behavior of a system (atom, molecule, etc.) subjected to extreme pressures [1–116], can, in first approximation, be simulated by placing it in a box of impenetrable walls, where the infinite potential is induced by neighboring particles of negative charge [25]. Under these conditions, the particle wave function must vanish at the walls, i.e. it ought to fulfill Dirichlet boundary conditions (DBC). However, this model only includes effects produced by the repulsive forces. To account for the existence of attractive forces between particles, such as Van der Waals forces, it has been proposed that the potential surface be finite (a box of penetrable walls). Spatial confinement induces changes in the observable properties of the systems, such as energy spectrum, transition frequencies and transition probabilities, as well as polarizabilty [1–189].

The confinement model has been extensively used to analyze the hydrogen atom enclosed by hard and soft spherical boxes [1–98], with confining boxes of diverse geometrical shapes [31,88–98], and it has also been applied to studies of the helium atom [99–122], many-electron atoms [50, 52,55,123–131], molecules [132–142] and the harmonic oscillator [143–171], among others.

The confinement model is also useful to systematically study effects on an atom or molecule trapped in a microscopic cavity or in fullerenes [72–78]. As mentioned above, some of the system observables undergo changes as a result of spatial confinement. The same situation is found at a nanoscopic scale in artificial systems constructed within semiconductors [79–87,172–188], such as two-dimensional quantum wells, quantum wires and quantum dots. Properties of a hydrogen-like impurity in a 2D quantum well have been investigated by several authors [172,173,185–188], who have concluded that particular features associated with the states, as well as the properties of an impurity, are determined, among other factors, by the size of the confining structure. Other applications of confined systems refer to: Metal properties [147,148], astrophysical spectroscopic data [40,146], phase transitions [155], matter embedded in electrical fields [68], nuclear models [164], etc. For a detailed list of references, several review articles [25, 48,54,95,125,127] are available.

2. METHODS USED IN VARIOUS STUDIES OF THE CONFINED HYDROGEN ATOM

2.1. First steps

Michels et al. [1] proposed a model to study ground state energy shifts and polarizability changes for the hydrogen atom when subjected to uniform compression. In their model, it is assumed that the nuclear charge is located at the centre of an impenetrable spherical box of radius r_0, thus the angular functions of the free atom are unaffected by the confinement, i.e., they remain spherical harmonic functions. The radial Schrödinger equation (in atomic units) for the confined hydrogen atom (CHA) is

$$\left(-\frac{1}{r^2}\frac{\mathrm{d}}{\mathrm{d}r}r^2\frac{\mathrm{d}}{\mathrm{d}r}+\frac{l(l+1)}{r^2}-\frac{2}{r}+2V_c(r)\right)R(r)=2ER(r), \tag{1}$$

where the confinement potential $V_c(r)$ is given by

$$V_c(r)=\begin{cases}0 & \text{if } r<r_0\\ \infty & \text{if } r\geq r_0.\end{cases} \tag{2}$$

Since the wave functions must vanish throughout $r \geq r_0$, we restrict our attention to the region $r < r_0$. The radial Schrödinger equation to be solved is

$$\left(-\frac{1}{r^2}\frac{\mathrm{d}}{\mathrm{d}r}r^2\frac{\mathrm{d}}{\mathrm{d}r}+\frac{l(l+1)}{r^2}-\frac{2}{r}\right)R(r)=2ER(r), \tag{3}$$

where DBC are imposed,

$$R(r_0)=0. \tag{4}$$

Michels et al. [1] proposed the following substitution

$$R(r)=\frac{1}{r}\mathrm{e}^{-r/a}f(r). \tag{5}$$

Equation (3) is thus transformed to

$$\left(\frac{\mathrm{d}^2}{\mathrm{d}r^2}-\frac{2}{a}\frac{\mathrm{d}}{\mathrm{d}r}-\frac{l(l+1)}{r^2}+\frac{2}{r}\right)f(r)=0, \tag{6}$$

where

$$a = \sqrt{-1/2E} \quad \text{or} \quad E = -\frac{1}{2a^2}. \tag{7}$$

For the free (unconfined) case, $\lim_{r\to\infty} R(r) = 0$ and $a = n$, with n a positive integer, resulting in the usual energy eigenvalue formula $E = -\frac{1}{2n^2}$. However, for the CHA the wave functions must vanish at a finite value of the radial variable r_0 (Equation (4)), therefore, a cannot be an integer. Michels et al. proposed a power series solution $f(r) = \sum_{l+1}^{\infty} b_s r^s$. The boundary conditions (4) can be written as $f(r_0) = \sum_{l+1}^{\infty} b_s r_0^s = 0$.

This equation determines the perturbed values of a and E. They restricted their attention to the ground state ($n = 1, l = 0$) only, and for large values of r_0.

They computed the ground state shift for $r_0 = 5, 6, 7$ and 8 a.u. and they also found the magnitude of the force $F = -\frac{\mathrm{d}E}{\mathrm{d}r_0}$ required to compress the hydrogen atom in a spherical box of radius r_0. The electronic pressure is defined as

$$p = \frac{F}{4\pi r_0^2} = -\frac{1}{4\pi r_0^2}\frac{\mathrm{d}E}{\mathrm{d}r_0}. \tag{8}$$

The polarizability α as a function of r_0 was computed by using Kirkwood's formula [57] that requires evaluation of $\langle r^2\rangle$. Michels et al. found a rising trend for the electron kinetic energy and a decreasing ratio $\Delta\alpha/\alpha$ as the pressure grows.

They defined $\Delta\alpha/\alpha$ as

$$\frac{\Delta\alpha}{\alpha} = \frac{2S^2(r_0) - (2/3)A(r_0)S(r_0)}{1 - 2S^2(r_0)},$$

where

$$S(r) = \mathrm{e}^{-r}\left(\frac{r^2}{3} + r + 1\right), \quad \text{and}$$

$$A(r) = \mathrm{e}^{-r}\left(\frac{3}{80}r^4 + \frac{3}{16}r^3 + \frac{3}{4}r^2 + \frac{9}{4}r + 3\right).$$

Also, they found an increasing pressure rate as r_0 is reduced.

In 1938, one year after Michels et al. [1] published their work, Sommerfeld and Welker [2] found the exact formal expression for the wave function. They also proposed that the CHA model was of astrophysical interest and the model has gained popularity since then. The radial Schrödinger equation (3) for $l = 0$ states can be written as

$$\frac{d^2R}{dr^2} + \frac{2}{r}\frac{dR}{dr} + 2\left(E + \frac{1}{r}\right)R = 0, \tag{9}$$

where the energy eigenvalues are given by Equation (7)

They proposed the following ansatz

$$R = e^{-\rho/2}F(\rho), \quad \rho = 2r/a, \tag{10}$$

where a is given by Equation (7). Equation (9) is then transformed to

$$\rho F'' + (2-\rho)F' + (a-1)F = 0. \tag{11}$$

This is Kummer's differential equation whose regular solution at the origin is the confluent hypergeometric function ${}_1F_1(-a+1, 2, \rho)$.

To find the energy eigenvalues they imposed the boundary conditions (Equation (4)), which is equivalent to finding the zeros of the confluent hypergeometric function ${}_1F_1$ at ρ_0, i.e. they must solve ${}_1F_1(-a+1, 2, \rho_0) = 0$.

Once the values of a (noninteger, in general) are determined, Equation (7) can be used to calculate the energy. However, the lack of algorithms as well as computational codes and, of course, even computers capable of executing them, made it impossible to obtain accurate results at the time that Sommerfeld and Welker published their work [2]. Instead, these authors solved the problem by using analytical expansions of the hypergeometric functions to estimate the energy levels. They computed the ground state energy for $r_0 = 2$, 3 and 4 a.u. and their results are shown in Table 1.

They also carried out a detailed investigation of the critical cage radius r_c, i.e., the radius at which the binding energy becomes zero as r_0 diminishes. The total energy for $r_0 < r_c$ is positive. The r_c value could be important when constructing the partition function of the atom and the pressure ionization of ground and excited states. They found that the critical radius is proportional to the zeros of the Bessel functions. Later, de Groot and ten Seldam [3] carried out a careful analysis of this topic.

de Groot and ten Seldam [3] were interested in computing energy level shifts for excited states. They treated the radial Schrödinger equation with a nonzero angular momentum (Equation (3)) subject to the boundary conditions given by Equation (4). They used the ansatz

$$R = e^{-\frac{1}{2}\rho}\rho^l F(\rho) \tag{12}$$

Table 1 Ground state energy eigenvalues (hartrees) obtained by Michels et al. [1], Sommerfeld and Welker [2] and de Groot and ten Seldam [3] as a function of the box radius r_0 (au)

r_0	$E(1s)$	$E(2s)$	$E(2p)$
2.0	−0.1250[b]		
3.0	−0.44[b]		
4.0	−0.48[b]		
5.0	−0.49659[a]		
6.0	−0.49928[a]		
7.0	−0.49986[a]	−0.0974[c]	−0.1058[c]
8.0	−0.49997[a]	−0.1055[c]	−0.1124[c]
10.0		−0.1162[c]	−0.1194[c]
15.0		−0.12451[c]	−0.12477[c]
20.0		−0.12499[c]	−0.12500[c]
∞	−0.50000[c]	−0.12500[c]	−0.12500[c]

[a] Reference [1].
[b] Reference [2].
[c] Reference [3].

to show that the function $F(\rho)$ satisfies the Kummer's differential equation whose regular solution at the origin is given by the confluent hypergeometric function ${}_1F_1$.

When $r_0 \to \infty$, the expansion of F becomes a polynomial with $a = n$, where n is a positive integer, the energies are given by the well known formula $E_n = -1/(2n^2)$. On the other hand, when r_0 is finite, the values of a are in general nonintegers, furthermore they can be imaginary when the energy value is positive. de Groot and ten Seldam [3] computed how, under compression, the hydrogen atom energies change for the 1s, 2s, and 3s states. They noted that for fixed values of l and E the wave functions depend on r or ρ. Whenever the free wave functions vanish, the corresponding node can be considered as the radius r_0 of the cage in which the hydrogen atom is confined. They divided the energy values into regions, where in each, energy computation is based on evaluation of the confluent hypergeometric functions by using tables, series expansions and interpolation methods. Their results, together with those of Michels et al. [1] and Sommerfeld and Welker [2], are summarized in Table 1.

de Groot and ten Seldam [3] also solved the problem for the region corresponding to total energy $E > 0$. Although their results were considered to be exact for some time, later calculations showed inaccuracies for some of them. These authors were the first to realize that for small values of r_0 the kinetic energy is higher than the Coulomb potential.

Sommerfeld and Welker [2] obtained the critical radius $r_c = 1.835$ au for the 1s state, whereas Dingle [4] obtained an improved value of $r_c = 1.8354$ au. de Groot and ten Seldam [3] discussed in detail the result found by

Sommerfeld and Welker [2] for the critical radius. The analytical expression for the critical radius can be found [46]:

$$r_c(n,l) = \frac{1}{8}(x_{2l+1,n-l})^2, \tag{13}$$

where $x_{p,i}$ denotes the ith zero of $J_p(z)$.

Boeyens [55], Varshni [23], Cohen and Burrow [41] and Tian et al. [56] also obtained the critical radius using different methods.

2.2. Perturbation theory

One of the most useful techniques in physics and chemistry is Time Independent Perturbation Theory (TP). Therefore, not surprisingly, it has been applied to the study of the CHA problem.

Dingle [4] and later Wigner [5] computed the CHA ground state by taking the Coulomb term as a perturbation. However, Wigner found that his expression diverges as $r_0 \rightarrow \infty$, giving rise to a problem that has been discussed by various authors [6–8,11].

de Groot and ten Seldam [3] recognized, for the first time, that for small box radii the CHA wave functions are very similar to those of a free particle in the box. Aguilera-Navarro et al. [9] solved the problem by introducing fifth order corrections within Perturbation Theory, where the free particle is taken as the unperturbed system and the Coulomb potential as the perturbation. They obtained, for the ground state,

$$E = \frac{\pi^2}{2r_0^2} - \frac{2.4377}{r_0} - \frac{1.0796}{\pi^2} - \frac{1.2112r_0}{\pi^4} - \frac{1.1928r_0^2}{\pi^6} - \frac{0.752r_0^3}{\pi^8}. \tag{14}$$

We can see immediately that his expression, although it diverges as $r_0 \rightarrow \infty$, works well for small values of r_0. Furthermore, this formula gives good results for a radius as large as 5 au, where the perturbation series convergence is not guaranteed. Gray and Gonda [10] noted that for box radii < 1.7 au the results of de Groot and ten Seldam [3] are inaccurate, whereas the values obtained with Equation (14) are better by comparison. At $r_0 = 2$ au, this equation predicts the exact energy value, where the exact wave function is the ordinary 2s function ($r \in [0, 2]$), which has a node at 2 au. However, the one obtained by means of second order perturbation theory is not the same 2s function. In this case, the energy value is exact, but some properties, for example, the polarizability, obtained through the perturbed wave function, do not yield exact values.

2.3. The method of linear variation functions

The linear variational method is a useful and accurate tool for solving problems in different fields, such as quantum chemistry, molecular physics and solid state physics, among others [200].

In this method the wave function is expanded as

$$\Psi = \sum_{i=1}^{N} C_i \phi_i, \tag{15}$$

where the $\{\phi_i\}$'s span a complete, linearly independent basis set that satisfies the boundary conditions of the problem. In some cases, the $\{\phi_i\}$'s are also orthonormal (albeit, they are not required to be). For a variational analysis we compute the energy functional $\int \Psi^* H \Psi d\tau$, which is minimized with respect to coefficients C_i, subjected to the constraint $\int \Psi^* \Psi d\tau = 1$, i.e.

$$\delta \left(\int \Psi^* H \Psi d\tau - \lambda \int \Psi^* \Psi d\tau \right) = 0. \tag{16}$$

This is equivalent to solving a set of homogeneous linear equations for the C_i's. In order for the homogeneous set of equations to have a nontrivial solution, the determinant of the C_i coefficients must vanish. This leads to the secular equation

$$|H_{ij} - E S_{ij}| = 0, \tag{17}$$

where the Hamiltonian and overlap matrix elements are

$$H_{ij} = \langle \phi_i | H | \phi_j \rangle \tag{18}$$

and

$$S_{ij} = \langle \phi_i | \phi_j \rangle, \tag{19}$$

respectively. Equation (17) is an Nth-order polynomial in E, which must be solved to obtain the N roots of E. The problem is usually treated as a generalized matrix eigenvalue problem, which is solved by standard numerical methods, thus yielding N energy eigenvalues and their corresponding eigenfunctions (eigenvectors). These values are upper limits to the energies of the N states of the system.

Goodfriend [12] developed a formalism in which the basis set does not satisfy the boundary conditions. His calculations were carried out only for s states. He constructed a trial wave function out of 1s, 2s and 3s functions corresponding to the free (unbounded) hydrogen atom.

$$\psi = C_1\phi_1 + C_2\phi_2 + C_3\phi_3, \tag{20}$$

where

$$\begin{aligned}
\phi_1 &= \frac{e^{-r}}{\sqrt{\pi}} \\
\phi_2 &= \frac{1}{4\sqrt{2\pi}}(2-r)e^{-r/2} \\
\phi_3 &= \frac{1}{81\sqrt{3\pi}}(27-18r+2r^2)e^{-r/3}.
\end{aligned} \tag{21}$$

Clearly, these functions do not fulfill the boundary conditions given in Equation (4), however, wave function ψ must satisfy the condition

$$\psi(r_0) = 0 = C_1\phi_1(r_0) + C_2\phi_2(r_0) + C_3\phi_3(r_0). \tag{22}$$

Goodfriend showed that the energy determinantal equation is given by

$$\begin{vmatrix}
H_{11}-ES_{11} & H_{12}-ES_{12} & H_{13}-ES_{13} & \phi_1(r_0) \\
H_{21}-ES_{21} & H_{22}-ES_{22} & H_{23}-ES_{23} & \phi_2(r_0) \\
H_{31}-ES_{31} & H_{32}-ES_{32} & H_{33}-ES_{33} & \phi_3(r_0) \\
\phi_1(r_0) & \phi_2(r_0) & \phi_3(r_0) & 0
\end{vmatrix} = 0. \tag{23}$$

The integrals involved in the matrix elements must be calculated between 0 and r_0. In Table 2 are shown the results obtained by this method for the 1s and 2s states. Despite using only three terms in the wave function expansion, the agreement with the results of de Groot and ten Seldam [3] is good in most cases. However, for the 2s state at $r_0 = 7$ au, the energy reported by Goodfriend [12] is about 50% higher than that obtained by de Groot and ten Seldam [3].

A different approach was proposed by Aquino [13], where the Hamiltonian matrix is diagonalized in the basis set of an isotropic, spherically confined harmonic oscillator (SICHO), where the CHA Hamiltonian is written as

$$H = \frac{1}{2}(p^2 + r^2) - \left(1/r + \frac{1}{2}r^2\right). \tag{24}$$

The first term is identified as the unperturbed Hamiltonian. This is the SICHO Hamiltonian that was accurately solved by Aquino [13]. The wave

Table 2 Energies (hartrees) obtained by Goodfriend [12] for levels 1s and 2s as a function of the box radius r_0 (au), where use is made of a linear combination of functions that do not satisfy boundary conditions, as compared with de Groot and ten Seldam [3]

r_0(au)	1s Energy	Ref. [3]
1.902	−0.0556	−0.0566
2	−0.1250	−0.1250
3	−0.4240	−0.4475
4	−0.4832	−0.4852
5	−0.4964	−0.49659
7	−0.4999	−0.49986
∞	−0.5000	−0.50000
r_0(au)	2s Energy	Ref. [3]
7	−0.0513	−0.0974
10	−0.1120	−0.1162
20	−0.1249	−0.12499
∞	−0.125	−0.12500

functions of the SICHO problem satisfy the boundary conditions (4) and form a complete orthonormal basis set. These are known with high accuracy. The matrix elements are obtained numerically and the energy eigenvalues correspond to solutions of Equation (17), in which S_{ij} is the identity matrix. Aquino used 7 functions in the expansion of the wave function. He computed the 1s–4s energy levels for $r_0 = 2, 4$ and 5 au, where the results remain within 1.5% of the most accurate figures [34,39]. One advantage of this method is that the accuracy is preserved as the box radius r_0 grows.

Recently, Ting-yun et al. [86] used a linear combination of B-splines to compute some energy values of the CHA for various states, although he aimed at studying the confined system off centre, as we will see in a later section.

2.4. Other variation methods

Gorecki and Byers Brown [14,15] proposed an approach based on the variational boundary perturbation theory. In this method the trial wave function for the confined system ψ is constructed as the product of the wave function for the free (unbounded) system ϕ, times a non-singular cut off function f, to ensure fulfillment of the boundary condition $\psi(r_0, \varphi, \theta) = 0$. The cut-off function clearly vanishes at r_0, $f(r_0) = 0$

$$\psi = \phi f. \tag{25}$$

They minimized the energy functional with respect to a class of functions $\{f\}$ that satisfy $f(r_0) = 0$ and finally were required to solve an integro-differential equation.

On the other hand, Marín and Cruz [16–18] used the direct variational method or Rayleigh–Ritz method with a trial function of the same form used by Gorecki and Byers Brown (Equation (25)). Marín and Cruz chose hydrogenic functions with a variational parameter as the exponent for the functions ϕ and the cut-off function simply as $f(r) = r - r_0$.

They proposed wave functions for arbitrary nodeless states of the form

$$\psi = N_{nl}(r - r_0)(2\alpha r)^l \, {}_1F_1(-n + l + 1, 2l + 2, 2\alpha r) \exp(-\alpha r), \tag{26}$$

where N_{nl} is a normalization constant, n is a positive integer (principal quantum number), α is a variational parameter and ${}_1F_1$ is the confluent hypergeometric function. They computed the energy functional as

$$E(\alpha) = \langle \psi | H | \psi \rangle, \tag{27}$$

which, minimized with respect to the variational parameter α, yields the optimal energies. Using this method they computed the energy of some nodeless states; 1s, 2p and 3d. Their results agree well with those reported previously by Ludeña [21] and Ley-Koo and Rubinstein [30]. In Ludeña's work [21] the Hamiltonian is diagonalized in a basis set of Slater-type functions that satisfy the boundary condition, whereas the calculations of Ley-Koo and Rubinstein [30] are based on a power series wave function expansion in the radial coordinate. The accuracy of the calculations of Marín and Cruz [16,18] is of the order of 1×10^{-2} to 1×10^{-3} hartrees, depending on the atomic state and the distance r_0; some of their results are shown in Table 3.

Marín and Cruz [16] also computed some properties of physical interest for the CHA, such as the Fermi contact term [30,34,35,38], diamagnetic screening constant [30,34,35], polarizability [30,34,35,44,45,57–61,63] and pressure [1,30,34,35]. In addition, they studied the hydrogen and helium atom in penetrable boxes, and the hydrogen atom between parallel hard walls [16].

Varshni [22] studied the CHA by means of the Rayleigh–Ritz method. He proposed a modification of wave function (26), introducing an additional variational parameter β. In his approach, the 1s, 2p and 3d CHA wave functions are written as

$$\begin{aligned} R_{10} &= (r_0 - r)(1 + \beta r)e^{-\alpha r} \\ R_{21} &= (r_0 - r)(1 + \beta r)re^{-\alpha r} \\ R_{32} &= (r_0 - r)(1 + \beta r)r^2e^{-\alpha r}. \end{aligned} \tag{28}$$

Table 3 Energy values obtained by different authors for the 1s, 2s and 3d states of the confined hydrogen atom as a function of r_0. The results obtained by Marín and Cruz [16] are shown in the second column, in the third column are those obtained by Varshni [22], and in the last column are the result of Goldman and Joslin [40]. The energies are in hartrees and the distances in au

r_0	Ref. [16]	Ref. [23]	Ref. [40]
1s			
0.1	949.803	938.853	937.986
0.5	29.794	29.515	29.496
1.0	4.7812	4.7497	4.7480
1.4	1.3002	1.2945	1.2942
2.0	−0.2500	−0.2500	−0.2500
3.0	−0.84509	−0.84509	−0.84793
2p			
0.6	50.401	49.935	49.874
0.8	27.155	26.910	26.879
1.0	16.611	16.464	16.446
1.4	7.6857	7.6214	7.6138
2.0	3.1791	3.1547	3.1520
3.0	0.96939	0.96309	0.96250
5.0	0.01547	0.01520	0.01519
7.0	−0.17482	−0.17482	−0.17496
2s			
1.0	30.234	29.979	29.935
1.5	12.692	12.587	12.570
2.0	6.7182	6.6640	6.6550
4.0	1.2532	1.2440	1.2427
6.0	0.36339	0.36098	0.36068
8.0	0.09281	0.09218	0.09212
12.0	−0.06250	−0.06250	−0.06250

In Table 3 we compare the results obtained by using other methods [16,40] with those obtained by Varshni [22] for selected values of the distance. As Varshni noted, the results of Marín and Cruz behave reasonably well in the region where the energy is negative, but not in the region where the energy becomes positive. It should be noted that by introducing a second variational parameter, the wave functions and energies are improved. As r_0 grows, parameter β approaches zero, obtaining the wave functions and energies of Marín and Cruz, but for small r_0, i.e. for regions where the energy is positive, it plays an important role. It is also known that some wave functions, although they give good energies, may yield other properties with a comparatively lower accuracy. Varshni also computed the Fermi contact term A [30,34,35,38]

$$A = (2/3) g\beta g_n \beta_n R_{n0}^2(0), \tag{29}$$

that occurs in the hyperfine splitting of the hydrogen atom by using his own wave functions and those of Marín and Cruz's [16]. He made an important improvement on this value with respect to those reported by Marín and Cruz.

2.5. The hypervirial method

Fernández and Castro [24–28,161] developed the hypervirial perturbative method (HPM) for confined quantum systems. It is founded on the hypervirial relationships [197,199] and on perturbation theory, representing a generalization of the treatments addressed by Swenson and Danforth [164] and Killingbeck [198]. By applying the HPM to the CHA, they obtained an analytical formula for the energies of different states:

$$E(r_0) = \pm 2/W^2, \tag{30}$$

where W, correct up to third order, is given by

$$W_{nc} \approx e_{cn} q_0^{-2} + A_0^{(2)} g q_0^2 + \frac{1}{2} A_1^{(2)} g^2 q_0^6 + \frac{1}{3} A_2^{(2)} g^3 q_0^{10}, \tag{31}$$

and

$$\begin{aligned} A_0^{(N)} &= \frac{1}{N+1} - \frac{N}{2(N+1)e_{cn}} \left[(N^2-1)/4 - t \right] A_0^{(N-2)}, \\ A_s^{(N)} &= \frac{2-s}{2(N+1)s e_{cn}} x_0^N A_{s-1}^{(-1)} - \frac{N}{2(N+1)e_{cn}} [(N^2-1)/4 - t] A_s^{(N-2)} \\ &\quad + \frac{2N+1}{2(N+1)e_{cn}} A_{s-1}^{(N-1)} - \frac{1}{e_{cn}} \sum_{j=1}^{s} \frac{A_{j-1}^{(-1)} A_{s-j}^{(N)}}{j}, \quad s > 0. \end{aligned} \tag{32}$$

In addition

$$\begin{aligned} t &= (2l+1)^2 - 1/4, \\ g &= -E/(2|E|), \\ q_0 &= (4r_0/W)^{1/2}, \\ e_{nc} &= j_{cn}^2/2, \end{aligned} \tag{33}$$

with j_{cn} being the nth zero of Bessel function J_c with $c = 2l + 1$.

In addition, this method is very useful for obtaining position expectation values. The accuracy in the ground state energy eigenvalues reported by Fernández and Castro was within 10^{-4} hartrees. It is well known that for impenetrable confinement, i.e., in cases where the wave functions fulfill DBC, the virial theorem is satisfied [29],

$$-r_0 \frac{\partial E}{\partial r_0} = 2\langle T \rangle + \langle V \rangle = 2E - \langle V \rangle. \tag{34}$$

Fernández and Castro made a careful analysis and generalized the virial theorem for systems subjected to sectionally defined potentials [29].

2.6. The exact solutions

Series solutions

One of the most utilized methods for solving the radial equation is by expanding the wave function as a power series of the radial coordinate. This approach is very appropriate because the power series [1–3,30–32,34, 35] correctly represents the confluent hypergeometric function.

Ley-Koo and Rubinstein [30] studied the CHA inside penetrable boxes. The Schrödinger equation to be solved is given by Equation (1), however, the confinement potential $V(r)$ is in this case more general than that of Equation (2). They used one which is vanishing inside the box ($r < r_0$) while outside it is constant (V_0). When V_0 becomes infinite, it is equivalent to an impenetrable box problem. In their work, they essentially aim at reproducing the measured hyperfine splitting value of a hydrogen atom embedded in alpha-quartz [37]. This problem was studied theoretically by Suryanrayana and Weil [38] by using the model of impenetrable boxes, albeit their result was not entirely satisfactory.

For the inner region ($r < r_0$) they propose a power series solution

$$R_{\nu l}(x) = A_{\nu l} x^l \phi_{\nu l}(x) = A_{\nu l} x^l \sum_{s=0}^{\infty} c_s^{(l)} x^l, \tag{35}$$

where

$$x = \frac{Zr}{a_o \nu}, \quad \text{and} \quad \nu^2 = \pm \frac{Z^2 e^2}{2a_0 E}, \tag{36}$$

a_0 is the Bohr radius, E is the energy of the system and $\pm$ stands for positive and negative energies, respectively, so that ν is always real.

For the region outside the box ($r \geq r_0$) they found that the solutions are of the form

$$R_{kl}(y) = B_{kl}\, y^{-l-1} \mathrm{e}^{-y}\, {}_1F_1(-l, 2l, 2y) \tag{37}$$

with

$$y = kr \quad \text{and} \quad k^2 = 2\mu(V_0 - E)/\hbar^2, \tag{38}$$

where μ is the reduced mass of the system. The confluent hypergeometric function in this case corresponds to a polynomial of degree l.

The energy values for different levels are obtained by imposing the continuity of the wave function and its derivative at $r = r_0$

$$\begin{aligned} R_{\nu l}(x_0) &= R_{kl}(y_0) \\ \left(\frac{Z}{a_0 \nu}\right) R'_{\nu l}(x_0) &= k R'_{kl}(y_0). \end{aligned} \tag{39}$$

Even for positive energies, they obtained the corresponding eigenvalues, which was a difficult task to perform because of the involvement of hypergeometric functions, and in fact it represents a feature previously mentioned by some authors [3,38]. In their procedure, Equations (39) are solved by choosing an angular momentum and an energy value, then finding the radius r_0 of the box where the hydrogen atom is embedded into.

Once the energies and wave functions are obtained with sufficient accuracy, some physical properties associated with the latter are calculated. Since the work of Michels et al. [1], it was recognized that it is essential to calculate the ground state to obtain these properties. For the CHA, some of the most important are the hyperfine splitting, given by the Fermi contact term A, Equation (29) [30,34,35,38] and the nuclear magnetic shielding [26], given by the diamagnetic screening constant [30,34,35],

$$\sigma = \frac{e^2}{3\mu c^2} \left\langle \frac{1}{r} \right\rangle, \tag{40}$$

whereas the polarizability [1] is given by Kirkwood's approximation [57]

$$\alpha = \frac{4}{9a_0} \langle r^2 \rangle^2, \tag{41}$$

and, by using the virial theorem, the pressure [1,30,34,35] in Equation (8) can thus be written as

$$p = -\frac{1}{4\pi r_0^2}\frac{\mathrm{d}E}{\mathrm{d}r_0} = \frac{1}{4\pi r_0^3}(2E - \langle V \rangle). \tag{42}$$

It has been proved that these properties depend only on the box radius r_0 for an atom within impenetrable boxes and that they vary monotonically as a function of it.

Ley-Koo and Rubinstein computed the energies and properties mentioned above for four barrier heights $V_0[e^2/2a_0] = 0, 1, 4, \infty$. Although for long time those calculations were considered exact, the numerical accuracy reported therein is only about 10^{-6} hartrees. They computed the 2s and 2p energy levels as a function of r_0. These states, when plotted as a function of r_0, appear to be very close to each other, where the 2p state remains below the 2s curve. The energy curves evolve as expected, growing monotonically as the box radius r_0 is reduced. When the spherical boxes become impenetrable, there appears to be no upper limit for the growth rate. By contrast, in the case of penetrable boxes, the energy eigenvalues are bounded from above, reaching the barrier height before ionization takes place. For a given barrier height, the electron remains in a bound state only for r_0 values larger than $r_I = r(E = V_0)$. For boxes of a given radius, the energy eigenvalues will be higher as V_0 increases, i.e., for less penetrable boxes. When $r_0 \to \infty$ the energy eigenvalues asymptotically attain those associated with the free (unbounded) hydrogen atom. Ley-Koo and Rubinstein noted that the energy curves for finite values of V_0 are concave upward for large radii, but become concave downward for radii close to r_I.

The variations of A and σ as a function of r_0 for different values of V_0 are similar. For large r_0's, A and σ asymptotically approach the values $A_H = 50.762$ mT and $\sigma = e^2/(3\mu a_0 c^2)$. As r_0 decreases, these quantities grow without limit for impenetrable boxes. By contrast, when the boxes become penetrable, these quantities show a similar behavior for large radii, though eventually, they attain a maximum for a radius slightly above r_I, diminishing very rapidly to zero at r_I. On the other hand, the polarizability exhibits a different behavior as a function of r_0: As we move toward ever smaller box radii r_0's, it sets out from the value $\alpha = 0.5927 \times 10^{-24}$ cm^3, which corresponds to that of the free hydrogen atom and decreases to zero upon reaching the limit of impenetrable boxes. On the other hand, for soft boxes, it diminishes all the way until reaching a radius close to r_I, then reversing drastically its trend afterwards, increasing very quickly without limit as $r_0 \to r_I$.

Ley-Koo and Rubinstein showed that for a penetrable box where $V_0 = 0$ and $r_0 = 4.003$ au, it is possible to obtain the experimental value of $A =$

51.811 mT for a hydrogen atom trapped in alpha-quartz. They suggested that, in a way, finite values of V_0 simulate the attractive Van der Waals forces.

Another approach was given by Killingbeck [32]. The radial Schrödinger equation (3) can be written as

$$\left(-\frac{1}{2}\frac{d^2}{dr^2} - r^{-1} + \frac{1}{2}l(l+1)r^{-2}\right) R = ER. \tag{43}$$

By making the substitution

$$R = r^{l+1}\phi(r)\exp(-r/(l+1)), \tag{44}$$

where

$$\phi = \sum_{n=0}^{\infty} A_n r^n = \sum_{n=0}^{\infty} T_n(r), \tag{45}$$

it can be shown that coefficients T_n fulfill the following recurrence relation:

$$\frac{1}{2}(n+2)(2l+n+3)T_{n+2} = (n+1)(l+1)^{-1} r T_{n+1} - \left[E + \frac{1}{2}(l+1)^{-2}\right] r^2 T_n. \tag{46}$$

By setting $T_0 = 1$ and the coefficients with negative sub-index to zero, as well as via selection of two energy values E_1 and E_2 where r is fixed to $r = r_0 = 2$ au, he was able to find the energy at which function R vanishes at $r = r_0$.

From the formula

$$\varepsilon = E_1 + (E_2 - E_1)/(1 - f), \tag{47}$$

with

$$f = \phi(E_2, r_0)/\phi(E_1, r_0). \tag{48}$$

Utilization of this fast and accurate method leads, after a few iterations, to results obtained within an accuracy of 10^{-8} hartrees. In Table 4 are shown Killingbeck's results in comparison with those of Friedman et al. [33] obtained by means of the finite element method.

One disadvantage encountered when numerically integrating the Schrödinger equation is that the wave function can be known only over a

Table 4 Energy eigenvalues for different states for the hydrogen atom confined by an impenetrable spherical box of radius $r_0 = 2$ au. The reported results were obtained by Killinbeck [32] and Friedman et al. [33] using the series method and the finite element method, respectively

State	Ref. [32]	Ref. [33]
1s	−0.12500000	−0.125
2s	3.32750916	3.3305
3s	9.31415047	10.3795
2p	1.57601879	1.576
3p	6.26900279	6.390
3d	3.32750916	3.328
4d	9.31415047	10.335
4f	5.34209438	5.3805
5g	7.64667494	7.781

discrete number of points defined on a mesh. On occasions, this number may be insufficient to compute the values of some matrix elements with the required accuracy. It is then necessary to repeat the calculations on a finer mesh. We note that in the same work, Killingbeck solved the hydrogen atom subject to Neumann boundary conditions, which gives rise to an interesting topic proposed by Wigner and Seitz [36] when analyzing the constitution of metallic sodium. We shall develop this topic in Section 4.

An alternative and very accurate way to solve the problem was given by Aquino [34,35], which is based on a procedure developed by Campoy and Palma [189,190] for free (unbounded) systems. This method has been successfully applied in the following contexts: The spherically confined harmonic oscillator [13], computation of the Einstein coefficients of the 1D asymmetrically confined harmonic oscillator [169], confined 2D hydrogen atom [185], and also in the study of free (unbounded) systems as the inversion frequencies of NH_3, in which the inversion potential is modeled by a 20th-degree polynomial [191], and in the Mitra potential [192].

In this method it is assumed that the wave function depends on a radial coordinate and energy,

$$R = R(r, E). \tag{49}$$

Two differential equations must then be solved simultaneously. One is the Schrödinger equation

$$\psi'' = 2[V(r) - E]\psi \tag{50}$$

and the other is a differential equation for the energy obtained from the Schrödinger equation,

$$\frac{\partial \psi''}{\partial E} = 2[V(r) - E]\left(\frac{\partial \psi}{\partial E}\right) - 2\psi, \tag{51}$$

where $V(r)$ is the potential to which the electron is subjected. The barrier is located at the position $r = r_0$, and the boundary conditions must be satisfied at this point for the exact energy.

$$\psi(r = r_0, E_{\text{excat}}) = 0. \tag{52}$$

Therefore, the problem of finding the Schrödinger equation energy eigenvalues is substituted by the problem of finding the zeros of the wave function ψ at r_0. One procedure is as follows: we give an initial guess value for E, then solve Equations (50) and (51) to obtain the values $\psi(r_0)$ and $\partial\psi(r_0)/\partial E$. To obtain an improved energy value, we use the Newton–Raphson method.

$$E_{i+1} = E_i - \psi(r_0, E_i)/(\partial\psi(r_0, E_i)/\partial E). \tag{53}$$

With this new value we iterate Equations (50) and (51) until the desired accuracy is achieved.

To apply this method, we write the radial Schrödinger equation (3) as

$$r^2 R'' = -2rR' + l(l+1)R - 2rR - 2Er^2 R, \tag{54}$$

then we expand the wave function at the origin in a Taylor series

$$R(r, E) = \sum_k (R^{(k)}(0, E)/k!)r^k = \sum_k T_k(r, E). \tag{55}$$

This expansion is similar to that used by Killingbeck [32]. After some algebraic manipulation, we obtain the recurrence relation for the coefficients T_p

$$T_{p+1} = -\frac{2r[ErT_{p-1} + T_p]}{[(p+1)(p+2) - l(l+1)]}. \tag{56}$$

Table 5 Energy eigenvalues for the ground state of the CHA as a function of the box radius obtained by Ley-Koo and Rubinstein [30] and Aquino [34]

r_0 (au)	Ref. [30]	Ref. [34]
0.53622	12.5000	12.50019917931
1.22195	1.18345	1.18344646955
2.0000	−0.1250	−0.12500000000
2.44558	−0.3200	−0.32000030292
3.04187	−0.42865	−0.42866962486
4.08671	−0.48535	−0.48533085511
5.80119	−0.4990	−0.49900149259
10.0000		−0.49999926328
20.0000		−0.50000000000

From Equation (55) we differentiate the wave function with respect to the energy

$$\partial R/\partial E = \sum_k \partial T_k/\partial E = \sum_k \dot{T}_k, \tag{57}$$

where $\dot{T}_k = \partial T_k/\partial E$ is obtained from Equation (56).

$$\dot{T}_{p+1} = -2r[Er\dot{T}_{p-1} + rT_{p-1} + \dot{T}_p][(p+1)(p+2) - l(l+1)]^{-1}. \tag{58}$$

By evaluating $T_p(r_0)$ and $\dot{T}_p(r_0)$, we can construct the wave function (Equation (55)) and its derivative (Equation (57)), yielding the improved energy (Equation (53)). After a few iterations, Aquino [34] obtained the energy eigenvalues with an accuracy of 10^{-12} hartrees. Recently, Aquino et al. [39] has improved the accuracy of the calculations up to 100^{-100} hartrees. Aquino's results [34] and those obtained by Ley-Koo and Rubinstein [30] are shown in Table 5. A few energy eigenvalues are shown in Figure 1.

Also, Aquino computed the Fermi contact term, diamagnetic screening constant, polarizability and the pressure as a function of the box radius, these quantities are graphically shown in Figure 2.

Use of confluent hypergeometric functions

As mentioned in an earlier section, solution of the CHA problem by means of the hypergeometric functions can be very difficult [1–4]. It is particularly difficult at the region of positive energies. In 1975 Suryanrayana and Weil [38] arrived at the solution found by de Groot and ten Seldam [3] and Dingle [4]. They solved the problem by finding the zeros of the confluent

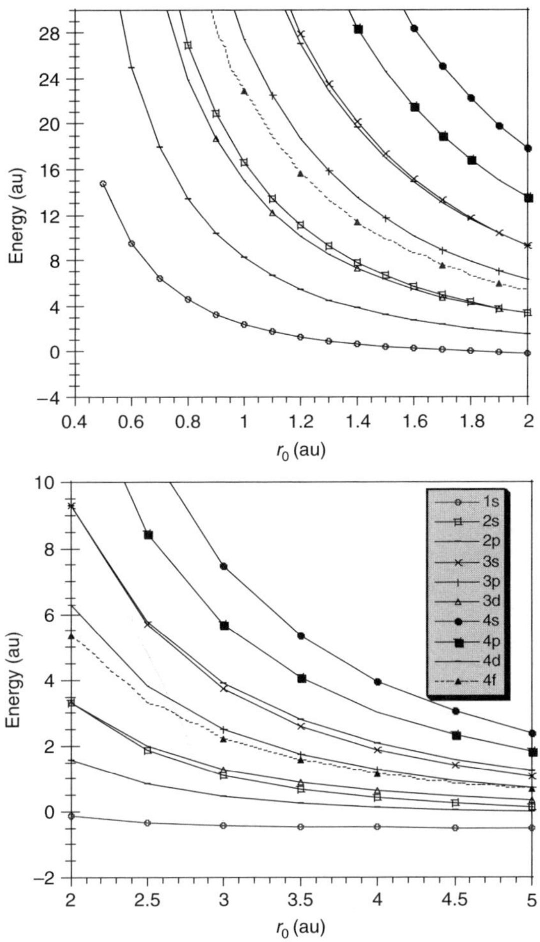

Figure 1 Ground and first nine excited state energies of the hydrogen atom confined by impenetrable walls as a function of the box radius. Observe that several level crossings occur at different radii.

hypergeometric function.

$$F = {}_1F_1(l + 1 - n, 2l + 2, \rho_0) = 0, \tag{59}$$

where

$$\rho = \frac{2r}{n} \quad \text{and} \quad E = -\frac{1}{2n^2}. \tag{60}$$

Their procedure is as follows: for a fixed value of the angular momentum l, a value for n (in general it is not an integer) is selected, then the function

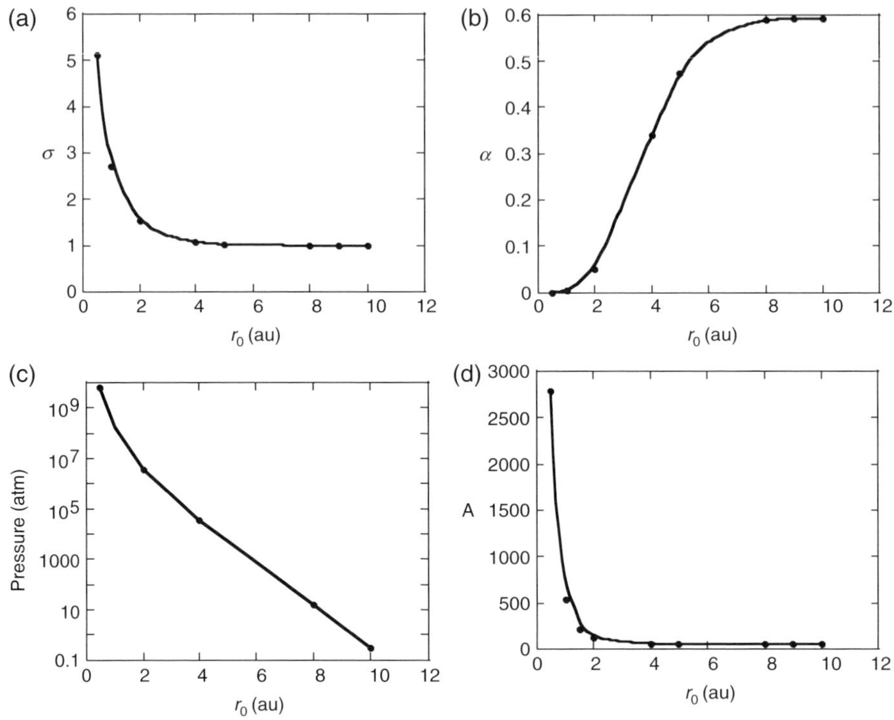

Figure 2 (2a) Magnetic screening constant $\sigma\,[e^2/3\mu a_0 c^2]$, (2b) polarizability in Kirkwood's approximation $\alpha\ [10^{-24}$ cm], (2d) hyperfine splitting constant A [mT] and (2c) pressure [atm] (for the 1s ground state) as a function of the confining radius r_0 (au).

F is evaluated from $\rho = 0$ until certain ρ_0 at which $F(\rho_0) = 0$, where the corresponding value of r is the box radius r_0. They also encountered some difficulties at the region of positive energies, which led to results being somewhat inaccurate.

Goldman and Joslin [40] also arrived at the same formal solution reported by de Groot and ten Seldam [3], Dingle [4] and Suryanrayana and Weil [38]. In addition, they found the energy eigenvalues by determination of the zeros of the confluent hypergeometric function (Equation (59)) To accomplish the latter they used an appropriate series representation for negative and positive energies, obtaining an accuracy of seven figures in their results, where 80–120 terms are included in the series. They studied the frequency and intensity shifts for absorption or stimulated emission, mainly for the principal lines in the Lyman and Balmer series of CHA. All of the emission frequencies are blue shifted since isotropic compression of the atom raises the energy levels. The effects on the intensities are more interesting: Goldman and Joslin [40] found that the relative intensities of the compressed system, as compared to those of the free atom, do not evolve monotonically as a function of the box radius. These relative intensities have a maximum greater than one for specific values of the box radius. The confined system intensities are closer to zero for very small cavity radii. For large boxes they approach those of the free atom, as expected.

They studied the lifetimes of the 2s, 2p, 3s, 3p and 3d states. For the 2p, 3s, 3p and 3d states an isotropic compression reduces their lifetimes as compared to those of the free atom. One very important result refers to the 2s state lifetime: for the free atom, the transition 2s $\rightarrow$ 1s is forbidden, which in the dipole approximation implies that the associated lifetime is infinite. By contrast, under compression, the 2p state energy lies lower than the 2s, whereas the 1s state remains below the 2p state, therefore, the transitions 2s $\rightarrow$ 2p and 2p $\rightarrow$ 1s are allowed. Under these conditions, the 2s state thus possesses a finite lifetime, which in fact becomes very sensitive in terms of the box size; for $(4 \leq r_0 \leq 8)$ au, Goldman and Joslin obtained a lifetime in the range $(7.5 \times 10^{-10}, 4.4 \times 10^{-7})$ s.

Very recently, Burrows and Cohen [41,42] have obtained formal solutions of the CHA [2–4,38,40] by combining algebraic and theoretical group methods. They obtained the energy eigenvalues by solving Equation (59) by means of the program Maple. Although their results are presented as being exact, those reported by Aquino [34] are superior by comparison. Furthermore, in a previous work, the same authors, Burrows and Cohen, performed calculations that yield more accurate results. Laughlin et al. [43] carried out a very accurate numerical calculation of the energy for the 1s, 2s and 2p states, where approximate algebraic solutions to the problem are also obtained.

Burrows and Cohen used their solutions found with Maple to compute the dipole polarizabilities and shielding factors of a general *n*s state, obtaining results that improve upon recent accurate calculations reported by Montgomery [44] and by Laughlin [43].

Recently, Aquino et al. [39] reviewed the solution of Equation (59) by using the computer algebra system Maple to evaluate the confluent hypergeometric functions, and Maple's root-finding procedure with careful handling of the precision of the involved variables. The obtained results were reported within 50 significant figures, shown in Table 6 where some of the ground state eigenvalues are given as a function of the box radius. Comparison of these results with the improved ones, obtained by the series method [34], led us to conclude that the energy eigenvalues can actually be determined within an accuracy of 10^{-100} hartrees.

2.7. Other approaches

Montgomery used Variational Perturbation Theory [46] to study the 1s, 2p and 3d states of the CHA, obtaining energies that are very close to those reported by Goldman and Joslin [40]. The WKB method was first used by

Table 6 Ground state 1s ($n = 1, l = 0$) energy of the hydrogen atom confined by an impenetrable spherical box of radius r_0 obtained by Aquino et al. [39]. Energies and distances are given in hartrees and bohrs, respectively. All figures are significative

r_0	Energy
0.5	14.74797003035027964459870376588061566847265618402933
0.7	6.46992612725126199619364245567140636265738478630981
0.9	3.26218953624011858388259606125630700497481876946612
1.0	2.37399086610366372719590654611349307963693745065997
1.5	0.43701806524725585076311008318713946334756309481263
2.0	−0.12500000000000000000000000000000000000000000000000
3.0	−0.42396728773345367391376953364592591279710293949447
4.0	−0.48326530207802202621468231192875311725590347878057
5.0	−0.49641700659145222132352437972159928232200449262657
6.0	−0.49927728637211503505666031511861895799541801925111
8.0	−0.49997510044574059549076692759149967927069874619390
9.0	−0.49999565421901901076894317532730537894873338002380
10.0	−0.49999926328152545254275469031816495470188999907569
20.0	−0.49999999999999355198935844427136657740142911588084

Dingle [4], whereas Patil [47] utilized the coalescence and inflexion properties of the wave function. Fröman et al. [48] based their analysis on the phase-integral method.

Patil and Varshni [49] used the analytical expressions for the limiting cases $r_0 \to 0$ and $r_0 \to \infty$ in conjunction with an interpolation procedure to obtain an expression valid for all values of r_0.

There exist methods that are fully numerical, e.g., the finite element method [33,51], the DFT method [52], and the Monte Carlo method [53]. As mentioned above, the CHA problem was approached through the finite element method by Friedman et al. [33]. Recently Guimaraes and Prudente [51] used a variational procedure that is based on the p-version of the finite element method. Their results are very close to the most accurate values published to date [34,39,43]. In addition, through this method it is possible to analyze all states.

Garza et al. [52] studied the problem by using the Khon-Sham equations, where their results are also close to the most accurate figures published hitherto. March and Tosi [53] used the path-integral Monte Carlo Method.

2.8. The off-centre spherically confined hydrogen atom

Most of studies have been devoted to the CHA analysis by placing the nucleus at the centre of the spherical box. However, the confined atom may conceivably be located anywhere inside the spherical box. For atoms trapped in fullerene cages, in the theory of endohedral compounds, there exist experimental and theoretical studies that show that the confined atom is not at the centre of M@C_{60} [72–78]. A similar problem has been studied in connection with the off-centre donor states in semiconductors, as for example the hydrogenic donor (D^0) or donor ions (D^-) confined in quantum dots or wells [79,80]. Researchers have utilized a variety of methods [79–87] to study the off-centre donor problem. These calculations are complicated due to lack of spherical symmetry, which results in a Schrödinger equation where the radial and angular variables are non-separable. Most authors have restricted their attention to computing ground state energies.

Let us suppose that a one-electron atom is located in an impenetrable spherical box. The origin of the coordinate system is at the centre of the sphere of radius r_0, and the atomic nucleus of charge Z is displaced a distance a from the origin, along the z axis. The Hamiltonian of the system now reads

$$H = -\frac{1}{2}\nabla^2 - \frac{Z}{|\vec{r} - \vec{a}|} + V_c(r), \tag{61}$$

with the boundary condition

$$\Psi(r = r_0, \theta, \phi) = 0, \tag{62}$$

where V_c is the confining potential (Equation (2)), while $\vec{r}$ and $\vec{a}$ refer to the electron and nucleus position, respectively.

Changa et al. [85] used the perturbation theory to estimate the energy eigenvalues of some low-lying electronic states. Their formula for the first order correction is a sum of terms of generalized hypergeometric functions. They also performed a numerical calculation based on the finite difference method.

The general conclusions of this problem are: the energy $E_{nl}(r_0, a)$ grows in a non-monotonic way as a increases, the equilibrium position of the nucleus is at the centre for all ns states and for $l > 0$ with higher $|m|$, while for low values of m the position at the centre is not favored.

By taking advantage of the rotational invariance of the system around the z-axis, Ting-yun et al. [86] used the ansatz

$$\Psi(r, \theta, \phi) = \psi(r, \theta)\frac{e^{im\phi}}{\sqrt{2\pi}}, \tag{63}$$

where θ is the angle between $\vec{r}$ and the z-axis. Upon substitution of Equation (63) into Equation (61) they obtained a Hamiltonian that depends on r and θ, the solutions of which are expressed as an expansion in B-splines functions,

$$\psi(r, \xi) = \sum_{ij} C_{ij} B_{i,k}(r) B_{j,k'}(\xi), \tag{64}$$

where $\xi = \cos(\theta)$.

Finally, the solutions of the Schrödinger equation are obtained by solving the generalized eigenvalue problem,

$$HC = ESC. \tag{65}$$

They only studied the states with $m = 0$, $r_0 = 2$ au, and $a = 0, 0.1, 0.5, 1.0, 1.5$ and 2 au. They computed the ground state and more than 20 excited states. Their calculations have around eight figures of accuracy.

Through their study it was found that the variation of energy levels $E_{nl}(r_0, a)$ is, in general, smooth and flat, except in a few particular cases. For $l = 0$ states, $E_{nl}(r_0, a)$ grows monotonically as a increases. For s states with n greater than 2, the wave function distribution is more concentrated near the boundary. For $l \neq 0$ the variation of $E_{nl}(r_0, a)$ is weaker. The energy curves decrease slightly, passing through a minimum, then growing again at a very low rate. Finally, they found the existence of anti-crossing levels at specific values of a.

3. SOFT SPHERICAL CONFINEMENT FOR HYDROGEN ATOM REVISITED

Simulation of confinement by penetrable boxes represents a more realistic physical model. A very simple approach was proposed by Marín and Cruz [18], where they used the Rayleigh–Ritz variational method via a trial wave function for the ground state, which consists of two piecewise functions, one for the inner region ($r < r_0$), and the other for the outer region ($r \geq r_0$). The trial wave function is defined as follows:

$$\begin{array}{ll} R_i = N_i(r_0 - \gamma r)\exp(-\alpha r), & r < r_0 \\ R_e = N_e r^{-1}\ \exp(-\beta r), & r > r_0 \end{array} \tag{66}$$

where α, β, γ are variational parameters and N_i, N_e are normalization constants. Only two variational parameters are independent, whereas the third one is obtained by imposing continuity of the logarithmic derivative of the wave function,

$$\beta = (\alpha r_0(1 - \gamma) + 2\gamma - 1)/(r_0(1 - \gamma)). \tag{67}$$

The results with that very simple wave function provide some insight only at a qualitative level.

In Section 2.6 this problem was discussed within an approach formulated by Ley-Koo and Rubinstein [30], in which the solution of the Schrödinger equation is obtained for the inner and outer regions. The equation required to define the energy is found by imposing continuity of the solutions and

Table 7 Ground state energy eigenvalues for the hydrogen atom confined by a penetrable spherical box as a function of r_0 and the barrier height $V_0 = 0.5$ hartrees. The reported results were obtained by Ley-Koo & Rubinstein [30] and by Aquino [35]

r_0 (au)	Ref. [30]	Ref. [35]
0.73918	0.4132	0.4132259339
0.83807	0.29585	0.2958597207
1.03054	0.0800	0.0799966714
1.25130	−0.1033	−0.103306791
1.53583	−0.2551	−0.255103218
1.96281	−0.37805	−0.378071403
2.46766	−0.4450	−0.444998290
3.13886	−0.4806	−0.480584480
4.14964	−0.4960	−0.496023853
4.57972	−0.4980	−0.498005999

their derivatives across r_0, which separates the interior and outer regions. This leads us to the logarithmic derivative equation of the wave function at the boundary $r = r_0$. The resulting equation must be evaluated numerically in order to find the energy eigenvalues, and although this procedure is in principle exact, the low accuracy attained in their results stems from an unsuitable way of computing the energy values. In Table 7 we show some energy eigenvalues for a barrier height $V_0 = 0.5$ hartrees.

This procedure is more accurate than that proposed by Marín and Cruz, however, its implementation is somewhat complicated.

This problem can be solved by the series method [35] described in Section 2.6. The radial Schrödinger equation of this problem is given by

$$\left(-\frac{1}{r^2}\frac{\mathrm{d}}{\mathrm{d}r}r^2\frac{\mathrm{d}}{\mathrm{d}r} + \frac{l(l+1)}{r^2} + 2V_c(r)\right)R(r) = 2ER(r), \tag{68a}$$

where the confining potential is the following:

$$V_c(r) = \begin{cases} -1/r, & \text{if } r < r_0 \\ V_0, & \text{if } r \geq r_0, \end{cases} \tag{68b}$$

in which V_0 is a positive constant. The Schrödinger equation is then solved inside the box ($r < r_0$) and outside of it ($r \geq r_0$).

The radial equation inside the box is

$$R'' + \frac{2}{r}R' - \left[\frac{l(l+1)}{r^2} - \frac{2}{r} - 2E\right]R = 0, \quad 0 \leq r \leq r_0, \tag{69}$$

where the solutions are of the form given by equation

$$R^i_{nl}(r, E) = \sum_{p=0}^{\infty} T_p(r, E) \tag{70}$$

and the functions $T_p(r, E)$ are those in Equation (56); the upper index i refers to the solution inside the box.

We focus our discussion on the solutions in the region outside the box, the radial Schrödinger equation is given by equation

$$\left(\frac{\mathrm{d}^2}{\mathrm{d}y^2} + \frac{2}{y}\frac{\mathrm{d}}{\mathrm{d}y} - \frac{l(l+1)}{y^2} - 1\right) R(y) = 0, \tag{71}$$

where

$$y = kr \quad \text{and} \quad k^2 = 2\mu(V_0 - E)/\hbar^2\,. \tag{72}$$

The solutions in this region were already discussed in a previous section; they are given by the Equation (37)

$$R_{kl}(y) = B_{kl}\, y^{-l-1} \mathrm{e}^{-y} f_l(y) \tag{73}$$

in which $f_l(y)$ is the confluent hypergeometric function, for the present case this function is reduced to a polynomial.

For example, for the s states ($l = 0$), $f_0(y) = 1$, $f'(y) = 0$; for p states ($l = 1$), $f_1(y) = 1 + y$, $f_1'(y) = 1$, etc.

For the s states the solutions outside the box are

$$R^e(r) = \frac{B\mathrm{e}^{-kr}}{r} \tag{74}$$

in which B is a normalization constant, and the super index e indicates the solution in the external region of the box.

In what follows we will concern ourselves with the s states. To determine the energy values we demand that the wave function and its derivative with respect to the variable r be continuous at r_0, thus arriving at the condition of continuity for the logarithmic derivative at r_0,

$$\frac{R^{i\prime}(r_0)}{R^i(r_0)} = \frac{R^{e\prime}(r_0)}{R^e(r_0)}. \tag{75}$$

Substituting the expression for the outer region solution it is found that

$$\frac{R^{i\prime}(r_0)}{R^i(r_0)} = -\left(k + \frac{1}{r_0}\right). \tag{76}$$

The problem of energy determination has therefore been reduced to finding the roots of this equation. In Table 7 we show some results obtained by this method, which are compared with those reported by Ley-Koo and Rubinstein [30].

With the present method an improvement on the energy eigenvalues and on the wave functions is attained. At present, work is in progress to calculate position expectation values.

4. TREATMENT OF THE CONFINED HYDROGEN ATOM UNDER NEUMANN BOUNDARY CONDITIONS

An approach whereby an electron is regarded as moving inside a spherical box [36], is commonly utilized in the analysis of alkaline metals. The box volume corresponds to the atomic volume, where the interaction of the electron with the remaining ones and with the nucleus, are taken into account. The solutions of the Schrödinger equation must fulfill Neumann boundary conditions (NBC) on the surface of the box.

The radial Schrödinger equation to be solved is given by Equation (1) in which the confining potential V_c is vanishing for every r,

$$\left(-\frac{1}{r^2}\frac{\mathrm{d}}{\mathrm{d}r}r^2\frac{\mathrm{d}}{\mathrm{d}r} + \frac{l(l+1)}{r^2} - \frac{2}{r}\right)R(r) = 2ER(r), \quad 0 < r < \infty. \tag{77a}$$

The solutions of this equation are subjected to the NBC

$$R'(r = r_0) = 0 \tag{77b}$$

in which the apostrophe indicates the derivative with respect to r. We will restrict ourselves to finding only bound state energy eigenvalues by the power series method [34,35].

The solutions for bound states are given by Equations (55) and (56), which are used to construct the wave function R and its derivative R'. In order to find the energy eigenvalues a guess value ε_0 is proposed at the outset of the procedure, to subsequently be improved upon within a Newton–Raphson scheme until a required precision is attained

$$\varepsilon_{i+1} = \varepsilon_i - \frac{R'(r_0, \varepsilon_i)}{\mathring{R}'(r_0, \varepsilon_i)}, \tag{78}$$

Table 8 Ground state energy ($l = 0$) for the hydrogen atom under Neumann boundary conditions as a function of r_0 obtained by Friedman et al. [33], Killinbeck [32] and Aquino [35]. Energies and distances are given in hartrees and bohrs, respectively

r_0	Ref. [33]	Ref. [32]	Ref. [35]
0.5	−3.055	−3.055 133 86	−3.055 133 860 168
1.0	−1.561	−1.561 462 18	−1.561 462 181 919
2.0	−0.829	−0.829 506 651	−0.829 506 651 041
3.0	−0.609	−0.609 196 267	−0.609 196 267 233
4.0	−0.529	−0.529 302 461	−0.529 302 460 977
5.0	−0.504	−0.506 158 374	−0.506 158 373 699
6.0	−0.495	−0.501 139 438	−0.501 139 438 570
7.0			−0.500 200 188 926
8.0			−0.500 034 174 169

where $\overset{\circ}{R}'(r_0, \varepsilon_i) = \partial R'/\partial \varepsilon$ is constructed numerically by means of the six point rule. In Table 8 energy eigenvalues obtained through this procedure are compared with those reported by Killinbeck [32] and Friedman [33]. As seen in that table, the results obtained by the finite element method [33] display a precision rate that progressively deteriorates for increasing values of r_0, while Killinbeck's results appear to preserve their precision.

5. THE HELIUM ATOM CONFINED IN SPHERICAL BOXES

Helium is the simplest many-electron atom. Unlike hydrogen, in the helium atom the electronic correlation can be studied, which is an essential property of the electronic structure for atoms and molecules.

The reported calculations on confined helium focus on studying the associated electronic properties under conditions of extreme pressure, in which the electron clouds, unlike those in free atoms, are forced to remain spatially restricted. A particularly important aspect refers to the systematic analysis of how energy and electronic correlation vary as a function of the confining cavity dimension.

Calculating electronic properties of confined atoms—as compared to free atoms—represents a more demanding task, therefore, it is not surprising that the first study on the confined helium atom was carried out 15 years after the pioneering work addressed by Michels et al. [1].

In what follows we will consider the problem of a helium atom confined in an impenetrable spherical box where the nucleus is placed at the centre.

In 1952 ten Seldam and de Groot [99] reported the first study of the confined helium atom. They performed a variational calculation via a trial wave function based on an ansatz proposed by Hylleraas in 1929, which is

Table 9 Ground state energy of the helium atom confined by an impenetrable spherical box of radius R obtained by ten Seldam and de Groot [99]. Distances and energies are given in bohrs and hartrees, respectively

R	∞	6.176	2.916	2.572
E	−2.902430	−2.901620	−2.84812	−2.76204

multiplied by cut-off functions to ensure fulfillment of DBC:

$$\psi(s,t,u) = \left(R - \frac{s-t}{2}\right)\left(R - \frac{s+t}{2}\right)(c_0 + c_1 u + c_2 t^2)e^{-ks}, \tag{79}$$

where R refers to the spherical box radius; c_0, c_1, c_2 and k are variational parameters; s, t and u denote the Hylleraas coordinates, defined as

$$\begin{aligned} s &= r_1 + r_2 \\ t &= -r_1 + r_2 \\ u &= r_{12}. \end{aligned} \tag{80}$$

The cut-off functions correspond to

$$(R - r_1) = \left(R - \frac{s-t}{2}\right) \quad \text{and} \quad (R - r_2) = \left(R - \frac{s+t}{2}\right). \tag{81}$$

In comparison to the free (unconfined) atom, inclusion of the cut-off functions increases both the number of terms in the trial wave and consequently the number of integrals to be evaluated. They calculated the ground state energy throughout a wide range of r values. For $R = 3.574$ bohrs they obtained an energy of −2.88852 hartrees, that is higher than the energy of the free (unconfined) case, −2.902430 hartrees, as expected, since the confinement within an impenetrable box raises the energy eigenvalue. Some of their results are summarized in Table 9. They also computed the pressure by differentiating the energy with respect to the volume

$$P = -\frac{\partial E}{\partial V} = -\left[\frac{1}{4\pi r^2}\frac{\partial E}{\partial r}\right]_{r=R}, \tag{82}$$

where $\partial E/\partial V$ is determined by means of a graphical derivative.

In a second work [100], they studied the polarization of the confined helium atom via the Kirkwood approach [57], finding a decreasing tendency for polarizability as the pressure grows.

Table 10 Ground state energy 1^1S of the helium atom confined by an impenetrable spherical box of radius R obtained by different authors. Energies and radial distances are given in hartrees and bohrs, respectively. The exact energy for the free (unconfined) system is −2.90372 hartrees

R (au)	Ref. [101]	Ref. [103]	Ref. [108]	Ref. [105]	Ref. [117]	Ref. [118]
2.0	−2.5977	−2.6026	−2.5028	−2.6051	−2.6040	−2.5998
2.2	−2.7074		−2.6947		−2.7145	−2.7088
2.4	−2.7760		−2.7902		−2.7836	−2.7765
2.6	−2.8194		−2.8358		−2.8271	−2.8191
2.8	−2.8472		−2.8570		−2.8472	−28462
3.0	−2.8652	−2.8708	−2.8684	−2.8727	−2.8548	−2.8636
3.5				−2.8935	−2.8935	−2.8851
4.0	−2.8956	−2.8988	−2.8764	−2.9003	−2.9004	−2.8931
4.5					−2.9026	−2.8963
5.0	−2.9004	−2.9020	−2.8764	−2.9032	−2.9034	−2.8978
6.0		−2.9024		−2.9035	−2.9037	−2.8990
7.0		−2.9025			−2.9037	
∞	−2.9024		−2.8764	2.9037	−2.9037	−2.8999

Table 11 Energy for the confined helium atom lowest triplet state 1^3S as a function of the box radius R obtained by Aquino et al. [117], Patil and Varshni [49] and Banerjee et al. [118]. Energies and distances are given in hartrees and bohrs, respectively. The exact energy for the free (unconfined) system is −2.17523 hartrees

R	Ref. [117]	Ref. [49]	Ref. [118]
2.0	0.5603	0.9809	0.5862
2.6	−0.9123		
2.8	−1.1723		
3.0	−1.3705	−1.1193	−1.3679
4.0	−1.8746	−1.7277	−1.8734
5.0	−2.0480	−1.9615	−2.0473
6.0	−2.1178	−2.0658	−2.1171
10.0	−2.1726	−2.1166	−2.1477
∞	−2.1752		

For spherical atoms, the polarizability in the Kirkwood approximation can be written as

$$\alpha = \frac{4}{9na_0}(\langle r_1^2\rangle + \langle r_2^2\rangle), \tag{83}$$

where

$$\langle r_i^2\rangle = \int \psi^* r_i^2 \psi \mathrm{d}V \Big/ \int \psi^* \psi \mathrm{d}V, \quad i = 1, 2. \tag{84}$$

Recently, Montgomery [110] obtained very good energies by performing calculations based on a wave function that is more general than that expressed in Equation (79), which includes 13 terms in the power expansion of s, t and u.

In the late 60's Gimarc [101] analyzed the confined helium atom problem by systematically studying the correlation energy in a two electron atom. The correlation energy is defined as the difference between the Hartree-Fock energy and the exact value,

$$E_{\text{corr}} = E_{\text{HF}} - E_{\text{exact}}. \tag{85}$$

The correlation energies for free (unconfined) H^-, He, Li^+ and Be^{++} were well known at that time, and so Gimarc wanted to analyze, in particular, how the correlation energy changes as a function of the box radius for the confined helium atom isoelectronic series. Gimarc performed a number of variational calculations based on the following wave functions:

$$\begin{aligned}
\psi_1 &= \left(1 - \frac{r_1}{R}\right)\left(1 - \frac{r_2}{R}\right) \mathrm{e}^{-\alpha r_1}\mathrm{e}^{-\alpha r_2}, \\
\psi_2 &= \left(1 - \frac{r_1}{R}\right)\left(1 - \frac{r_2}{R}\right) (1 + \beta r_{12})\mathrm{e}^{-\alpha r_1}\mathrm{e}^{-\alpha r_2}, \\
\psi_3 &= \left(1 - \frac{r_1}{R}\right)\left(1 - \frac{r_2}{R}\right) (\mathrm{e}^{-\alpha_1 r_1} + b\mathrm{e}^{-\alpha_2 r_1})(\mathrm{e}^{-\alpha_1 r_2} + b\mathrm{e}^{-\alpha_2 r_2}), \\
\psi_4 &= [(\mathrm{e}^{-\alpha r_1}\mathrm{e}^{-\alpha r_2})(1 + c_1 r_{12}) + c_2(r_1^2\mathrm{e}^{-\alpha r_1}\mathrm{e}^{-\alpha r_2} + \mathrm{e}^{-\alpha r_1} r_2^2 \mathrm{e}^{-\alpha r_2} \\
&\quad - 2r_1\mathrm{e}^{-\alpha r_1} r_2 \mathrm{e}^{-\alpha r_2})] \left(1 - \frac{r_1}{R}\right)\left(1 - \frac{r_2}{R}\right), \\
\psi_5 &= \left(1 - \frac{r_1}{R}\right)\left(1 - \frac{r_2}{R}\right) (\mathrm{e}^{-\alpha r_1} + \mathrm{d}r_1^2\mathrm{e}^{-\alpha r_1})(\mathrm{e}^{-\alpha r_2} + \mathrm{d}r_2^2\mathrm{e}^{-\alpha r_2}),
\end{aligned} \tag{86}$$

where $\alpha, \alpha_1, \alpha_2, \beta, c_1, c_2$ and d are variational parameters. The double-zeta function ψ_3 and the wave function ψ_5 are excellent approximations to the Hartree-Fock wave functions suggested by Green et al. [193]. Wave function ψ_4 is the one used by ten Seldam and de Groot, expressed in terms of the Slater type orbitals. Gimar found that the correlation energy is relatively insensitive to significant changes in R.

In 1978, Ludeña [102] carried out a Hartree-Fock calculation by using a wave function consisting of a single Slater determinant for the closed-shell atoms, whereas he used a linear combination of the Slater determinants for the open-shell atoms. Each Slater-type orbital times a cut-off function of the form $(1 - r/R)$ to satisfy the boundary conditions. Ludeña studied pressure effects on the electronic structure of the He, Li, Be, B, C and Ne neutral atoms. The energies he obtained for the confined helium atom are slightly lower than those Gimarc obtained, especially for box radii in the range $R > 1.6$ au.

One year later, Ludeña and Gregory [103] reported CI calculations by using a basis of 6s, 4p and 4d orbitals, spanned by 41 terms. Their calculations yield a lower energy than those obtained by ten Seldan and de Groot, and by Gimarc through a 3-term Hylleras function ψ_3. They apparently did not run into numerical problems when performing calculations for radius < 1.0 au, as reported by Gimarc, who experienced a loss of precision in his calculations for box radii at this particular range, thus restricting his analysis when attempting to increase the confinement regime by reducing the cavity dimension. They also found that the correlation energy changes to a relatively low extent as a function of R, increasing from 0.0409 hartrees for $R = 9$ bohrs to 0.0471 hartrees for $R = 0.5$ bohrs. They thus concluded that the correlation energy grows slightly when the box radius is reduced.

Gorecki and Byers Brown [104] performed the first energy calculation of the helium atom confined in a penetrable box. They argued that the hard box model gives results agreeing with observations only at a qualitative level, and that it overestimates pressure effects through a factor in the range from 2 to 5. They proposed using a finite barrier of height U_0, instead of an infinite barrier, because the electron wave functions do not approach zero very rapidly, rather, they extend over toward the vicinity of other atoms. On the other hand, U_0 can be used as a parameter to fit some results of a particular experiment, which in turn could be used as a predictive tool for possible outcomes in other experiments. They utilized the Hartree-Fock method, where inside the box the potential energy consists of the nucleus-electron attraction and the electron–electron repulsion. Outside the box the potential is constant with height U_0.

$$V = -\frac{2}{r_1}\chi(r_1) - \frac{2}{r_2}\chi(r_2) + \frac{\chi(r_1)\chi(r_2)}{r_{12}} + U_0\left(2 - \chi(r_1) - \chi(r_2)\right) \tag{87}$$

where

$$\chi(x) = \begin{cases} 1 & \text{if } x \in [0, R] \\ 0 & \text{if } x \notin [0, R] \end{cases} \tag{88}$$

$\vec{r}_1$ and $\vec{r}_2$ denote the electron positions and R refers to the confining sphere radius. For the ground state, they considered an uncorrelated two electron wave function of the form

$$\psi(\vec{r}_1, \vec{r}_2) = \phi(\vec{r}_1)\phi(\vec{r}_2)/4\pi, \tag{89}$$

where ϕ is an orbital function to be determined, which satisfies the Hartree-Fock equation

$$\frac{1}{r^2}\frac{\mathrm{d}}{\mathrm{d}r}\left(r^2\frac{\mathrm{d}\phi}{\mathrm{d}r}\right) + 2\left(\varepsilon - U(r)\right)\phi = 0, \tag{90}$$

where

$$U(r) = \begin{cases} U_0, & \text{if } r \geq R \\ -\frac{2}{r} + \frac{1}{4\pi}\int \frac{\phi^2(r_1)}{r_{12}}\mathrm{d}^3 r_1, & \text{if } r \leq R \end{cases} \tag{91}$$

and ε is the orbital energy.

They computed the total energy for a variety of box sizes and different values of U_0 by numerically solving the Hartree-Fock equation. In order to test the accuracy accomplished by this method, they computed the energy for the case of an impenetrable box ($U_0 = \infty$), finding slightly lower energies than those reported by Ludeña [102]. They also obtained the pressure numerically.

In a first report, Marín and Cruz [16] studied the helium atom confined in an impenetrable spherical box where they used the direct variational method to optimize the energy value. The Hamiltonian for a spherically confined helium atom within a hard box is given by

$$H = -\frac{1}{2}\nabla_1^2 - \frac{1}{2}\nabla_2^2 - \frac{2}{r_1} - \frac{2}{r_2} + \frac{1}{r_{12}} + V(r_1, r_2), \tag{92}$$

where

$$V(r_1, r_2) = \begin{cases} 0, & \text{if } r_1, r_2 < R \\ \infty, & \text{if } r_1, r_2 \geq R \end{cases} \tag{93}$$

and the trial ground state wave function they utilized is

$$\psi = \mathrm{e}^{-\alpha r_1}\mathrm{e}^{-\alpha r_2}(R - r_1)(R - r_2). \tag{94}$$

Their results yield energies that are higher than those obtained by Ludeña [102] through a fraction of 0.02 hartrees. In a second paper [18] they implemented soft boundaries, and performed ground state energy calculations for H^-, He, Li^+ and Be^{2+}. The two-electron Hamiltonian for these systems is given as

$$H = -\frac{1}{2}\nabla_1^2 - \frac{1}{2}\nabla_2^2 + V(r_1, r_2), \tag{95}$$

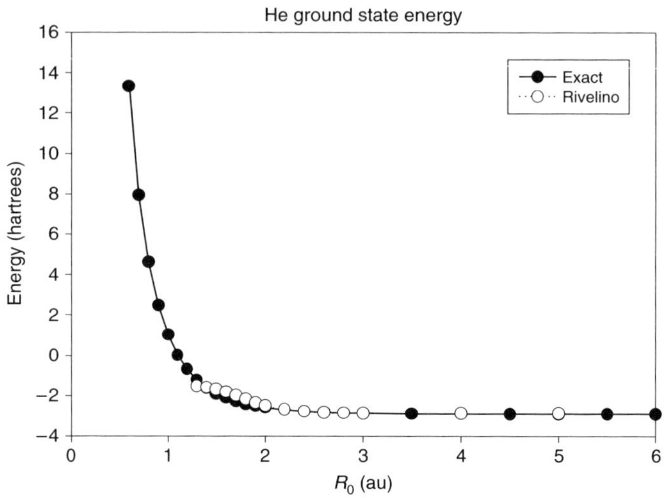

Figure 3 Helium ground state energy computed by two different methods. The Rivelino and Vianna calculations [108] vs the results of reference [117] considered here as the exact ones.

where

$$V(r_1, r_2) = \begin{cases} -\dfrac{Z}{r_1} - \dfrac{Z}{r_2} + \dfrac{1}{r_{12}}, & \text{if } r_1, r_2 < R \\ U_0, & \text{if } r_1, r_2 \geq R \end{cases} \tag{96}$$

and use is made of the direct variational method and a stepwise wave function is defined. For the region inside the box they used

$$\chi_i(r_1, r_2) = A(R - \alpha r_1)(R - \alpha r_2)\mathrm{e}^{-\alpha(r_1+r_2)} \tag{97}$$

and for the outer region they chose

$$\chi_e(r_1, 4_2) = B\mathrm{e}^{-\beta(r_1+r_2)}/(r_1 r_2), \tag{98}$$

where α, β are variational parameters, and A, B are constants. They carried out calculations with $U_0 = 0, 10, \infty$ hartrees, finding that for relatively large boxes and $U_0 = 10$ hartrees the energies become slightly lower in comparison to those for $U_0 = \infty$, a feature they attribute to the rigidity of the utilized wave function.

Joslin and Goldman [105] in 1992 studied this problem by using the Diffusive Quantum Monte Carlo Methods. By resorting to the hard spherical box model, they performed calculations, not only on the ground state of helium atom, but also for H^- and Li^+. In this method the Schrödinger equation is transformed into a one similar to the diffusion equation, where the simulation is represented by the motion of imaginary particles called walkers. In their calculations they included the trial wave function

$$\psi_0 = \psi_1(r_1, r_2, r_{12})\psi_2(r_1)\psi_2(r_2), \tag{99}$$

where

$$\psi_1(r_1, r_2, r_{12}) = \exp\left(-Z(r_1 + r_2) + \frac{r_{12}}{2(1 + br_{12})}\right) \tag{100}$$

and $\psi_2(r) = j_0(kr) = \sin(kr)/kr, k = \pi/R$, which is the solution of a free particle problem in an impenetrable spherical box. Since it vanishes at $r = R$, the trial wave function fulfills DBC.

$$\psi(r_1 = R) = \psi(r_2 = R) = 0.$$

The reason to include this trial wave function in their calculation was to increase the efficiency of the method, ensuring that the walkers remain for longer periods of time in spatial regions where $|\psi|^2$ is largest. The energies attained by this method are lower by approximately 0.002 hartrees than those obtained by Ludeña and Gregory [103] with their CI method. In the simulations they used a number of configurations ranging in [4000, 10000], where the processing times in the calculations were around 1000 hours of CPU on a workstation with 8 processors. For a long time their results were considered as the exact ones.

Rivelino and Vianna [108] studied quantum confined systems with many electrons, they constructed configurations with wave functions that do not satisfy any particular boundary conditions, which constitutes a generalization of the method proposed by Goodfriend [12]. They applied such a method to analyze the helium atom trapped in an impenetrable spherical box. Their description is based on wave functions constructed as a linear superposition of CI functions, where *ad hoc* boundary conditions are implicitly implemented by imposing some restrictions on the expansion coefficients. For the helium calculation they used 2 CI functions constructed with the atomic basis set 6-311G (for free atoms) which does not satisfy boundary conditions. Their results are slightly better than those obtained by Gimarc [101] and slightly higher compared with those of Ludeña & Gregory [103]. However, the behavior of the energy curve for radii smaller than 1.5 bohrs is somewhat peculiar, as shown in Figure 3. The energy curve obtained with this method remains below the most precise values reported to date, which shows an uncharacteristic and most probably incorrect behavior toward small radii. However, as they themselves point out, their aim was to develop a formalism to study electronic correlation in many electron atoms rather than obtaining the best energy.

Varshni [116] used a model potential within an independent electron scheme to study the spectrum of the helium atom under high pressure. Through the model potential, he was able to obtain excited state energies correctly, however, not the ground state. He used the model potential to simulate an electron-He^+ core interaction [116,196]

$$V(r) = -\frac{1}{r} + \frac{1}{r}(1 + \beta r)e^{-2\beta r}, \tag{101}$$

where β is a parameter that is optimized to reproduce as closely as possible the ^{1}S and ^{3}S experimental energies of the free helium atom. For the ^{1}P and ^{3}P states, he used a two parameter potential

$$V(r) = -\frac{\alpha}{r} + \frac{1}{r}(1 + \beta r)e^{-2\beta r}. \tag{102}$$

The Schrödinger equation is solved numerically by Numerov's method with a logarithmic mesh. He computed the energies of twelve excited states for large values of the box size R and a variety of transition wavelengths involving different levels, where, as R decreases, the transition lines get blue-shifted. Based on this result he suggested an application for studying bubbles of helium implanted in a variety of materials, high pressure helium plasmas and helium white dwarfs.

In 2003, Aquino, Flores-Riveros and Rivas-Silva [115] approached the problem via a wave function expansion in terms of generalized Hylleraas (GH) basis sets, where the Hamiltonian is expressed in Hylleraas coordinates

$$\begin{aligned} H = & -\frac{1}{2}\left(\frac{\partial^2}{\partial r_1^2} + \frac{2}{r_1}\frac{\partial}{\partial r_1} + \frac{\partial^2}{\partial r_{12}^2} + \frac{2}{r_{12}}\frac{\partial}{\partial r_{12}} + 2\hat{\mathbf{r}_1}\cdot\hat{\mathbf{r}_{12}}\frac{\partial^2}{\partial r_1 \partial r_{12}}\right) \\ & -\frac{1}{2}\left(\frac{\partial^2}{\partial r_2^2} + \frac{2}{r_2}\frac{\partial}{\partial r_2} + \frac{\partial^2}{\partial r_{12}^2} + \frac{2}{r_{12}}\frac{\partial}{\partial r_{12}} + 2\hat{\mathbf{r}_2}\cdot\hat{\mathbf{r}_{12}}\frac{\partial^2}{\partial r_1 \partial r_{12}}\right) \\ & -\frac{Z}{r_1} - \frac{Z}{r_2} + \frac{1}{r_{12}}, \end{aligned} \tag{103}$$

where $\hat{\mathbf{r}_1}$ and $\hat{\mathbf{r}_2}$ are unit vectors from the nucleus to electrons 1 and 2 respectively, and r_{12} denotes the electron–electron distance.

The trial wave function in terms of GH bases including the cut-off factors is given as

$$\psi = \left(1 - \frac{r_1}{R}\right)\left(1 - \frac{r_2}{R}\right)\sum_k^N c_k(1 + P_{12})r_1^{n_k} r_2^{m_k} r_{12}^{l_k} \times \exp(-\alpha_k r_1 - \beta_k r_2 - \gamma_k r_{12}), \tag{104}$$

where the c_k coefficients and the $\{\alpha, \beta, \gamma\}$ exponents are variational parameters, while P_{12} refers to the two particle exchange operator. In that work they used a total number of terms N=1, 5 and 10 where the associated expansions are spanned according to the condition $n+m+l \leq 2$. As expected, the lowest energies are obtained through a 10-term expansion, which in comparison to those obtained by Joslin and Goldman, turn out to be slightly higher. They also obtained the ionization radius, the atomic pressure and the expectation value $\langle r_{12} \rangle$ for the confined helium atom.

Aquino et al. [117] extended their calculations by using $N = 20, 30$ and 40 generalized Hylleraas wave functions spanned according to $n + m + l \leq 5$. Their ground state energies, calculated by means of the 40-term expansion over the box radius $R \geq 3.5$ bohrs, are in fact slightly lower than those obtained by Josling and Goldman, which were long held as the exact figures. Aquino et al. [117] were also the first in obtaining the lowest triplet state energy 1^3S, by using 20- 30- and 40-term generalized Hylleraas functions for this calculation. Their results are shown in Tables 10 and 11.

They found that, as the box radius is reduced, the singlet-triplet energy splitting increases, as can be seen in Figure 4.

In addition, they carried out a study of the singlet 1^1S and triplet 1^3S states via a DFT method by solving the spin-polarized Kohn-Sham equations. They used the local-density approximation (LDA) to the exchange-correlation functional, the Dirac functional was used for the exchange contribution and the Perdew–Wang parametrization for the correlation energy. Also, the self interaction correction (SIC) was considered in order to improve the energies. As expected, the calculations including the SIC are closest to the energies obtained with the generalized Hylleraas functions. As a general feature, it was found that the energies obtained through the LDA approach plus SIC overestimate the variationally calculated figures. However, for the ground state, the excess energy fraction occurs in a uniform fashion, that in principle could be corrected by an *ad hoc* procedure to parametrize the correlation functional. They concluded that the self-interaction correction proposed by Perdew and Zwunger works well for the open shell state 1^3S of the free helium atom, but not when the system is under confinement. They also found a reduced numerical precision for DFT calculations performed at box radii < 2.2 bohrs.

Patil and Varshni [49] calculated the ground and a few excited state energies by replacing the electron–electron interaction by an effective

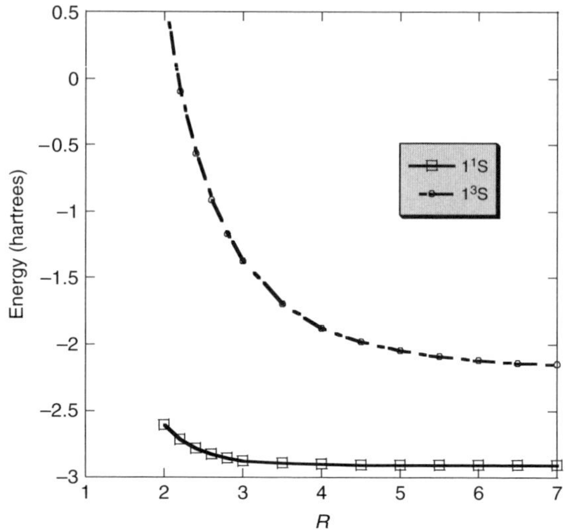

Figure 4 Helium ground state 1^1 S and the lowest triplet energy 1^3 S. The energy splitting between these two states grows as R diminishes.

screening nuclear charge which is determined by interpolating the expressions for small and large box sizes. With this approach they were able to calculate energies for the ground and some excited states for confined H^-, He, and Li. For Helium, they replaced the ground state potential

$$V = -Z\left(\frac{1}{r_1} + \frac{1}{r_2}\right) + \frac{1}{r_{12}} \tag{105}$$

by

$$V = -Z\left(\frac{1}{r_1} + \frac{1}{r_2}\right) + S(R)\left(\frac{1}{r_1} + \frac{1}{r_2}\right), \quad \text{if } r_1 < R, r_2 < R \tag{106}$$

where $S(R)$ represents the amount of screening as a function of the confinement. They found that the form of this function must be the following

$$S(R) = 0.296 + 0.071\mathrm{e}^{-(Z-5/6)R}, \tag{107}$$

which attains the appropriate limits when $R \to 0$ and $R \to \infty$. Their energies are about 0.002 hartrees higher than those of Joslin and Goldman [40].

The effective Hamiltonian utilized for the excited states is of the form

$$H = \frac{1}{2}p_1^2 - \frac{Z}{r_1} + \frac{1}{2}p_2^2 - \frac{Z'}{r_2}, \quad \text{for } r_i < R, i = 1, 2 \tag{108}$$

where Z' is the screening charge experienced by the outer electron, whose value is determined by imposing that the energy attains the correct value as $R \to \infty$. When obtaining the lowest triplet state energies they found a reduced numerical accuracy as R diminishes.

Recently, Banerjee, Kamal and Chowdhury [118] computed confined helium atom energies for the ground and three low-lying excited states using a two parameter wave function that represents a generalization of an ansatz proposed by Le Sech [195] for free atomic systems.

$$\psi(r_1, r_2) = \phi(r_1, r_2)\Omega(r_1, r_2, r_{12}) \left(1 - \frac{r_2^2}{R^2}\right)\left(1 - \frac{r_2^2}{R^2}\right), \tag{109}$$

where $\phi(r_1, r_2)$ includes only the wave function of an electron in the field of the nucleus and $\Omega(r_1, r_2, r_{12})$ is the correlation function. They used the quadratic form for the cut-off function $(1 - \frac{r^2}{R^2})$ which satisfies DBC and warrants fulfillment of the electron–nucleus cusp condition.

For the ground state

$$\phi(r_1, r_2) = A\{1s(r_1)1s(r_2)\}[\alpha(1)\beta(2) - \alpha(2)\beta(1)], \tag{110}$$

where A is a normalization constant, function $1s(r)$ is a ground state hydrogenic orbital and α, β are spin-up and spin-down, respectively. The correlation function was chosen as

$$\Omega(r_1, r_2, r_{12}) = \cosh(\lambda r_1)\cosh(\lambda r_2)\left(1 + \frac{1}{2}r_{12}e^{ar_{12}}\right), \tag{111}$$

where a and λ are variational parameters.

For the excited states

$$\phi(r_1, r_2) = A[1s(r_1)nl(r_2) \pm 1s(r_2)nl(r_1)]\chi_s(r_1, r_2), \tag{112}$$

where A is a normalization constant and χ_s is the spinorial part of the wave function. The correlation function $\Omega(r_1, r_2, r_{12})$ for these states is

$$\Omega(r_1, r_2, r_{12}) = (\cosh(\lambda r_1) + \cosh(\lambda r_2))\left(1 + \frac{1}{2}r_{12}e^{-\alpha r_{12}}\right). \tag{113}$$

The ground state energy values compare very well with those obtained by Aquino et al. through a 40-term expansion of generalized Hylleraas functions (see Tables 10 and 11).

The traditional way to obtain the critical cage radius [55,101,102,115,123] is by separate calculation of the confined helium and confined helium ion ground states, as a function of the box radius. The point at which the curves cross each other is defined as the critical cage size. In this approach, a different Hamiltonian is defined for each system, where both of these are subjected to the same confinement regime, usually given by hard wall boxes. In 2006 Díaz-García and Cruz [121] proposed an alternative way to obtain the critical cage radius for helium ionization, where the energy evolution of the two electron atom confined by a penetrable spherical box is analyzed. They expressed the total energy in such a way that the first and second ionization potentials can be easily identified.

In 1999 Whitkop [120] achieved a novel contribution upon studying the helium atom confined in an impenetrable spherical box, where the nucleus is placed off the centre. He performed a variational calculation by using a linear combination of normalized Gaussian-type orbitals (GTO's) times a cut-off function, thus ensuring the wave function fulfilment of boundary conditions on the sphere surface. He computed the helium atom ground state energy with a confining sphere or radius $R = 1$ au, where two basis sets are used: four s orbitals for the first, and four s plus one p(z) orbitals for the second. In both calculations he found that the energy increases as separation D between the nucleus and the sphere centre gets larger. In Figure 5 this behavior is illustrated.

6. OTHER APPROACHES

In 2002, Mukherjee, Karwoski and Diercksen [113] and Saha et al. [114] studied the helium atom and two-electron ions embedded in an overall charge neutral environment like that prevailing in a plasma. The confining potential is taken as a screening of the Coulomb potential between charges, represented by a short range Yukawa-type potential defined over the entire space. This procedure was suggested by Lam and Varshni [194] twenty years ago. The Hamiltonian with the Debye screening potential is given by

$$H = -\frac{1}{2}\nabla_1^2 - \frac{1}{2}\nabla_2^2 - \frac{Ze^{-\mu r_1}}{r_1} - \frac{Ze^{-\mu r_2}}{r_2} + \frac{Ze^{-\mu r_{12}}}{r_{12}}, \tag{114}$$

where Z is the nuclear charge and μ is the Debye shielding parameter

$$\mu = \sqrt{\frac{4\pi(1+Z)n}{kT}}, \tag{115}$$

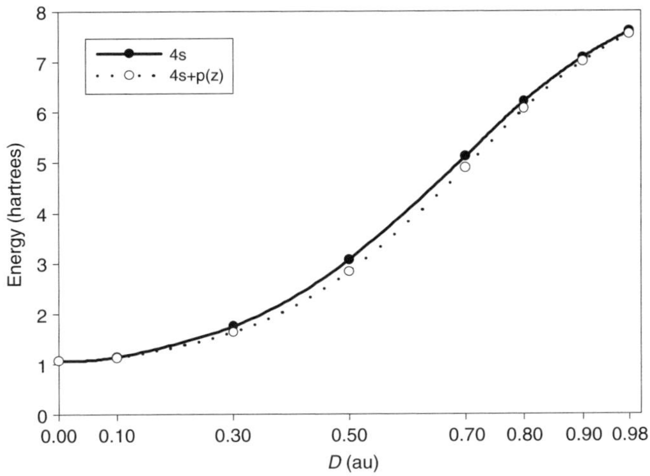

Figure 5 Helium ground state energy as a function of distance D between the nucleus and the spherical box centre obtained by means of two basis sets; four s orbitals for the first, and four s plus one p(z) for the second.

n is the density number of electrons in the plasma and T is the absolute temperature. The Debye shielding parameter is related to the Debye radius r_D as $\mu = 1/r_D$.

Mukherjee et al. [113] used a 55-term CI wave function to compute energies for the He atom, whereas Saha et al. [114] used a 75-term wave function that contain correlation factors. They also performed calculations on ground state energies for He to Ar^{16+} for a variety of values of μ from zero (no screening) to an upper limit in which the system becomes unstable.

They found that when the screening constant gets larger the energy increases, becoming less bounded and eventually leading to an unstable system. They also found that the one- and two-electron expectation values are a smooth function of μ even for those values in which the energy is positive. The values $\langle 1/r \rangle$ and $\langle 1/r_{12} \rangle$ decrease continuously as μ grows, those of $\langle r \rangle$, $\langle r^2 \rangle$, $\langle r_{12} \rangle$and $\langle r_{12}^2 \rangle$ grow with μ. The instability in the system occurs due to an increase in the screening parameter μ, because μ increases the absolute values of the kinetic and potential energies decrease, the latter does it at a faster rate and eventually the system becomes unstable.

Sabin and Sabin [109] studied confinement effects on the mean excitation energy of atoms and molecules. In particular, they analyzed the effect of confinement on I_0, the mean excitation energy for helium. I_0 is the material parameter that characterizes the ability of the target to absorb energy, independent of the nature of the incident particle. They use the jellium model, in which the ellipsoids possess impenetrable boundaries and are filled up by a uniform electron gas, whose number is precisely the number of electrons in the atom or molecule under consideration. For an N-electron atom confined in a spherical cavity of impenetrable walls of radius R, they arrive at the following expression for I_0

$$I_0 = \hbar \left(\frac{e^2 N}{m R^3} \right)^{1/2} . \tag{116}$$

They concluded that when the box radius increases the jellium density decreases and also the mean excitation energy diminishes, i.e., the stopping cross section grows as the atom becomes strongly confined. Before concluding this review it must be mentioned that the helium atom has been studied by considering other, non-spherical geometrical confining boxes and through potentials different from those described in this work. Examples of the former are: helium atom confined by spheroidal boxes [107] and within paraboloidal walls [106]. As example of the latter the helium atom confined by a harmonic oscillator potential [122] can be mentioned.

7. CONCLUSIONS

In the present work we have reviewed the main methods used in the study of one and two-electron atoms confined in spherical boxes, and have seen that the physical properties of atomic systems analyzed here may undergo drastic changes when under compression.

Michels et al. [1] showed that the effect of external pressure on a hydrogen atom can be reasonably approached by placing it inside spherical boxes, whereas later works showed the formal solution of the problem [2,3]. However, at the time those studies were addressed, no analytical solution was proposed for the energy eigenvalues of the CHA. An alternative approach followed by other researchers was the use of approximate methods such as perturbation theory and the variational method, among others. However, with the advent of computers in the modern age, along with the development of sophisticated programming software, it has been possible to numerically obtain energy eigenvalues ever more accurately for a hydrogen atom confined in spherical boxes. At present, the most accurate values can be obtained within an error of 1×10^{-100} hartrees for the hydrogen atom confined in an impenetrable spherical box.

A few of the most studied physical properties in the CHA are: the Fermi contact term A, the nuclear magnetic shielding σ, the pressure P and the polarizability α. In the case of impenetrable confinement, the first three quantities grow boundlessly as $r \to 0$, while the polarizability approaches zero. By contrast, when soft (penetrable) boxes are involved A, σ and P become vanishingly small as $r \to 0$, while the polarizability increases very rapidly and no upper bound is apparently found.

In this work we have also presented some of the most relevant methods utilized in various studies of the confined hydrogen with the nucleus located off-centre and when the atom is confined by imposing Neumann boundary conditions. Likewise, the most important methods applied in the analysis of a helium atom confined in spherical boxes of penetrable and impenetrable boxes, and by placing the nucleus off-centre in such cavities. Unlike the CHA problem, the accuracy attained in a confined helium atom energy calculation is around 1×10^{-5} hartrees. At present, only a few of the low-lying state energies for the helium atom have been obtained with the latter precision.

It is well known that confinement simulated via impenetrable boxes overestimates the effects on the atomic systems analyzed here. For example, transitions to the continuum and related properties are beyond the scope of such models. Even though resorting to cavities of soft or penetrable walls would provide a more realistic model, systems under impenetrable confinement represent the most broadly used models and the predictions resulting thereof are in general qualitatively correct.

Goldman and Joslin [40] showed that when a hydrogen atom is subjected to extreme pressures, as occurs in stars and giant planets, the number of emission lines and their corresponding intensities vary as a function of the pressure exerted on it. It is thus natural to expect that high pressures will have similar effects on many-electron atoms and molecules, which may even compound the identification of the associated spectral lines when attempting to determine their chemical composition. For this reason it is necessary to resort to more realistic potentials in the calculation of energies. To this end, answering some of the following questions may be essential: How can these confinement potentials be obtained from first principles? How can we estimate the involved parameters? Also, in which situation is a particular potential more appropriate? Of course, for heavy atoms the answer to such fundamental questions would require the extension of non-relativistic confinement models to fully relativistic ones [131], including electron correlation effects.

Nevertheless, important progress can still be made by introducing some refinements and improvements on existing models, which would lead us to a better understanding and a deeper insight into the electronic structure of confined atoms. New experimental results will undoubtedly guide us in the countless tasks that must be performed in the near future.

ACKNOWLEDGMENTS

I would like to thank H. E. Montgomery Jr., A. Flores-Riveros and an anonymous referee for their comments and suggestions during this work.

REFERENCES

[1] A. Michels, J. de Boer, A. Bijl, Physica 4 (1937) 981.
[2] A. Sommerfeld, H. Welker, Ann. Physik [5] 32 (1938) 56.

[3] S.R. de Groot, C.A. ten Seldam, Physica XII (1946) 669.
[4] R.B. Dingle, Proc. Camb. Phil. Soc. 49 (1953) 103.
[5] E.P. Wigner, Phys. Rev. 94 (1954) 77.
[6] R.E. Trees, Phys. Rev. 102 (1956) 1553.
[7] B.F. Gray, J. Chem. Phys. 36 (1962) 1801.
[8] B.F. Gray, J. Chem. Phys. 55 (1971) 2848.
[9] V.C. Aguilera-Navarro, W.M. Kloet, A.H. Zimermann, Rev. Brasil Fis. 1 (1971) 55.
[10] B.F. Gray, I. Gonda, J. Chem. Phys. 62 (1975) 2007.
[11] L.S. Cederbaum, K. Schönhammer, Phys. Rev. A 12 (1975) 2257.
[12] P.L. Goodfriend, J. Phys. B: At. Mol. Opt. Phys. 23 (1990) 1373.
[13] N. Aquino, J. Phys. A: Math. Gen. 30 (1997) 2403.
[14] J. Gorecki, W. Byers Brown, J. Phys. B: At. Mol. Opt. Phys. 20 (1987) 5953.
[15] J. Gorecki, W. Byers Brown, J. Phys. B: At. Mol. Opt. Phys. 22 (1989) 2659.
[16] J.L. Marín, S.A. Cruz, J. Phys. B: At. Mol. Opt. Phys. 24 (1991) 2899.
[17] J.L. Marín, S.A. Cruz, Am. J. Phys. 59 (1991) 931.
[18] J.L. Marín, S.A. Cruz, J. Phys. B: At. Mol. Opt. Phys. 25 (1992) 4365.
[19] J.A. Weil, J. Chem. Phys. 71 (1979) 2803.
[20] K.R. Browstein, Phys. Rev. Lett. 71 (1993) 1427.
[21] E.V. Ludeña, J. Chem. Phys. 66 (1977) 468.
[22] Y.P. Varshni, J. Phys. B: At. Mol. Opt. Phys. 30 (1997) L589.
[23] Y.P. Varshni, J. Phys. B: At. Mol. Opt. Phys. 31 (1998) 2848.
[24] F.M. Fernández, E.A. Castro, Int. J. Quantum Chem. 19 (1981) 533.
[25] F.M. Fernández, E.A. Castro, Kinam 4 (1982) 193.
[26] F.M. Fernández, E.A. Castro, J. Math. Phys. 23 (1982) 1103.
[27] F.M. Fernández, E.A. Castro, J. Chem. Phys. 76 (1982) 2506.
[28] F.M. Fernández, E.A. Castro, Int. J. Quantum Chem. XXIV (1983) 169.
[29] F.M. Fernández, E.A. Castro, J. Chem. Phys. 75 (1981) 2908.
[30] E. Ley-Koo, S. Rubinstein, J. Chem. Phys. 71 (1979) 351.
[31] E. Ley-Koo, S. Rubinstein, J. Chem. Phys. 73 (1980) 887.
[32] J. Killlinbeck, Phys. Lett. 84a (1981) 95.
[33] M. Friedman, A. Rabinovitch, R. Thieberger, J. Comput. Phys. 33 (1979) 359.
[34] N. Aquino, Int. J. Quantum Chem. 54 (1995) 107.
[35] N. Aquino, Doctoral thesis, Universidad Autónoma Metropolitana Iztapalapa, México, 1997 (unpublished).
[36] E. Wigner, F. Seitz, Phys. Rev. 43 (1933) 804.
[37] B.D. Perlson, J.A. Weil, J. Mag. Reson. 15 (1974) 594.
[38] D. Suryanrayana, J.A. Weil, J. Chem. Phys. 64 (1976) 510.
[39] N. Aquino, G. Campoy, H.E. Montgomery, Int. J. Quantum Chem. 107 (2007) 1548.
[40] S. Goldman, C. Joslin, J. Phys. Chem. 96 (1992) 6021.
[41] B.L. Burrows, M. Cohen, Int. J. Quantum Chem. 106 (2006) 478.
[42] B.L. Burrows, M. Cohen, Phys. Rev. A 72 (2005) 032508.
[43] C. Laughlin, B.L. Burrows, M. Cohen, J. Phys. B: At. Mol. Opt. Phys. 35 (2002) 701.
[44] H.E. Montgomery, Chem. Phys. Lett. 352 (2002) 529.
[45] C. Laughlin, J. Phys. B: At. Mol. Opt. Phys. 37 (2004) 4085.
[46] H.E. Montgomery, Int. J. Mol. Sci 2 (2001) 103.
[47] S.H. Patil, J. Phys. B: At. Mol. Opt. Phys. 35 (2002) 255.
[48] P.O. Fröman, S. Yngve, N. Fröman, J. Mat. Phys. 28 (1987) 1813.
[49] S.H. Patil, Y.P. Varshni, Can. J. Phys. 82 (2004) 647.
[50] C. Zicovich-Wilson, J. Planelles, W. Jaskólski, Int. J. Quantum Chem. 50 (1994) 429.
[51] M.N. Guimaraes, F.V. Prudente, J. Phys. B: At. Mol. Opt. Phys. 38 (2005) 2811.
[52] J. Garza, R. Vargas, A. Vela, Phys. Rev. E 58 (1998) 3949.
[53] N.H. March, M.P. Tosi, Il Nuovo Cimento 18 (1996) 1061.
[54] W. Jaskólski, Phys. Rep. 271 (1996) 1.
[55] J.C.A. Boeyens, J. Chem. Soc. Faraday Trans. 90 (1994) 3377.

[56] X. Tian, C. Zhuang-Qi, O. Yong-Cheng, S. Qi-Shun, Z. Guo-Long, Chinese J. Physics 15 (2006) 1172.
[57] J.G. Kirkwood, Phys. Z 33 (1932) 57.
[58] A.D. Buckinngham, K.P. Lawley, Mol. Phys. 3 (1960) 219.
[59] P.W. Fowler, Mol. Phys. 53 (1984) 865.
[60] K.D. Sen, J. Garza, R. Vargas, N. Aquino, Phys. Lett. A 295 (2002) 299.
[61] K.D. Sen, B. Mayer, P.C. Schmidt, J. Garza, R. Vargas, A. Vela, Int. J. Quantum Chem. 90 (2002) 491.
[62] A. Banerjee, K.D. Sen, J. Garza, R. Vargas, J. Chem. Phys. 116 (2002) 4054.
[63] R. Dutt, A. Mukherjee, Y.P. Varshni, Phys. Lett. A 280 (2001) 318.
[64] W. Wilcox, Am. J. Phys. 57 (1989) 526.
[65] M.L. Glasser, Am. J. Phys. 71 (2003) 574.
[66] D. Djajaputra, B.R. Cooper, Eur. J. Phys. 21 (2000) 261.
[67] M.E. Changa, A.V. Scherbinin, V.I. Pupishev, Int. J. Quantum Chem. 96 (2004) 167.
[68] M. Friedman, A. Rabinovitch, R. Thieberger, J. Comp. Phys. 33 (1979) 359.
[69] V.I. Pupyshev, A.V. Scherbinin, Phys. Lett. A 299 (2002) 371.
[70] E. Drigo, R.M. Ricotta, Phys. Lett. A 299 (2002) 137.
[71] X.J. Meng, X.R. Zhu, M.F. Tian, Chinese Phys. Lett. 22 (2005) 310.
[72] F. Weiss, J.L. Elkind, S.C. O'brien, R.F. Curl, R.E. Smalley, J. Am. Chem. Soc. 110 (1988) 4464.
[73] L.S. Wang, J.M. Alford, Y. Chai, M. Diener, R.E. Smalley, Z. Phys. D 26 (1993) S297.
[74] P. Decleva, G. De Alti, G. Fronzoni, M. Stener, J. Phys. B: At. Mol. Opt. Phys. 32 (1999) 4523.
[75] L.S. Wang, J.M. Alford, Y. Chai, M. Diener, J. Zhang, S.M. McClure, T. Guo, G.E. Scuseria, R.E. Smalley, Chem. Phys. Lett. 207 (1993) 354.
[76] C.G. Joslin, J. Yang, C.G. Gray, S. Goldman, J.D. Doll, Chem. Phys. Lett. 208 (1993) 86.
[77] E. Broclawik, A. Eilmes, J. Chem. Phys. 108 (1998) 3498.
[78] J. Hernández-Rojas, J. Breton, J.M. Gomez Llorente, J. Chem. Phys. 104 (1996) 1179.
[79] J.L. Zhu, X. Chen, Phys. Rev. B 50 (1994) 4497.
[80] N. Porras-Montenegro, S.T. Pérez-Merchencano, A. Latgé, J. Appl. Phys. 74 (1993) 7624.
[81] J.L. Zhu, Phys. Rev. B 39 (1989) 8780.
[82] J.L. Zhu, J.J. Xiong, B.L. Gu, Phys. Rev. B 41 (1990) 6001.
[83] J. Silva-Valencia, N. Porras-Montenegro, J. Appl. Phys. 81 (1997) 901.
[84] J.J. Diamond, P.L. Goodfriend, S. Tsonchev, J. Phys. B: At. Mol. Opt. Phys. 24 (1991) 3669.
[85] M.E. Changa, A.V. Scherbinin, V.I. Pupyshev, J. Phys. B: At. Mol. Opt. Phys. 33 (2000) 421.
[86] S. Ting-yun, Q. Hao-xue, L. Bai-wen, J. Phys. B: At. Mol. Opt. Phys. 33 (2000) L349.
[87] J.L. Movilla, J. Planelles, Phys. Rev. B 71 (2005) 075319.
[88] E. Ley-Koo, S.A. Cruz, J. Chem. Phys. 74 (1981) 4603.
[89] S.H. Patil, J. Phys. B: At. Mol. Opt. Phys. 34 (2001) 1049.
[90] S. Satpathy, Phys. Rev. B 28 (1983) 4585.
[91] Z. Liu, D.L. Lin, Phys. Rev. B 28 (1983) 4413.
[92] A.F. Kovalenko, E.N. Sovyak, M.F. Golovko, Phys. Stat. Sol. b 155 (1989) 549.
[93] J.D. Levine, Phys. Rev. 140 (1965) 586.
[94] E. Ley-Koo, R.M.G. García-Castellan, J. Phys. A: Mat. Gen. 24 (1991) 1481.
[95] D.S. Krähmer, W. Schleich, V. Yakovlev, J. Phys. A: Mat. Gen. 31 (1998) 4493.
[96] E. Ley-Koo, S. Mateos-Cortés, Am. J. Phys. 61 (1993) 246.
[97] E. Ley-Koo, S. Mateos-Cortés, Int. J. Quantum Chem. 46 (1993) 609.
[98] A.F. Kovalenko, M.F. Holovko, J. Phys. B: At. Mol. Opt. Phys. 25 (1992) L233.
[99] C.A. ten Seldam, S.R. de Groot, Physica 18 (1952) 891.
[100] C.A. ten Seldam, S.R. de Groot, Physica 18 (1952) 905.
[101] B.M. Gimarc, J. Chem. Phys. 47 (1967) 5110.
[102] E.V. Ludena, J. Chem. Phys. 69 (1978) 1770.
[103] E.V. Ludeña, M. Gregori, J. Chem. Phys. 71 (1979) 2235.
[104] J. Gorecki, W. Byers-Brown, J. Phys. B: At. Mol. Opt. Phys 21 (1988) 403.
[105] C. Joslin, S. Goldman, J. Phys. B: At. Mol. Opt. Phys. 25 (1992) 1965.
[106] E. Ley-Koo, A. Flores-Flores, Int. J. Quantum Chem. 66 (1998) 123.

[107] A. Corella-Madueño, R.A. Rosas, J.L. Marín, R. Riera, Int. J. Quantum Chem. 77 (2000) 509.
[108] R. Rivelino, J.D.M. Vianna, J. Phys. B: At. Mol. Opt. Phys. 34 (2001) L645.
[109] P.B. Sabin, J.R. Sabin, Int. J. Quantum Chem. 82 (2001) 277.
[110] H.E. Montgomery, unpublished data.
[111] D. Bielinska-Waz, J. Karwowski, G.H.F. Diercksen, J. Phys. B: At. Mol. Opt. Phys. 34 (2001) 1987.
[112] T. Sako, S. Yamamoto, G.H.F. Diercksen, J. Phys. B: At. Mol. Opt. Phys. 36 (2003) 1681.
[113] P.K. Mukherjee, J. Karwowski, G.H.F. Diercksen, Chem. Phys. Lett. 363 (2002) 323.
[114] B. Saha, T.K. Mukherjee, P.K. Mukherjee, G.H.F. Diercksen, Theor. Chem. Acc. 108 (2002) 305.
[115] N. Aquino, A. Flores-Riveros, J.F. Rivas-Silva, Phys. Lett. A 307 (2003) 326.
[116] Y.P. Varshni, Eur. Phys. J. D 22 (2003) 229.
[117] N. Aquino, J. Garza, A. Flores-Riveros, J.F. Rivas-Silva, K.D. Sen, J. Chem. Phys. 124 (2006) 054311.
[118] A. Banerjee, C. Kamal, A. Chowdhury, Phys. Lett. A 350 (2006) 121.
[119] X. Wen-Fang, Chin. Phys. Lett. 23 (2006) 1742.
[120] P.G. Whitkop, Int. J. Quantum Chem. 73 (1999) 459.
[121] C. Díaz-García, S.A. Cruz, Phys. Lett. A 353 (2006) 332.
[122] D. Bielinska-Waz, J. Karwowski, G.F.H. Diercksen, J. Phys. B: At. Mol. Opt. Phys. 34 (2001) 1987.
[123] J. Garza, R. Vargas, N. Aquino, K.D. Sen, J. Chem. Sci. 117 (2005) 379;
K.D. Sen, J. Garza, R. Vargas, A. Vela, Chem. Phys. Lett. 325 (2000) 29.
[124] J. Garza, R. Vargas, A. Vela, K.D. Sen, J. Mol. Strcut. (Theochem) 501 (2000) 183.
[125] A.L. Buchachenko, J. Phys. Chem. B 105 (2001) 5839.
[126] J.P. Conerade, V.K. Dolmatov, P.A. Lakshmi, J. Phys. B: At. Mol. Opt. Phys. 33 (2000) 251.
[127] V.K. Dolmatov, A.S. Baltenkov, J.P. Connerade, S.T. Manson, Rad. Phys. Chem. 70 (2004) 417; and references therein.
[128] R.F.W. Bader, M.A. Austen, J. Chem. Phys. 107 (1997) 4271.
[129] P.K. Chattaraj, U. Sarkar, J. Phys. Chem. A 107 (2003) 4877.
[130] A. Borgoo, D.J. Tozer, P. Geerlings, F. De Proft, Phys. Chem. Chem. Phys. 10 (2008) 1406.
[131] M. Gruchowski, R. Szmytkowski, J. Phys. A: Mat. Gen. 37 (2004) 7783.
[132] T.L. Cotrell, Trans. Faraday Soc. 47 (1951) 337.
[133] K.K. Singh, Physica 30 (1964) 211.
[134] R. Lesar, D.R. Hersbach, J. Phys. Chem. 85 (1981) 2798.
[135] R. Lesar, D.R. Hersbach, J. Phys. Chem. 87 (1983) 5202.
[136] J. Gorecki, W.J. Byers-Brown, J. Chem. Phys. 89 (1988) 2138.
[137] S. Mateos-Cortés, E. Ley-Koo, S.A. Cruz, Int. J. Quantum Chem. 86 (2002) 376.
[138] T. Pang, Phys. Rev. A 49 (1994) 1709.
[139] V.I. Pupyshev, V.V. Bobrikov, Int. J. Quantum Chem. 100 (2004) 528.
[140] V.V. Bobrikov, V.I. Pupyshev, Russian Chem. Bull. 54 (2005) 1.
[141] J.M.H. Lo, M. Klobuskowski, D. Bielinska-Waz, E.W.S. Schreiner, G.H.F. Diercksen, J. Phys. B: At. Mol. Phys. 39 (2006) 2385.
[142] D. Bielinska-Waz, G.H.F. Diercksen, M. Klobuskowski, Chem. Phys. Lett. 349 (2001) 215.
[143] F.C. Auluck, Proc. Nat. Inst. Sci. India 7 (1941) 133.
[144] D.S. Kothari, F.C. Auluck, Sci. Culture 6 (1940) 370.
[145] F.C. Auluck, Proc. Nat. Inst. Sci. India 8 (1942) 147.
[146] S. Chandrasekhar, Astrophys. J. 97 (1943) 263.
[147] E.M. Corson, I. Kaplan, Phys. Rev 71 (1947) 130.
[148] R.B. Dingle, Proc. R. Soc. London Ser. A 212 (1952) 47.
[149] F.C. Auluck, Proc. Nat. Inst. Sci. India 7 (1941) 383.
[150] F.C. Auluck, D.S. Kothari, Proc. Cambridge Philos. Soc. 41 (1945) 175.
[151] J.S. Baijal, K.K. Singh, Progr. Theoret. Phys. (Kyoto) 14 (1955) 214.
[152] T.E. Hull, R.S. Julius, Can. J. Phys. 34 (1956) 914.
[153] S. Sengupta, S. Gjosh, Phys. Rev. 115 (1959) 1681.
[154] K.K. Singh, Physica 30 (1964) 211.

[155] B. Suryan, Phys. Rev. 71 (1947) 741.
[156] P. Dean, Proc. Cambridge Phil. Soc. 62 (1966) 277.
[157] R. Vawter, Phys. Rev. 174 (1968) 749.
[158] R. Vawter, J. Math. Phys. 14 (1973) 1864.
[159] A. Consortini, B.R. Frieden, Nuovo Cimento 35B (1976) 153.
[160] F.C. Rotbart, J. Phys. A: Math. Gen. 11 (1978) 2363.
[161] F.M. Fernández, E.A. Castro, Int. J. Quantum Chem. 19 (1981) 521.
[162] V.C. Aguilera-Navarro, J.F. Gomes, A.H. Zimerman, E. Ley-Koo, J. Phys. A: Math. Gen. 16 (1983) 2943.
[163] F.M. Fernández, E.A. Castro, Int. J. Quantum Chem. 20 (1981) 623.
[164] R.J. Swenson, S.H. Danforth, J. Chem. Phys. 57 (1972) 1734.
[165] V.C. Aguilera-Navarro, E. Ley-Koo, A.H. Zimerman, J. Phys. A: Math. Gen. 13 (1980) 3585.
[166] F.M. Fernández, E.A. Castro, Phys. Rev. A 24 (1981) 2883.
[167] G.A. Arteca, S.A. Maluendes, F.M. Fernández, E.A. Castro, Int. J. Quantum Chem. 24 (1983) 169.
[168] A. Sinha, R. Roydchoudhury, Int. J. Quantum Chem. 73 (1999) 497.
[169] G. Campoy, N. Aquino, V.D. Granados, J. Phys A: Math. Gen. 35 (2002) 4903.
[170] N. Aquino, E. Castaño, G. Campoy, V.D. Granados, Eur. J. Phys. 22 (2001) 645.
[171] H.E. Montgomery Jr., G. Campoy, N. Aquino, arXiv:0803.4029v1[math-phy], 28 Mar 2008.
[172] S. Chaudhury, Phys. Rev. B 28 (1983) 4480.
[173] M.W. Lin, J.J. Quinn, Phys. Rev. B 31 (1985) 2348.
[174] D.M. Larsen, S.Y. McCann, Phys. Rev. B 45 (1992) 3485.
[175] D.M. Larsen, S.Y. McCann, Phys. Rev. B 46 (1992) 3966.
[176] J.W. Brown, H.N. Spector, J. Appl. Phys. 59 (1986) 1179.
[177] J.W. Brown, H.N. Spector, J. Appl. Phys. 35 (1987) 3009.
[178] S.I. Tsonchev, P.L. Goodfriend, J. Phys. B: At. Mol. Opt. Phys. 25 (1992) 4685.
[179] J.L. Zhu, J.H. Zhao, J.J. Xiong, J. Phys. Condens. Mat. 6 (1994) 5097.
[180] J.L. Zhu, X. Chen, J. Phys. Condens. Mat. 6 (1994) L123.
[181] M. El-Said, Superlattices and Microstruct. 23 (1998) 1237.
[182] T. Garm, J. Phys. Condens. Mat. 8 (1996) 5723.
[183] J.L. Zhu, J.Z. Yu, Z.Q. Li, Y. Kawazoe, J. Phys. Condens. Mat. 8 (1996) 7857.
[184] G. Bastard, Phys. Rev. B 24 (1981) 4714.
[185] N. Aquino, G. Campoy, A. Flores-Riveros, Int. J. Quantum. Chem. 103 (2005) 267.
[186] N. Aquino, E. Castaño, Rev. Mex. Fís. 44 (1998) 628.
[187] N. Aquino, E. Castaño, Rev. Mex. Fís. E 51 (2005) 126.
[188] L. Chaos-Cador, E. Ley-Koo, Int. J. Quantum Chem. 103 (2005) 369.
[189] A. Palma, G. Campoy, Phys. Lett. 121A (1987) 221.
[190] G. Campoy, A. Palma, Int. J. Quantum. Chem. Symp. S20 (1986) 33.
[191] N. Aquino, G. Campoy, H. Yee-Madeira, Chem. Phys. Lett. 296 (1998) 111.
[192] J.F. Rivas-Silva, G. Campoy, A. Palma, Int. J Quantum Chem. XL (1991) 405.
[193] L.C. Green, M.M. Mulder, M.N. Lewis, J.W. Woll, Phys. Rev. 93 (1954) 757.
[194] C.S. Lam, Y.P. Varshni, Phys. Rev. A 27 (1983) 418.
[195] C. Le Sech, Chem. Phys. Lett. 200 (1992) 369.
[196] Y.P. Varshni, Phys. Rev. A 38 (1988) 1595.
[197] J.O. Hirschfelder, J. Chem. Phys. 33 (1960) 1462.
[198] J. Killingbeck, Phys. Lett. A 65 (1978) 87.
[199] T.L. Cottrell, S. Paterson, Philos. Mag. 42 (1951) 391.
[200] R.E. Christoffersen, Basic Principles and Techniques of Molecular Quantum Mechanics, Springer-Verlag, New York, 1989, pp. 363–368;
I.N. Levine, Quantum Chemistry, Prentice Hall, New Jersey, 1991, pp. 202–205;
J.P. Lowe, Quantum Chemisty, Academic Press, New York, 1978, pp. 157–166;
W. Kauzmann, Quantum Chemisty An Introduction, Academic Press, New York, 1957, pp. 125–127.

CHAPTER 5

Exact Solutions for Confined Model Systems Using Kummer Functions

B.L. Burrows[a] and **M. Cohen**[b]

Contents

1. INTRODUCTION

The traditional methods of solution of many of the soluble problems of non-relativistic quantum mechanics employ a wide variety of analytical and algebraic methods, and their closed-form eigensolutions are usually expressed in terms of many different higher mathematical functions. However, most of these diverse functions can also be expressed quite conveniently in terms

[a] Mathematics Section, Faculty of Computing, Engineering and Technology, Staffordshire University, Beaconside, Stafford, ST18, 0DP UK
[b] Department of Physical Chemistry, The Hebrew University of Jerusalem, Jerusalem 91904, Israel

E-mail addresses: B.L.Burrows@staffs.ac.uk (B.L. Burrows), maurice@fh.huji.ac.il (M. Cohen).

Advances in Quantum Chemistry, Vol. 57
ISSN: 0065-3276, DOI: 10.1016/S0065-3276(09)00609-1

of the regular Kummer (confluent hypergeometric) function $M(a,b,x)$. (In this work, we generally follow the notations and definitions of Abramowitz and Stegun [1]). The eigenfunctions assume particularly simple forms under standard physical (boundary) conditions.

A second linearly independent solution of Kummer's equation, conventionally denoted $U(a,b,x)$, is not regular at the origin and can often be excluded on physical grounds. However, as we have shown recently [2–4], when a normal hydrogen atom is *confined* inside a finite sphere or spherical shell, its *radial* eigenfunctions may involve $U(a,b,x)$ or a combination of two linearly independent solutions of Kummer's equation, chosen so as to satisfy the boundary conditions which reflect the particular confinement considered. We have felt it necessary to address this choice in Section 2 of the present work.

Much of the recent work on confined atoms has focussed on the possibility of deriving exact (analytical) solutions, less on the physical origin of the confinement process itself, and we will follow this course here. Our justification is the belief that a model problem amenable to analytical solution is of general interest *per se*, as it is expected to furnish a good zero-order model for some truly physical system.

We begin by observing that the standard treatment of the normal unconfined hydrogen atom (UHA), solves Schrödinger's equation for a *model* Hamiltonian with a central-field potential (we use the usual atomic units), specifically

$$H = -\frac{1}{2}\nabla^2 + V(r), \qquad V(r) = -\frac{1}{r}, \quad 0 \le r < \infty. \tag{1}$$

Here, r denotes the *scalar* distance of the electron measured from the centre of mass (assumed to be at rest) of the complete system; in the infinite proton mass approximation this corresponds to the electron–proton distance. The resulting *reduced* Schrödinger equation for the *relative* electronic motion is found to be soluble in several different *orthogonal* coordinate systems. In particular, in spherical polar coordinates (r, θ, Φ) referred to the centre of mass, the *natural* unconfined ranges of these variables are

$$0 \le r < \infty, \quad 0 \le \theta < \pi, \quad 0 \le \Phi < 2\pi \tag{2}$$

and any *central* potential such as the Coulombic $V(r)$ of Equation (1) leads to solutions which may be expressed conveniently in the form of products of a *harmonic* function (a single-valued solution of the three-dimensional Laplace equation), and a purely *radial* function. As we will show in Section 3, the K-dimensional (K-d) generalisation of such central-field problems may be solved analogously using *hyperradial* coordinates: $\{r; \theta_i, i = 1, 2, \ldots, K-1\}$. In principle, both radial and angular confinement may be thought of as resulting from external environmental effects (such as various applied fields); models which seek to mimic such effects frequently restrict the range of one or more of the variables. For example, the motion of a normal H-atom confined to the plane $\theta = \pi/2$ may be viewed as a 2-dimensional H-atom model, which might be confined further by restricting the range of Φ or r (two examples are treated by Chaos-Cador et al. [5] along similar lines.)

We have been interested mainly in *radial* confinement of the 3-d H-atom [2–4], with Hamiltonian as in Equation (1), with the range of r *restricted*:

$$L < r < R \tag{3}$$

but with θ, Φ unrestricted as in Equation (2). There are actually three distinct models of *radial* confinement here, since L may be zero or nonzero, while R may be finite or infinite; the normal UHA corresponds to $L = 0$, $R \to \infty$, and the most popular models take R to be finite (exterior confinement). We may treat many unconfined and radially confined central-field K-d models together as follows. A reduced *radial* equation for any system with central potential $V(r)$ (r is the K-d hyperradius) is obtained from the *partial* separation

$$\psi = r^l Y_{l\mathbf{m}} f(r), \qquad \mathbf{m} = (m_1, m_2, \ldots, m_k), \tag{4}$$

where the K-d *hyperspherical harmonic* function $r^l Y_{l\mathbf{m}}$ satisfies the K-d Laplace equation

$$\nabla_K^2 r^l Y_{l\mathbf{m}} = 0, \qquad \nabla_K^2 = \sum_{i=1}^{K} \frac{\partial^2}{\partial x_i^2}. \tag{5}$$

In the UHA 3-d example, the angular variables θ and Φ remain unconfined and the total angular momentum quantum number l appearing in the harmonic functions (4) is a natural integer, and the *partial* separation of variables we have employed here leads naturally to an eigenequation for the radial function $f_l(r)$:

$$H_l f_l = E_l f_l, \qquad H_l = -\frac{1}{2}D^2 - \frac{l+1}{r}D + V(r), \tag{6}$$

where $D = D_r = \mathrm{d}/\mathrm{d}r$, and E_l denotes a discrete or continuous energy.

The analogous H_l for the K-d case (see Section 3 and the Appendix) has the form

$$H_l = -\frac{1}{2}\left(D^2 + \frac{2l+K-1}{r}D\right) + V(r). \tag{7}$$

For convenience, we rewrite the radial equation in the form

$$[rD^2 + cD + 2r(E - V(r))]f_l(r) = 0, \tag{8}$$

where

$$D = \frac{\mathrm{d}}{\mathrm{d}r}, \qquad c = (2l + K - 1), \quad l = 0, 1, 2 \ldots . \tag{9}$$

This will be brought into the form of the canonical Kummer equation,

$$L(a, b, x)F(x) = 0, \qquad L = xD^2 + (b - x)D - a, \quad D = \frac{\mathrm{d}}{\mathrm{d}x} \tag{10}$$

using transformations which depend on the detailed forms of the potentials for each case. We treat several examples, including both unconfined and various confinement models for hydrogenic atoms, harmonic oscillators and constant potentials. These models lead to particular *boundary conditions* which the appropriate solutions of Kummer's equation must satisfy in each case. In the following section, we give a brief survey of the Kummer functions which will be used in the later sections. In many physical models it is the exceptional solutions of the Kummer equation, or solutions with particular asymptotic behaviour that are important, so we have included these less well known cases in our analysis.

2. SOLUTIONS OF KUMMER'S EQUATION

2.1. The general theory

The Kummer Equation (10) is a linear second-order differential equation, with two *independent solutions* $X(a, b, x)$ and $Y(a, b, x)$, say. The *functional* forms of these depend on the values of the equation parameters a and b, and the general solution is of the form

$$F(x) = pX(a, b, x) + qY(a, b, x) \tag{11}$$

where p and q are suitably chosen constants. Here we consider the conventional choices of X and Y, together with the special forms used in the present work.

The standard approach (Piaggio [6]) is to examine a series solution to the Kummer equation so that we look for a solution in the form

$$y = x^c \sum_{n=0}^{\infty} a_n x^n, \quad a_0 \neq 0. \tag{12}$$

Substituting this into the Kummer equation we obtain

$$L(a, b, x)y = \sum_{n=0}^{\infty} a_n\{(n + c + b - 1)(n + c)x^{n+c-1} - (n + c + a)x^{n+c}\} = 0. \tag{13}$$

The lowest power is x^{c-1} and since $a_0 \neq 0$ we obtain the *indicial* equation

$$c(c + b - 1) = 0 \Rightarrow c = 0, \quad \text{or} \quad c = 1 - b. \tag{14}$$

The recurrence relation for the coefficients is

$$(n + c + 1)(n + c + b)a_{n+1}(c) = (n + c + a)a_n(c). \tag{15}$$

Unless $b = 1$ the roots of the indicial equation are distinct and provided the ratio $a_{n+1}(c)/a_n(c)$ does not become infinite then this analysis leads to *two independent solutions.* (A careful calculation is also required if the ratio becomes indeterminate). We will refer to these solutions as the *unexceptional* cases.

For $c = 0$ we obtain the standard KummerM function:

$$M(a, b, x) = 1 + \frac{a}{b}x + \frac{a(a+1)}{b(b+1)}\frac{x^2}{2!} + \cdots + \frac{(a)_{(n)}}{(b)_{(n)}}\frac{x^n}{n!} + \cdots, \tag{16}$$

where the coefficients are written in terms of the Pochhammer symbol

$$(a)_{(n)} = \frac{\Gamma(a+n)}{\Gamma(a)}, \quad (a)_{(0)} = 1. \tag{17}$$

We note that when $a = -m, m = 0, 1, 2\ldots$ the infinite series terminates and we obtain a polynomial solution. *Exceptional* cases occur when $b = -m, m = 0, 1, 2\ldots$ when some of the coefficients become infinite.

The KummerM function has many properties that can be used to relate these functions for varying a and b. One of the most useful properties is

$$\begin{aligned} xM(a, b, x) = {} & (a - b)M(a - 1, b, x) + (b - 2a)M(a, b, x) \\ & + aM(a + 1, b, x). \end{aligned} \tag{18}$$

This is easily derived by examining the coefficient of a general power of x in the series expansions of the functions and using the Pochhammer symbol. There is a comprehensive list of similar properties in Abramowitz and Stegun [1], which may be derived in the same way. In addition to these, there are derivative properties which depend on the result:

$$\frac{\mathrm{d}M(a,b,x)}{\mathrm{d}x} = \frac{\mathrm{d}}{\mathrm{d}x}\left\{1 + \frac{a}{b}x + \frac{a(a+1)}{b(b+1)}\frac{x^2}{2!} + \cdots\right\}$$
$$= \frac{a}{b}M(a+1, b+1, x). \tag{19}$$

For $c = 1 - b$ we obtain

$$N(a,b,x) = x^{1-b}\left[1 + \frac{(a-b+1)}{(2-b)}x + \frac{(a-b+1)(a-b+2)}{(2-b)(3-b)}\frac{x^2}{2!} + \cdots\right]$$
$$= x^{1-b}\sum_{n=0}^{\infty}\frac{(a-b+1)_{(n)}}{(2-b)_{(n)}}\frac{x^n}{n!} = x^{1-b}M(1+a-b, 2-b, x). \tag{20}$$

We note that exceptional cases, when the coefficients are infinite, are for $b = m, m = 2, 3, \ldots$ and that for $b = 1$, $M(a,b,x) = N(a,b,x)$. For $b > 1$ we have a singularity at $x = 0$ and this is often used to eliminate the possibility of using $N(a,b,x)$ in a particular solution where the behavior at the origin is required to be uniform.

The KummerM function is convergent for all x and finite at $x = 0$ while $N(a,b,x)$ is a valid second independent solution (apart from the singularity at $x = 0$) but it is sometimes convenient to express them in other ways. For example, if we write the solution in the form $y = \exp(x)w(x)$, then the resulting equation for $w(x)$ is

$$[xD^2 + (b+x)D - (a-b)]w(x) = 0. \tag{21}$$

This may be brought into the precise form of the Kummer equation by the transformation $x \to -x$, $D \to -D$, so we have formally

$$L(b-a, b, x)w(-x) = 0 \Rightarrow w(x) = M(b-a, b, -x), \quad \text{or}$$
$$w(x) = N(b-a, b, -x). \tag{22}$$

These procedures lead to two additional forms (AS 13.1.14-15) and the set of four standard solutions

$$\{M(a,b,x), N(a,b,x), \mathrm{e}^x M(b-a,b,-x), \mathrm{e}^x N(b-a,b,-x)\}. \tag{23}$$

This is not an independent set of solutions since there are only two independent solutions of the Kummer equation and the Wronskians for each pair of these forms are listed in Abramowitz and Stegun equation (13.1.21) [1].

2.2. Exceptional solutions

In all exceptional cases we can follow the procedures outlined in Piaggio [6] and there are essentially two separate cases.The first is when $b = 1$, so that we obtain just one solution with $c = 0$. To obtain another solution we may first *define* the coefficients to satisfy the recurrence relation (15) for a general c, and if we then substitute the resulting series, $Q(c)$, into the Kummer differential equation (with $a_o = 1$) we obtain

$$L(a, 1, x)Q(c) = c^2. \tag{24}$$

This verifies that for $c = 0$, when $Q(0) = M(a, 1, x)$, we have a solution. But, since c appears quadratically on the right-hand side of Equation (24), partial differentiation with respect to c leads to

$$L(a, 1, x)\frac{\partial Q}{\partial c} = 2c \Rightarrow L(a, 1, x)\frac{\partial Q(0)}{\partial c} = 0. \tag{25}$$

Consequently we may construct a second, independent solution from

$$\lim_{c\to 0} \frac{\partial Q(c)}{\partial c} = \ln(x)M(a, 1, x) + \lim_{c\to 0} x^c \frac{\partial}{\partial c} \sum_{n=0}^{\infty} a_n(c)x^n, \tag{26}$$

where $a_n(c)$ are defined from the recurrence relation (15) with $a_0 = 1$.

The second case is when one of the denominators in a standard series solution becomes infinite. We can examine all possibilities by considering the form

$$Q = x^{c+s}M(d, s - m, x), \tag{27}$$

where s is a small parameter and the infinities occur in the series when $s = 0$ and $n \geq m$. For $M(a, b, x)$ this corresponds to choosing $c = 0, d = a$ so that $b = -m$ and for $N(a, b, x)$ choosing $c = 1 - m, d = 1 + a - m$ so that $b = m$. In either case there is a singularity in the mth term when $s = 0$. We may form the recurrence relations starting at $a_0 = s$ instead of $a_0 = 1$ and the corresponding solution is then $P = sx^{c+s}M(d, s - m, x) = sQ$, so that $\lim_{s\to 0} P$ is finite and consequently the singularity at $s = 0$ has been removed. Using the differential operator we have

$$L(d, -m, x)P = L(d, s - m, x)P - s^2DQ = 0 - s^2\frac{\mathrm{d}Q}{\mathrm{d}x}. \tag{28}$$

As before, since we have a factor of s^2 we may deduce

$$\lim_{s\to 0} L(d,-m,x)P = 0, \quad \text{and} \quad \lim_{s\to 0} L(d,-m,x)\frac{\partial P}{\partial s} = 0. \tag{29}$$

The order of these limits is important and the functions P and $\frac{\partial P}{\partial s}$ are calculated *first* in terms of the coefficients $a_n(c,s)$ and then the limit as $s \to 0$ is applied. Carrying this out for P, before taking the final limit, we have the form

$$P = x^{c+s}[su(x,s) + v(x,s)], \tag{30}$$

where neither su nor v is infinite at $s = 0$. Thus taking the limit as $s \to 0$ we obtain

$$X = x^c v(x,0), \tag{31}$$

a series solution in x. Applying this procedure to $\frac{\partial P}{\partial s}$

$$x^{c+s} ln(x)[su(x,s) + v(x,s)] + x^{c+s}\frac{\partial}{\partial s}[su(x,s) + v(x,s)]. \tag{32}$$

Taking the limit as $s \to 0$ we obtain the independent solution

$$Y = x^c[ln(x)v(x,0) + w(x)], \tag{33}$$

where $w(x)$ is another series. The solutions are independent because of the term in $ln(x)$ which ensures that the Wronskian of X and Y is not identically zero for all x.

As an example of this procedure we may consider the problem

$$L(3,3,x)y = 0. \tag{34}$$

From the definition

$$N(3,3,x) = x^{-2}M(1,-1,x) \tag{35}$$

we obtain an invalid solution since the coefficient of x^0 is infinite. Using the procedure outlined above we have

$$sM(1,s-1,x) = s + \frac{sx}{s-1} + \frac{x^2}{s-1} + \frac{x^3}{(s-1)(s+1)} + \cdots \tag{36}$$

which is not singular at $s = 0$. We may then calculate X and Y as above:

$$X = \lim_{s \to 0} x^{-2+s} \left\{ s + \frac{sx}{s-1} + \frac{x^2}{s-1} + \frac{x^3}{(s-1)(s+1)} + \cdots \right\}$$
$$= -\left(1 + x + \frac{x^2}{2} \cdots\right) = -e^x \tag{37}$$

and

$$Y = \lim_{s \to 0} x^{-2+s} \left\{ ln(x) \left[s + \frac{sx}{s-1} + \frac{x^2}{s-1} + \frac{x^3}{(s-1)(s+1)} + \cdots \right] \right.$$
$$\left. + \left[1 - \frac{x}{(s-1)^2} - \frac{1}{(s-1)^2} x^2 + \cdots \right] \right\} = ln(x)X + (1 - x - x^2 + \cdots), \tag{38}$$

which provide *two* independent solutions. In addition from the earlier analysis, $M(3, 3, x)$ is a solution, but in this case we have $M(3, 3, x) = -X$. A similar result is to be expected since any other solution must be a combination of the two independent solutions X and Y.

We have therefore considered all possibilities and obtained two independent solutions for all cases. The techniques for finding solutions in the exceptional cases rely on limits for which the order of operations is important. For example the case $a = b = -1$ is an exceptional case since $M(-1, -1, x)$ has an indeterminate coefficient. In general we have

$$e^x = M(b, b, x) \Rightarrow \lim_{b \to -1} M(b, b, x) = \lim_{b \to -1} e^x = e^x. \tag{39}$$

But this can also be treated by the limit process

$$M(-1, -1, x) = \lim_{b \to -1} \lim_{a \to -1} M(a, b, x) = 1 + x. \tag{40}$$

Thus we have found two independent solutions by two different limit processes. In general any solution may be expressed in terms of $M(a, b, x)$ and $N(a, b, x)$, but it may be convenient to consider other ways of writing solutions.

2.3. The KummerU function

The general solution can be written in many ways and it is frequently expressed in terms of the KummerM function, $M(a, b, x)$ and the KummerU function, which is defined by

$$U(a,b,x) = \alpha M(a,b,x) - \beta N(a,b,x),$$
$$\alpha = \frac{\pi}{\sin(\pi b)\Gamma(1+a-b)\Gamma(b)}, \qquad \beta = \frac{\pi}{\sin(\pi b)\Gamma(a)\Gamma(2-b)} \tag{41}$$

and we assume that the limiting forms are used for the cases where the coefficients are infinite. (See [1,7,8] for equivalent definitions in slightly different notations). From the results

$$\Gamma(2-b)\sin(\pi b)\Gamma(b-1) = -\pi,$$
$$\Gamma(1-m-b)\sin(\pi b)\Gamma(b+m) = (-1)^m\pi$$
$$\Gamma(-m) = \infty \quad m = 0, 1, 2, \ldots \tag{42}$$

we can construct examples where $\beta = 0$. For example, with b a positive integer and $a = -m$, $m = 0\ldots$ we have $\beta = 0$ and α finite. In these cases, $M(a,b,x)$ and $U(a,b,x)$ are multiples of each other and *do not form two independent solutions,* but we may use $M(a,b,x)$ and $N(a,b,x)$ in these situations. One advantage of $U(a,b,x)$ is that it satisfies a set of recurrence relations and properties similar to $M(a,b,x)$, for example

$$\frac{dU(a,b,x)}{dx} = -aU(a+1,b+1,x) \tag{43}$$

(see Abramowitz and Stegun [1] for a comprehensive list of the properties of both $U(a,b,x)$ and $M(a,b,x)$).

Another major advantage of using $U(a,b,x)$ is when applying the boundary conditions to the solution of the Kummer equation. Suppose we require a solution that vanishes at $x = R \to \infty$. The general solution may be written

$$y = pM(a,b,x) + qU(a,b,x) \tag{44}$$

and we need to examine the asymptotic form of $M(a,b,x)$ to establish the appropriate constants p and q. We may write the coefficient of x^n in $M(a,b,x)$ as

$$\frac{\Gamma(a+n)\Gamma(b)}{\Gamma(n+1)\Gamma(b+n)\Gamma(a)}. \tag{45}$$

Using Stirling type asymptotic results ([1]: Eq. 6.1.46)

$$\frac{\Gamma(a+n)}{\Gamma(b+n)} \sim n^{a-b} \quad n \to \infty \tag{46}$$

and

$$\frac{\Gamma(n+1)}{\Gamma(b-a+n+1)} \sim n^{a-b} \quad n \to \infty. \tag{47}$$

Thus for large n ($n > N_1$, N_1 a large integer) $M(a, b, x)$ is dominated by the terms in the sum

$$\frac{\Gamma(b)}{\Gamma(a)} \sum_{n>N_1} \frac{x^n}{\Gamma(b-a+n+1)}. \tag{48}$$

Writing $n \approx m + a - b$ for some large $m > N_1$ we have asymptotically

$$M(a, b, x) \sim \frac{\Gamma(b)}{\Gamma(a)} \sum_{m>N_1} \frac{x^{m+a-b}}{\Gamma(m+1)} \sim \frac{\Gamma(b)}{\Gamma(a)} \exp(x) x^{a-b}. \tag{49}$$

This asymptotic result is for large $x > 0$ and is given in Abramowitz and Stegun [1]. (For a more detailed derivation see Morse and Feshbach [8]). Thus for large R, in order to satisfy the boundary condition at $x = R$ we need to choose

$$\frac{p}{q} = -\frac{U(a, b, R)}{M(a, b, R)}. \tag{50}$$

If we use the asymptotic form given above for $M(a, b, x)$ in the definition (41) of $U(a, b, x)$ we obtain

$$U(a, b, x) \sim \left\{ \alpha \frac{\Gamma(b)}{\Gamma(a)} - \beta \frac{\Gamma(2-b)}{\Gamma(1+a-b)} \right\} \exp(x) x^{a-b} \tag{51}$$

as $x \to \infty$. From this we may deduce that

$$\lim_{R \to \infty} \frac{U(a, b, R)}{M(a, b, R)} = 0 \tag{52}$$

so that the appropriate solution for R infinite is simply $U(a, b, x)$. To describe the asymptotic behaviour of $U(a, b, x)$ more precisely we may consider the rearranged Kummer equation

$$\left[D^2 + \left(\frac{b}{x} - 1 \right) D - \frac{a}{x} \right] y = 0, \qquad y = x^{-r} \sum_{n=0}^{N} A_n x^{-n}, \quad A_0 = 1. \tag{53}$$

The leading term ($n = 0$) gives

$$\frac{r - a}{x^{r+1}} + O(x^{-(r+2)}) = 0 \Rightarrow r = a \tag{54}$$

so that

$$U(a, b, x) \sim x^{-a}, \quad x \to \infty. \tag{55}$$

This result may also be found in Abramowitz and Stegun [1], where it is also extended to negative x and to more terms in the expansion so that

$$U(a, b, x) \sim x^{-a} \sum_{n=0}^{N} \frac{(a)_n (1 + a - b)(n)}{n!} (-x)^{-n}. \tag{56}$$

Note that the solution which vanishes at ∞ may also be obtained by writing the general solution in terms of $M(a, b, x)$ and $N(a, b, x)$, with $q = 1 - p$ chosen so that

$$\lim_{R \to \infty} pM(a, b, R) + (1 - p)N(a, b, R) = 0 \tag{57}$$

and this will be necessary for the case where $U(a, b, x)$ is not independent of $M(a, b, x)$.

A similar analysis may be carried out to deal with the case of $R \to -\infty$, but it may be more convenient to use an alternative form for the solution. The analysis given above to construct different solutions in the form $\exp(x)w(x)$ can be used to generate the four solutions

$$\{U(a, b, x), x^{1-b}U(1 + a - b, 2 - b, x), e^x U(b - a, b, -x), \\ e^x x^{1-b} U(1 - a, 2 - b, -x)\}. \tag{58}$$

The Wronskians of these functions have been calculated in [1] and the appropriate choice of independent solutions can be made for each boundary condition imposed.

3. K-DIMENSIONAL SYSTEMS

3.1. Unconfined systems

The *reduced* K-dimensional radial equation for spherically symmetric potentials $V(r)$ is given by

$$
\left(-\frac{1}{2}\left(D^2+\frac{K-1}{r}D\right)+\frac{l(l+K-2)}{2r^2}+V(r)\right)\Psi = E\Psi,
$$
$$
D=\frac{\mathrm{d}}{\mathrm{d}r} \tag{59}
$$

(see the Appendix for a brief derivation). Writing $\Psi = r^p\psi$ and taking $p = l$ we obtain

$$
[rD^2 + (2l + K - 1)D]\psi = 2r[V(r) - E]\psi. \tag{60}
$$

For the example of the hydrogen atom we have $V(r) = -1/r$ and we write $\psi = \exp(-\beta r)\phi$ so that we obtain

$$
\begin{aligned}
\{rD^2 + [(2l + K - 1) - 2\beta r]D - \beta(2l + K - 1)\}\phi \\
= (-2 - 2Er - \beta^2 r)\phi.
\end{aligned} \tag{61}
$$

Choosing

$$
E = -\frac{\beta^2}{2}, \qquad u = 2\beta r, \qquad c = \left(l + \frac{K-1}{2}\right) - \frac{1}{\beta} \tag{62}
$$

leads to

$$
\{uD_u^2 + [(2l + K - 1) - u]D_u - c\}\phi = 0, \quad D_u = \frac{\mathrm{d}}{\mathrm{d}u}. \tag{63}
$$

This is the canonical Kummer equation so we may write

$$
\phi = aM(c, 2l + K - 1, 2\beta r) + bN(c, 2l + K - 1, 2\beta r). \tag{64}
$$

For general $K > 1$, if we require a solution which is finite everywhere, has a finite derivative at the origin and truncates so that $\Psi \to 0$ as $r \to \infty$, then $b = 0$, $c = -n$, $n = 0, 1, 2 \ldots$. Thus up to normalization we have

$$
\phi = M(-n, 2l + K - 1, 2\beta r) \tag{65}
$$

with

$$
\beta = \frac{1}{n + l + \frac{K-1}{2}}, \qquad E = -\frac{\beta^2}{2}. \tag{66}
$$

(The special case $K = 1$ is treated below.)

We may treat the harmonic oscillator, with potential $V(r) = r^2/2$, in a similar manner. Writing $\psi = \exp(-r^2/2)\phi$ in (60) we obtain

$$\{rD^2 + [(2l + K - 1) - 2r^2]D + r(2E - 2l - K)\}\phi = 0 \tag{67}$$

and a further transformation $u = r^2$ gives

$$\left\{uD_u^2 + \left[\left(l + \frac{K}{2}\right) - u\right]D_u - \frac{1}{2}\left(l + \frac{K}{2} - E\right)\right\}\phi = 0, \quad D_u = \frac{\mathrm{d}}{\mathrm{d}u}. \tag{68}$$

This is the standard Kummer equation with solutions

$$\phi = aM(c, d, r^2) + bN(c, d, r^2), \qquad c = \frac{1}{2}\left(l + \frac{K}{2} - E\right), d = l + \frac{K}{2}. \tag{69}$$

In general, for a solution which is well behaved at the origin we have $b = 0$. If we also require that $M(c, d, r^2)$ truncates so that $\Psi \to 0$ as $r \to \infty$, then we obtain

$$\phi = M(-n, d, r^2), \quad n = 0, 1, 2, \ldots, \qquad E = 2n + l + \frac{K}{2}. \tag{70}$$

Although these two models have been treated using slightly different transformations the generic analysis is identical. The essential transformations are of the form $\psi = \exp(-f(r))\phi$, where $f(r)$ is chosen to describe the correct asymptotic form for ψ and leads to the Kummer equation for ϕ. The different transformations used for the hydrogen and harmonic oscillator problems lead to E being related differently to the parameters of the Kummer functions.

The case of a constant potential, $V(r) = V_0$, can be treated analogously. We may write $\psi = \exp(-\mu r)\phi$, where $\mu^2 = 2(V_0 - E)$, $\mu > 0$ and obtain

$$\{rD^2 + [(2l + K - 1) - 2\mu r]D - \mu(2l + K - 1)\}\phi = 0 \tag{71}$$

and a further transformation $u = 2\mu r$ leads to the standard Kummer equation

$$\left\{uD_u^2 + [(2l + K - 1) - u]D_u - \frac{1}{2}(2l + K - 1)\right\}\phi = 0 \tag{72}$$

so that we have

$$\begin{aligned} \phi &= aM(c,d,2\mu r) + bN(c,d,2\mu r), \quad d = 2l + K - 1, \\ c &= \frac{1}{2}(2l + K - 1). \end{aligned} \tag{73}$$

For $K > 1$ and a solution well behaved at the origin we require $b = 0$. If, in addition, we require a solution finite as $r \to \infty$, since $l \geq 0$ and $K > 1$, then we may deduce that $c > 0$ and the KummerM function does not truncate. We can get a solution in $r < R$ however by solving for μ and hence E from the equations

$$M(c,d,2\mu R) = 0, \qquad 2E = 2V_0 - \mu^2. \tag{74}$$

This is an example where the potential is *confined* to an interval, such as $0 < r < R$, and we now turn to an analysis of these restrictions.

3.2. Confinement

The main purpose of this article is to review the solutions of the radial equation when the region of space considered is *confined* and, in such a region, the potential $V(r)$ takes on a particularly simple form so that exact solutions may be found in terms of Kummer functions. The examples considered here are when $V(r) = -1/r$, $V(r) = r^2/2$ or $V(r) = V_0$, a constant potential; of course, an elementary scale change enables the examples $V(r) = -Z/r$ and $V(r) = \omega^2 r^2/2$ to be treated analogously.

The solutions obtained are not completely independent of the rest of the space and this manifests itself through the *boundary conditions* For an *unconfined* system we mean that the radial coordinate is completely unrestricted but *natural* boundary conditions still exist. For example, in the usual treatments we require the wavefunctions to be finite at $r = 0$ and to be asymptotically zero as $r \to \infty$. For *confined* systems one or both of these boundary conditions may be changed.

The usual treatment of confinement is where the charge $\hat{\rho} = |\Psi(r)|^2$ is essentially zero on the sphere (or hypersphere) $r = R$ and the system may therefore be isolated. For example, if we envisage a large repulsive potential in $r \geq R$, then we may *model* this as an infinite barrier on $r = R$ and consequently $\Psi(R) = 0$. This model isolates the wavefunction completely in $r \leq R$ and, for the examples considered here, leads to the general form

$$\Psi = [aM(c,d,f(r)) + bN(c,d,f(r))]r^l \exp(-g(r)). \tag{75}$$

For $d > 1$, in order that the solution remains finite at $r = 0$ we require $b = 0$, and for Ψ to vanish on the sphere we need to choose the energy so that

$$M(c, d, f(R)) = 0. \tag{76}$$

We describe this type of confinement as an *exterior* confinement, that is one caused by systems outside the confined space. There are other types of confinement, however. Another exterior confinement is one where the *dimension* of the space is confined. As an example, we consider the three dimensional problem

$$\left(-\frac{1}{2}\nabla^2 - \left(\frac{1}{z} + V(x, y)\right)\right)\Psi = \epsilon\Psi, \tag{77}$$

where V is an unspecified potential which confines the electron to the interval on the z axis $0 < z < R$. Equation (77) is separable and, with $\Psi = \psi(z)\omega(x, y)$, we have the confining equation

$$-\frac{1}{2}\left(\frac{\partial^2\omega}{\partial x^2} + \frac{\partial^2\omega}{\partial y^2}\right) - V\omega = \lambda\omega. \tag{78}$$

Then, for *known* λ, we obtain the one-dimensional hydrogen-like atom, where z ($z \geq 0$) is the radial coordinate:

$$-\frac{1}{2}\frac{\mathrm{d}^2\psi}{\mathrm{d}z^2} - \frac{1}{z}\psi = (\epsilon - \lambda)\psi = E\psi. \tag{79}$$

Let $\psi = \exp(-\alpha z)\phi$, then we obtain

$$-\frac{1}{2}\left(\frac{\mathrm{d}^2\phi}{\mathrm{d}z^2} - 2\alpha\frac{\mathrm{d}\phi}{\mathrm{d}z} + \alpha^2\phi\right) - \frac{1}{z}\phi - E\phi = 0. \tag{80}$$

Choosing

$$E = -\frac{\alpha^2}{2}, \qquad u = 2\alpha z, \tag{81}$$

we obtain

$$u\frac{\mathrm{d}^2\phi}{\mathrm{d}z^2} - u\frac{\mathrm{d}\phi}{\mathrm{d}u} + \frac{1}{\alpha}\phi = 0. \tag{82}$$

This is a Kummer equation, and we can choose the solutions to match the boundary conditions. For a solution finite at the origin we cannot use the invalid KummerM function but

$$N\left(-\frac{1}{\alpha}, 0, 2\alpha z\right) = 2\alpha z M\left(1-\frac{1}{\alpha}, 2, 2\alpha z\right). \tag{83}$$

We can find α and hence E, with the solution zero at $z = R$ by solving the implicit equation

$$M\left(1-\frac{1}{\alpha}, 2, 2\alpha R\right) = 0. \tag{84}$$

For R infinite the series must truncate so that

$$1-\frac{1}{\alpha} = -n. \tag{85}$$

where n is a non-negative integer, and this leads to

$$E = -\frac{1}{2}\frac{1}{(n+1)^2}. \tag{86}$$

This type of model may be used to confine a system to a lower dimension, and a particularly well known example of such a model is two-dimensional confined hydrogen (treated later), which has been applied to the theory of quantum dots so that the atoms are essentially confined to a plane.

There is another type of confinement, which we may describe as *interior* confinement, where the region near the origin cannot be treated by a simple model, and may have to be treated numerically, but for large enough r, $r > r_0$ say, $V(r)$ is sufficiently simple for our Kummer models to be applied; essentially this is a type of *asymptotic* model. Again the effect of the inner region affects the solution in the outer region through boundary conditions, and in some cases a prescribed energy. Examples of this are presented below, and often the appropriate choice of the wavefunction is

$$\Psi = \exp(-\alpha r)U(c, d, f(r)), \tag{87}$$

where the use of the KummerU function ensures that the wavefunction is asymptotically zero; since $r > r_0$ the behaviour at the origin is not a constraint. In the following subsections we consider some particular cases of this analysis.

4. CONFINED HYDROGEN ATOM IN 3-D AND 2-D

4.1. Three dimensional hydrogen atom

If we now specialize to the transformations of the 3-d hydrogen, then after separation the radial equation is

$$\left[-\frac{1}{2}\left(D^2+\frac{2}{r}D\right)+\frac{l(l+1)}{2r^2}-\frac{1}{r}\right]\Psi=E\Psi, \qquad D=\frac{\mathrm{d}}{\mathrm{d}r}. \tag{88}$$

As shown earlier the transformation $\Psi = r^l \exp(-\beta r)\phi$ leads to the general solution

$$\begin{aligned}\phi &= aM(c,d,2\beta r)+bN(c,d,2\beta r), \qquad c=(l+1)-\frac{1}{\beta}, d=(2l+2),\\ E &= -\frac{\beta^2}{2}.\end{aligned} \tag{89}$$

The solution that is finite at the origin and truncates is obtained from $M(-n, 2l+2, 2\beta r)$, $c=-n$, with n a non-negative integer and consequently

$$E=-\frac{1}{2}\frac{1}{(n+l+1)^2}. \tag{90}$$

As an example of interior confinement, we consider the related model

$$\left[-\frac{1}{2}\left(D^2+\frac{2}{r}D\right)+\frac{l(l+1)}{2r^2}-\frac{1}{r}+V(r)\right]\Psi=E\Psi, \quad D=\frac{\mathrm{d}}{\mathrm{d}r}, \tag{91}$$

where $V(r)$ is a short-range potential, which is either, exactly, or essentially zero in the region $r > r_0$. Thus interior confinement may be used as a model for the *asymptotic* solution to this equation. In this case the behaviour at the origin is irrelevant, but we require a solution which decays to zero as $r \to \infty$ for a *given* energy E. The appropriate solution in $r > r_0$ is $r^l \exp(-\beta r)U(c, 2l+2, 2\beta r)$, where c and β are found from the known value of E:

$$E=-\frac{\beta^2}{2}, \quad \beta>0, \qquad c=(l+1)-\frac{1}{\beta}. \tag{92}$$

We next consider an example of exterior confinement where, at $r = R$, a potential or a field is applied that prevents charge from leaving the region

$r < R$. We therefore have the charge density, $\hat{\rho}$, satisfying

$$\frac{\partial \hat{\rho}(R)}{\partial r} = 0, \qquad \rho\hat{(r)} = |\Psi(r)|^2. \tag{93}$$

For real wavefunctions this implies

$$2\frac{\partial \Psi(R)}{\partial r}\Psi(R) = 0. \tag{94}$$

One model of this would be an infinite spherical potential at $r = R$ so that the boundary condition is simply $\Psi(R) = 0$. This model has been considered by many authors [9–13] and recently we have found the solution in terms of the Kummer functions [2] and extended the theory to the first-order polarizabilities of such systems [4]. In this case, since we require the solution to be finite at the origin, the energies are found from

$$M(c, 2l + 2, 2\beta R) = 0, \qquad E = -\frac{\beta^2}{2}. \tag{95}$$

If this is the appropriate physical model for the confinement all the energies may be found from this equation. However, if the physics of the confining barrier is more general, and we only require that the flow of charge is zero, then in addition to these solutions we need to solve for the set of energies from

$$\frac{\partial \Psi(R)}{\partial r} = 0. \tag{96}$$

This gives rise to a parallel set of stationary states for the confined hydrogen-like atom, which coincide with the original set as $R \to \infty$. As an example we consider the s-state ($l = 0$) for the 3-d hydrogen solutions, confined in $r < R$, so that the rate of flow of charge at $r = R$ is

$$\frac{\partial \rho}{\partial r} = \frac{\partial |\psi|^2}{\partial r} = 2\psi\frac{\partial \psi}{\partial r}. \tag{97}$$

For confinement Equation (97) implies that *either* $\psi(R)$ or $\frac{d\psi(R)}{dr}$ vanishes, which lead to one of the following equations for β:

$$M(c, 2, 2\beta R) = 0, \quad c = 1 - \frac{1}{\beta} \tag{98}$$

Table 1 Spherically confined 3-d hydrogen. (For a given R in column 1 the energies obtained by solving $\psi(R) = 0$ and $\mathrm{d}\psi(R)/\mathrm{d}r = 0$ are given in columns 2 and 3)

R	$\psi(R) = 0$	$\frac{\mathrm{d}\psi(R)}{\mathrm{d}r} = 0$
1.85	−0.012946	−0.887016
1.9	−0.054055	−0.866752
2	−0.125	−0.829507
2.2	−0.232033	−0.766136
2.5	−0.334910	−0.692516
3	−0.423967	−0.609196
3.5	−0.464357	−0.558546
4	−0.483265	−0.592302
5	−0.496417	−0.506158
6	−0.499277	−0.501139
7	−0.499863	−0.500200
10	−0.499993	−0.500001

or

$$cM(c+1, 3, 2\beta R) = M(c, 2, 2\beta R), \quad c = 1 - \frac{1}{\beta}. \tag{99}$$

Equation (99) makes use of the derivative property (Equation (19)) of the KummerM functions.

In Table 1 we present results of two distinct models of exterior confinement and compare the ground state energies obtained by setting either $\Psi(R) = 0$ or $\frac{\mathrm{d}\Psi(R)}{\mathrm{d}r} = 0$.

Note that the eigenvalues converge to -0.5 as $R \to \infty$; this is to be expected since the unconfined hydrogen wavefunction has both the wavefunction and its derivative zero at infinity. If the *process* of confinement is undefined (perhaps using electric or magnetic fields) then both may be valid. For an *infinite* barrier confinement only the former is valid. In both of these models it is often assumed that $\Psi = \partial\Psi/\partial r = 0$ in $r > R$ so that there is a mathematical discontinuity in either $\partial\Psi/\partial r$ or Ψ. The *physical potential* would contain an *interface potential* that connects the two regions, and for small R this may have a comparable effect to the Coulomb potential. A more realistic model would be obtained from boundary conditions describing the matching of $\partial\Psi/\partial r$ and Ψ to the wavefunctions in $r > R$. This is discussed further in Section 6 and leads to a consistent limit as the barrier becomes infinite.

As an example of both exterior and interior confinement we consider the Schrödinger equation for the 3-d hydrogen atom in the interval $L < r < R$, where the parameters in the Kummer functions are used to match boundary conditions at $r = L$ and $r = R$. The general solution in this region is given by

$$\Psi(r) = r^l \exp(-\beta r) f(c, d, 2\beta r), \quad d = (2l + 2), \tag{100}$$

where

$$f(c, d, 2\beta r) = aM(c, d, 2\beta r) + bN(c, d, 2\beta r). \tag{101}$$

This general form contains three free parameters (a, b, β) which can be used to match this solution to the interior and exterior regions. One example is the case where the confinement can be modeled so that the electron is completely confined in $L < r < R$ and, consequently,

$$\Psi(L) = 0 \Rightarrow aM(c, d, 2\beta L) + bN(c, d, 2\beta L) = 0 \tag{102}$$

and

$$\Psi(R) = 0 \Rightarrow aM(c, d, 2\beta R) + bN(c, d, 2\beta R) = 0. \tag{103}$$

To obtain non-zero solutions for a and b we need to solve

$$M(c, d, 2\beta L)N(c, d, 2\beta R) - M(c, d, 2\beta R)N(c, d, 2\beta L) = 0. \tag{104}$$

This is an equation for β, and once this parameter has been calculated the ratio of a/b may be determined. This problem has been considered in [3], where stationary state energies have been calculated for various L, R and l. Note that *in principle* we may use $U(c, d, 2\beta r)$ instead of $N(c, d, 2\beta r)$, leading to different values for a and b but an equivalent solution. However in some cases it is impractical to use $U(c, d, 2\beta r)$ since it is not universally independent of $M(c, d, 2\beta r)$. If $c = -n$, n a integer, and d is a positive integer then $U(c, d, 2\beta r)$ is a multiple of $M(c, d, 2\beta r)$ and for values of β where $c \approx -n$ the numerical computations in (104) become unstable [3]. However, from the analysis in Section 2 we see that it is always possible to choose an appropriate second solution to the Kummer equation.

Another case is where the internal potential approaches $-1/r$ asymptotically and we may take a model where $r > L$, $V(r) = -1/r$ and the energy is prescribed from the internal region. In this case β is prescribed and the two remaining parameters, a and b, are fixed by matching the solution to the known wavefunction in $r \leq R$. If we denote the wavefunction in the inner region by $\hat{\Psi}$ we choose a and b so that Ψ and its derivative match with $\hat{\Psi}$ at $r = R$.

4.2. Two dimensional hydrogen atom

Consider the equation

$$\left(-\frac{1}{2}\nabla^2 - \left(\frac{1}{\rho} + V(z)\right)\right)\Psi = \epsilon\Psi. \tag{105}$$

This separates in cylindrical polar coordinates (ρ, θ, z) so $\Psi = \psi(\rho)u(\theta)\omega(z)$ and we may regard $V(z)$ as the known confining potential. Thus

$$-\frac{1}{2}\frac{\mathrm{d}^2\omega}{\mathrm{d}z^2} - V(z)\omega = \lambda\omega. \tag{106}$$

Thus with the known λ and $E = \epsilon - \lambda$

$$(T_1 + T_2)\psi u = E\psi u, \qquad T_1 = -\frac{1}{2}\left(D^2 + \frac{1}{\rho}D\right) - \frac{1}{\rho}, \qquad D = \frac{\mathrm{d}}{\mathrm{d}\rho}, \\ T_2 = -\frac{1}{2\rho^2}\frac{\mathrm{d}^2}{\mathrm{d}\theta^2}. \tag{107}$$

Separating further

$$T_2 u = \frac{\mu^2}{2\rho^2}u \Rightarrow u = \mathrm{e}^{\mathrm{i}\mu\theta} \tag{108}$$

where we choose the θ dependence to be periodic. Finally

$$\left(T_1 + \frac{\mu^2}{2\rho^2}\right)\psi = E\psi \tag{109}$$

and we take $\psi = \rho^l \exp(-\beta\rho)\phi = \rho^l\phi_1$ $(l \geq 0)$ to obtain

$$\left[D^2 + \frac{(2l+1)}{\rho}D + \frac{2}{\rho} + 2E\right]\phi_1 = 0 \tag{110}$$

with $l = |\mu|$. This is identical with the equation for 2-d hydrogen and consequently we may use the 2-d hydrogen as a model for the confinement of an electron to a two-dimensional plane. The equation for ϕ may be written

$$\left(uD_u^2 + (2l + 1 - u)D_u - \left(l + \frac{1}{2} - \frac{1}{\beta}\right)\right)\phi = 0 \tag{111}$$

with $E = -\beta^2/2$ and $u = 2\beta\rho$, which is again the standard Kummer equation. For a solution finite at the origin

$$\phi = M\left(l + \frac{1}{2} - \frac{1}{\beta}, 2l + 1, 2\rho\beta\right) \tag{112}$$

which truncates when $l+\frac{1}{2}-\frac{1}{\beta} = -n, n$ a non-negative integer. which leads to

$$E = -\frac{1}{2}\frac{1}{(n + l + 1/2)^2}. \tag{113}$$

This may be regarded as a *partially* confined solution; although it is confined to a plane there are no further confinements. Further radial confinement, analogous to those described above for the hydrogen models may be either interior or exterior or both, but with the following scenarios.

1. Interior confinement: given an interior potential asymptotic to $-1/\rho$ with prescribed energy (determined from the inner region) the solution in the exterior region may be modeled as

$$\psi = \rho^l \exp(-\beta\rho)U(c, 2l + 1, 2\beta\rho), \quad c = l + \frac{1}{2} - \frac{1}{\beta}. \tag{114}$$

2. Exterior confinement: given a region $\rho > R$ which is completely isolated so that

$$\frac{\mathrm{d}|\psi(R)|^2}{\mathrm{d}\rho} = 0 \tag{115}$$

and such that the wavefunction is finite at the origin we need either
(a)

$$\psi(R) = 0 \Rightarrow M(c, 2l + 1, 2\beta R) = 0 \tag{116}$$

or
(b)

$$\frac{\mathrm{d}\psi(R)}{\mathrm{d}\rho} = 0 \Rightarrow \frac{\mathrm{d}}{\mathrm{d}\rho}\{\rho^l \exp(-\beta\rho)M(c, 2l + 1, 2\beta R)\} = 0. \tag{117}$$

3. Interior and Exterior confinement: here we take the model where $L < \rho < R$ and the general solution is

$$\psi = \rho^l \exp(-\beta\rho)\{aM(c, 2l + 1, 2\beta\rho) + bN(c, 2l + 1, 2\beta\rho)\} \tag{118}$$

where a, b and β are free parameters which may be determined from the boundary conditions.

5. TWO-DIMENSIONAL HARMONIC OSCILLATOR

The theory of confined harmonic oscillators parallels the theory for constrained hydrogenic systems. Here we consider the details when a 3-d harmonic oscillator model is constrained to two dimensions This is analogous to hydrogen, and after separation in cylindrical coordinates the ρ equation is

$$\left[-\frac{1}{2}\left(D^2+\frac{1}{\rho}D\right)+\frac{\rho^2}{2}+\frac{\nu^2}{2\rho^2}\right]\Psi = E\Psi. \tag{119}$$

If we make the transformation $u = \rho^2$ we obtain

$$\left(u\frac{\mathrm{d}^2}{\mathrm{d}u^2}+\frac{\mathrm{d}}{\mathrm{d}u}-\frac{u}{4}-\frac{\nu^2}{4u}\right)\Psi = -\frac{E}{2}\Psi. \tag{120}$$

A further transformation $\Psi = u^s\psi$ with $2s = |\nu|$ leads to

$$u\frac{\mathrm{d}^2\psi}{\mathrm{d}u^2}+(2s+1)\frac{\mathrm{d}\psi}{\mathrm{d}u}-\frac{u}{4}\psi = -\frac{E}{2}\psi. \tag{121}$$

Finally using $\psi = \exp(-u/2)\phi$ we arrive at the Kummer equation

$$\left[uD^2+(2s+1-u)D-\frac{1}{2}(2s+1-E)\right]\phi = 0. \tag{122}$$

The partially confined solution, finite at the origin and truncated for the correct behaviour at infinity, is

$$\phi = M(-n, |\nu|+1, \rho^2), \qquad E = 2n+|\nu|+1. \tag{123}$$

Furthermore, if instead we wish to confine at some $\rho \leq R$ then we need to solve

$$M(c, |\nu|+1, R^2) = 0, \quad c = \frac{1}{2}(|\nu|+1-E). \tag{124}$$

6. THE CONSTANT POTENTIAL

Here we consider two explicit examples of the constant potential. One widely used model is that of the 3-d square well, where the interval is restricted to $0 < r < R$, and at $r = R$ the solution is matched to a solution in the region

$r > R$ with a different potential. For this model we require the solution to be finite at the origin and consequently the solution may be written in the form $\psi = \exp(-\mu r)\phi$, where

$$\phi = aM(l + 1, 2l + 2, 2\mu r). \tag{125}$$

There are *two* free constants, μ and a, which may be chosen to match the wavefunction and its derivative at $r = R$. Once μ is calculated the energy is determined from $2E = 2V_0 - \mu^2$. The particular case where the potential is infinite in $r > R$ requires that we choose μ as in Equation (74). This model is a case of exterior confinement.

An example of an interior confinement is a 3-d constant potential in $R < r < \infty$, which is matched at $r = R$ to a confined hydrogen potential. Since we require the solution to be asymptotically zero as $r \to \infty$ the form of the solution for the constant potential is

$$\psi_+ = a\exp(-\mu r)U(l + 1, 2l + 2, 2\mu r). \tag{126}$$

In the inner region we have the standard confined hydrogen solution

$$\psi_- = bM\left(1 - \frac{1}{\alpha}, 2, 2\alpha r\right)\exp(-\alpha r), \qquad E = -\frac{\alpha^2}{2}, \tag{127}$$

where for simplicity of presentation we have chosen $l = 0$. Since E is prescribed from the inner region we have a relationship between μ and α:

$$\mu^2 = 2V_0 + \alpha^2. \tag{128}$$

At $r = R$ we require

$$\psi_+ = \psi_-, \qquad \frac{d\psi_+}{dr} = \frac{d\psi_-}{dr} \Rightarrow \psi_+\frac{d\psi_-}{dr} - \psi_-\frac{d\psi_+}{dr} = 0. \tag{129}$$

This equation may be solved for α and the energy determined. Note that the ratio a/b may then be found from the matching conditions. In Table 2 we illustrate this model, taking $R = 2$ and various values of V_0, solving for E with α in the range $0 \leq \alpha \leq 1$. For large V_0 the numerical process needs to be carried out to a high precision to obtain the solution. As $V_0 \to \infty$ we have $\mu \to \infty$ and

$$\lim_{\mu\to\infty} \psi_+(2) = \lim_{\mu\to\infty} \frac{d\psi_+(2)}{dr} = 0. \tag{130}$$

Table 2 Interior confinement for a spherically constant potential. (For a given value of V_0 in column 1 the energy E is calculated from Equation (129)).

V_0	E
0	−0.431219
2	−0.32248
4	−0.285579
10	−0.240385
20	−0.212047
30	−0.198137
50	−0.183297
100	−0.167423
500	−0.144708
700	−0.135067

Thus for a model with an infinite barrier we cannot solve (129) exactly for *both* $\psi_-(2) = 0$ and $\mathrm{d}\psi_-(2)/\mathrm{d}r = 0$. The standard model [10] uses just $\psi_-(2) = 0$ and we obtain, up to normalization,

$$\psi_- = M(-1, 2, r)\mathrm{e}^{-r/2} \tag{131}$$

for the radially confined 3-d 1s state of hydrogen, but $d\psi_-(2)/dr \neq 0$. However an alternative model is to choose

$$\frac{\mathrm{d}\psi_-(2)}{\mathrm{d}r} = 0 \tag{132}$$

and this leads to a solution with $E = -0.829507$ (see Table 1) where $\alpha > 1$. In any physical situation it is necessary to examine the interface potential to determine the appropriate model of confinement, and perhaps a more realistic model is the ground state obtained from some large finite V_0.

7. SUMMARY

We have examined *exact* solutions, in the form of Kummer functions, that may be used to describe various forms of confinement to specific spatial regions. The definition of confinement has been enlarged from that used conventionally, and in all cases the physical aspects of the confinement are modeled by *boundary conditions*. These boundary conditions in some cases include prescribed energy values, but are mainly conditions on the wavefunction and its derivative on the boundary of the confined region. In this way the main aspects of the physics are modeled by the exact solutions found. However, for any particular physical problem it is necessary to define the given conditions precisely.

ACKNOWLEDGEMENTS

One of the authors (BLB) would like to express his appreciation of support from the EPSRC (grant number EP/F005199/1) during the preparation of this article.

APPENDIX

The K-dimensional radial equation is

$$\left[-\frac{1}{2}\left(D^2+\frac{K-1}{r}D\right)+\frac{l(l+K-2)}{2r^2}+V(r)\right]\psi=E\psi. \tag{A.1}$$

To illustrate this in 4 dimensions consider the hyperspherical co-ordinates:

$$\begin{aligned} x_1 &= r\cos(\theta_1)\sin(\theta_2)\sin(\theta_3)\\ x_2 &= r\sin(\theta_1)\sin(\theta_2)\sin(\theta_3)\\ x_3 &= r\cos(\theta_2)\sin(\theta_3)\\ x_4 &= r\cos(\theta_3). \end{aligned} \tag{A.2}$$

To find ∇^2 we need to calculate

$$h_i=\left|\frac{\partial \mathbf{r}}{\partial u_i}\right|, \quad u_i=\{r,\theta_j, j=1,2,3\}. \tag{A.3}$$

This notation describes the order of the 4 hyperspherical co-ordinates. We have

$$h_1=1, \quad h_2=r\sin(\theta_2)\sin(\theta_3), \quad h_3=r\sin(\theta_3), \quad h_4=r \tag{A.4}$$

and

$$\nabla^2=\frac{1}{J}\sum_{i=1}^{4}\frac{\partial}{\partial u_i}\frac{J}{h_i^2}\frac{\partial}{\partial u_i}, \quad J=\prod h_i. \tag{A.5}$$

This can be calculated as

$$\left(\frac{\partial^2}{\partial r^2}+\frac{3}{r}\frac{\partial}{\partial r}\right)+\frac{L^2}{r^2}, \tag{A.6}$$

where

$$L^2 = \left(\frac{1}{\sin^2(\theta_3)} \frac{\partial}{\partial \theta_3} \sin^2(\theta_3) \frac{\partial}{\partial \theta_3} \right) + \frac{L_2^2}{\sin^2(\theta_3)}$$

and

$$L_2^2 = \frac{1}{\sin(\theta_2)} \frac{\partial}{\partial \theta_2} \sin(\theta_2) \frac{\partial}{\partial \theta_2} + \frac{1}{\sin^2(\theta_2)} \frac{\partial^2}{\partial \theta_1^2}. \tag{A.7}$$

The operator L_2^2 is the total angular momentum operator in 3-d, but more importantly it is independent of θ_3 so that it commutes with L^2, the angular momentum operator in 4-d. It can be shown [14] that the algebraic properties of the angular momentum in 4-d mirror those in 3-d, and in particular we can always choose the zero eigenvalue for L_2^2 in order to solve the eigenvalue problem for L^2: $(L^2 - \lambda)\phi(\theta_3, \theta_1, \theta_2) = 0$. This may be separated with $\phi(\theta_3, \theta_1, \theta_2) = f(\theta_3)g(\theta_1, \theta_2)$ and written

$$g(\theta_1, \theta_2) \left\{ \left[\frac{\partial}{\partial \theta_3} \sin^2(\theta_3) \frac{\partial}{\partial \theta_3} \right] - \lambda \sin^2(\theta_3) \right\} f(\theta_3) + f(\theta_3) L_2^2 g(\theta_1, \theta_2) = 0. \tag{A.8}$$

This can be done for *any* eigenvalue λ of L^2. To solve for λ we may choose $g(\theta_1, \theta_2)$ to be an eigenfunction of L_2^2 with eigenvalue zero. Thus we may find λ from the eigenvalue equation

$$\left\{ \left[\frac{1}{\sin^2(\theta_3)} \frac{\partial}{\partial \theta_3} \sin^2(\theta_3) \frac{\partial}{\partial \theta_3} \right] - \lambda \right\} f(\theta_3) = 0. \tag{A.9}$$

This technique is used by Merzbacher [15] for 3-d and can be generalized to K-dimensions. Note that this procedure only gives *one* eigenfunction with eigenvalue λ, but this is degenerate. In order to find the eigenvalues of L^2 in K-dimensions we need to choose zero eigenvalues, so that we only need to solve the generalized eigenproblem for L^2:

$$\left\{ \left[\frac{1}{\sin^{K-2}(\theta)} \frac{\mathrm{d}}{\mathrm{d}\theta} \sin^{K-2}(\theta) \frac{\mathrm{d}}{\mathrm{d}\theta} \right] - \lambda \right\} y = 0. \tag{A.10}$$

Using the substitution $x = \cos(\theta)$ we obtain

$$\left\{ (1 - x^2) \frac{\mathrm{d}^2}{\mathrm{d}x^2} - (K - 1)x \frac{\mathrm{d}}{\mathrm{d}x} - \lambda \right\} y = 0. \tag{A.11}$$

A series solution in the form

$$y = \sum_{n=0}^{\infty} a_n x^n \tag{A.12}$$

results in a two-term recurrence:

$$a_{n+2} = \frac{\lambda + (K-1)n + n(n-1)}{(n+1)(n+2)} a_n \tag{A.13}$$

and the series diverges unless it is truncated. Truncation at $n = l$ leads to

$$-\lambda = l(l-1) + l(l+K-1) = l(l+K-2). \tag{A.14}$$

REFERENCES

[1] M. Abramowitz, I.A. Stegun, Handbook of Mathematical Functions, Dover, NewYork, 1965.
[2] B.L. Burrows, M. Cohen, Int. J. Quant. Chem. 106 (2006) 478.
[3] B.L. Burrows, M. Cohen, Mol. Phys. 106 (2008) 267.
[4] B.L. Burrows, M. Cohen, Phys. Rev. A 72 (2005) 032508-1.
[5] L. Chaos-Cador, E. Ley-Koo, Int. J. Quant. Chem. 103 (2005) 369.
[6] H.T.H. Piaggio, An Elementary Treatise on Differential Equations, G.Bell and Sons, London, 1962.
[7] N.N. Lebedev, in: R.A. Silverman (Ed.), Special Functions and Their Applications, Prentice Hall, Englewood Cliffs, NJ, 1965.
[8] P.M. Morse, H. Feshbach, Methods of Theoretical Physics, McGraw-Hill, New York, 1953.
[9] E. Ley-Koo, S.J. Rubinstein, J. Chem. Phys. 71 (1979) 351.
[10] P.W. Fowler, Mol. Phys. 53 (1984) 865.
[11] H.E. Montgomery, Chem. Phys. Lett. 352 (2002) 529.
[12] C. Laughlin, J. Phys. B 37 (2004) 4085.
[13] C. Laughlin, B.L. Burrows, M. Cohen, J. Phys. B 35 (2005) 701.
[14] B.G. Adams, Algebraic Approach to Simple Quantum Systems, Springer-Verlag, Berlin, 1994.
[15] E. Merzbacher, Quantum Mechanics, Wiley, New York, 1998.

CHAPTER 6

Perturbation Theory for a Hydrogen-like Atom Confined Within an Impenetrable Spherical Cavity

Cecil Laughlin [a]

Contents

1. INTRODUCTION

A simple cavity model of a compressed hydrogen atom was first introduced by Michels et al. [1] more than 70 years ago and it was further developed

[a] School of Mathematical Sciences, University of Nottingham, Nottingham NG7 2RD, UK

E-mail address: Cecil.Laughlin@nottingham.ac.uk.

Advances in Quantum Chemistry, Vol. 57
ISSN: 0065-3276, DOI: 10.1016/S0065-3276(09)00610-8

by Sommerfeld and Welker [2] and De Groot and Ten Seldam [3]. The main interest in these early papers was to study the effects of high pressure on the atom and, in particular, the ground state polarisability was estimated for a series of cavity radii [1]. Subsequent earlier work was reviewed by Ley-Koo and Rubinstein [4] and by Fröman et al. [5].

In this cavity model the atom is confined at the centre of a spherical 'box', of finite radius R, with an impenetrable wall. This is, of course, a highly idealised model and its limitations were clearly recognised [3]. However, as pointed out in Fowler's extensive work on confined 1-electron problems [6], it does serve as a framework for discussing properties of localised systems.

More recently, there has been renewed interest in confined atomic systems in several areas of research; for reviews and references see Jaskólski [7], Sako and Diercksen [8,9] and Dolmatov et al. [10], as well as papers here in the present volume. In addition to the spherical box model, the hydrogen atom has also been studied under various types of confinement (see, for example, Ley-Koo and Rubinstein [4], Fröman et al. [5], Connerade et al. [11] and Saha et al. [12]), and off-centre investigations of the spherical cavity model have been performed as well [13].

It is also worth mentioning that numerical solutions of the Schrödinger equation frequently enclose the atom in a spherical box of finite radius; for example, discrete variable methods, finite elements methods and variational methods which employ expansions in terms of functions of finite support, such as B-splines, all assume that the wave function vanishes for $r \geq R$, which is exactly the situation we deal with here. For such solutions to give an accurate description of the unconfined system it is, of course, necessary to choose R sufficiently large that there is negligible difference between the confined and unconfined atoms.

The present contribution is concerned with perturbation theory treatments of spherically confined 1-electron systems. It is part review of earlier work, in particular for the 'large-box' problem, and part new, in particular for the 'small-box' problem.

The energies and properties of a confined atom smoothly approach the unconfined atom values as the box radius R becomes very large, so it is of interest to know the behaviour of these small departures from the free atom values for large R. In practice, it is desirable to have reliable analytical estimates of the departures, rather than calculate them directly as the difference of two, relatively large, quantities, a process which becomes increasingly prone to cancellation, and other, errors as R increases. The problem here is to treat a (small) perturbation of the boundary conditions; a more traditional approach, in which the perturbation would be in the potential, cannot be used because, although the corrections to the properties are small, the perturbation itself is infinite for $r \geq R$.

We can, however, employ Rayleigh–Schrödinger perturbation theory (RSPT) for the small-box problem, treating the electron-nucleus Coulomb interaction as the perturbation. This results in expansions for energies and wave functions, and also for properties such as oscillator strengths and polarisabilities, in powers of R. RSPT, with the Coulomb attraction treated as the perturbation, fails to converge in the limit $R \to \infty$ (Wigner [14]), so in this work we give estimates of the radii of convergence of our perturbation expansions. Earlier work on a perturbation series for the ground state of a confined hydrogen atom has been performed by Aguilera-Navarro et al. [15] and summarised by Gray and Gonda [16]. More recently, Fernández [17] converted the Schrödinger equation for the hydrogen-like atom into the Schrödinger equation for a harmonic oscillator and then developed a perturbation theory expansion for the energies.

The plan of the paper is as follows. We first review the long-range analytical forms of the energies, oscillator strengths, ground-state dipole polarisabilities and nuclear shielding factors of a spherically confined hydrogen-like atom for large confining radii R. First-order corrections to energies and wave functions, due to a perturbation of the boundary condition at $r = R$ to satisfy a Dirichlet condition, are presented which can subsequently be employed to develop first-order corrections to properties (other than energies). We next consider the small-box problem and develop RSPT expansions in powers of R, presenting results up to order 15 (31 in the case of energy). The radii of convergence of these perturbation expansions are also investigated. Finally, we discuss the usually more difficult problem of the application of perturbation theory to intermediate box sizes. Here, use is made of known exact solutions at some discrete values of R.

2. ASYMPTOTIC EXPANSIONS FOR LARGE BOX RADII

2.1. The non-relativistic Schrödinger equation for a confined hydrogen-like atom

We consider a hydrogen-like atom, with nuclear charge Z, enclosed in a spherical well, of radius R, with an impenetrable wall. The nucleus is assumed fixed at the centre of the well and we note that, for finite R, it is not therefore possible to separate out the translational motion of the centre of mass of the system. Pupyshev [18] proved that, for the ground state, the energy is a minimum when the nucleus is at $r = 0$. In a non-relativistic approximation the Schrödinger equation for the electronic motion is[1]

$$\left[-\frac{1}{2}\nabla^2 + V^{(R)}(r)\right]\Psi_{nl}(\mathbf{r}) = E_{nl}\Psi_{nl}(\mathbf{r}), \tag{1}$$

[1] Unless otherwise stated, atomic units are used throughout this paper.

where

$$V^{(R)}(r) = \begin{cases} -\dfrac{Z}{r}, & 0 \le r < R, \\ \infty, & r \ge R. \end{cases} \tag{2}$$

The eigenfunctions $\Psi_{nl}(\mathbf{r})$ and associated eigenenergies E_{nl} are, of course, functions of R but we do not usually indicate this dependence explicitly (on occasions where necessary we add the superscript (R), or (∞), where (∞) indicates the unconfined atom). Due to the spherical symmetry of the problem, $\Psi_{nl}(\mathbf{r})$ has the form

$$\Psi_{nl}(\mathbf{r}) = \frac{1}{r}\psi_{nl}(r)Y_{lm}(\hat{\mathbf{r}}), \tag{3}$$

where $Y_{lm}(\hat{\mathbf{r}})$ is a spherical harmonic and $\psi_{nl}(r)$ is the solution of the differential equation

$$\left[\frac{\mathrm{d}^2}{\mathrm{d}r^2} - \frac{l(l+1)}{r^2} + \frac{2Z}{r} + 2E_{nl}\right]\psi_{nl}(r) = 0 \tag{4}$$

which satisfies the Dirichlet boundary conditions

$$\psi_{nl}(0) = \psi_{nl}(R) = 0. \tag{5}$$

Setting

$$\psi_{nl}(r) = r^{l+1}F(r)\mathrm{e}^{-\alpha_{nl}r}, \tag{6}$$

where

$$E_{nl} = -\frac{1}{2}\alpha_{nl}^2, \tag{7}$$

transforms (4) to

$$\frac{\mathrm{d}^2F}{\mathrm{d}r^2} + 2\left(\frac{l+1}{r} - \alpha_{nl}\right)\frac{\mathrm{d}F}{\mathrm{d}r} - 2\frac{(l+1)\alpha_{nl} - Z}{r}F = 0. \tag{8}$$

Next, introduction of the new independent variable

$$x = 2\alpha_{nl}r \tag{9}$$

transforms (8) into the second-order differential equation

$$xF'' + (2l + 2 - x)F' + (N - l - 1)F = 0, \tag{10}$$

where

$$N = \frac{Z}{\alpha_{nl}}, \tag{11}$$

for the function $F(x)$, with the prime denoting differentiation with respect to x.

As has been known since the early studies of a spherically confined hydrogen atom [2,3], the required solution of Equation (10) is a confluent hypergeometric function, or Kummer M function,

$$F(a, b, x) = \sum_{j=0}^{\infty} \frac{(a)_j x^j}{(b)_j j!}, \tag{12}$$

with

$$a = l + 1 - N, \quad b = 2l + 2 \tag{13}$$

and

$$(a)_j = \prod_{k=0}^{j-1}(a + k), \quad (a)_0 = 1, \tag{14}$$

which satisfies the boundary condition

$$F(a, b, X) = 0, \tag{15}$$

where

$$X = 2\alpha_{nl} R. \tag{16}$$

For the unconfined atom (R infinite), N has to be a positive integer for normalisable solutions, $N = n = 1, 2, 3, \ldots$, and the series for $F(a, b, x)$ terminates to give a polynomial $L_{n+l}^{2l+1}(x)$, the associated Laguerre polynomial of degree $n - l - 1$. For finite R, N is not, in general, an integer. However, as the confining radius R decreases from an infinite value, N increases from its unconfined value n and, at specific values of R, it attains integer values, again giving polynomial solutions $L_{n+i+l}^{2l+1}(x)$, $i = 1, 2, 3, \ldots$.

These solutions occur when the confining radius is at the $(n-l)$th node of $L^{2l+1}_{n+i+l}(x)$. We retain n as a label for all values of R, i.e. $\psi_{nl}(r)$ has $n-l-1$ nodes in the interval $(0, R)$.

2.2. Energies

De Groot and Ten Seldam [3] first employed the solution (12) to calculate energies for 1s, 2s and 2p levels as a function of the confining radius R by choosing fixed values of the energy and then calculating the zeros of the confluent hypergeometric function F to give the corresponding R values. This approach was more recently extended by Goldman and Joslin [19] to calculate energy levels E_{nl} for $n = 1-6$ and $l = 0-5$ to 7 significant figures for a range of values of R from 0.1 up to 50. Their procedure was to employ a root-finding method to calculate values of α_{nl} at which the function $F(a, b, 2\alpha_{nl}R)$ is zero, thus satisfying the boundary condition at $r = R$, and providing E_{nl} via Equation (7). A very similar procedure was followed more recently by Burrows and Cohen [20], whose solutions were presented as exact, using the algebraic manipulation package MAPLE to evaluate $F(a, b, 2\alpha_{nl}R)$, to calculate E_{ns} and $E_{n+1,\mathrm{p}}$ for $n = 1-5$ and for specimen values of R between 1 and 20. Aquino et al. [21] have, very recently, published extremely accurate energies (50 significant digits!) for a few of the lowest s and p states using two different procedures: evaluation of the zeros of the hypergeometric function by MAPLE algorithms, and a solution in series method.

Although the works reported in [19–21] apparently give highly accurate energies E_{nl}, their procedures may not be appropriate for all n, l and R. For example, when $n-l$ is large, the highly oscillatory nature of the confluent hypergeometric function can lead to problems with its accurate evaluation. In addition, it is, of course, valuable to have a simple analytical estimate of the departure of E_{nl} from $E^{(\infty)}_{nl}$, the unconfined atom value, which can be calculated easily and efficiently. It is known that this departure is exponentially small for large box radii R [22], i.e. R large compared to the radius of the unconfined atom orbital. A staightforward confirmation of the exponential dependence of $E_{nl} - E^{(\infty)}_{nl}$ on R can easily be obtained using the trial function

$$\phi_t(r) = \psi^{(\infty)}_{nl}(r) - \psi^{(\infty)}_{nl}(R), \quad 0 \le r \le R, \tag{17}$$

in the energy ansatz

$$E_t = \frac{\langle \phi_t | H | \phi_t \rangle}{\langle \phi_t | \phi_t \rangle}. \tag{18}$$

The error in E_t depends on the factor $\left| \psi^{(\infty)}_{nl}(R) \right|^2$, viz $\mathrm{e}^{-2ZR/n}$.

To derive an explicit expression for $E_{nl} - E_{nl}^{(\infty)}$ we proceed as follows. Since, for large R, the energy eigenvalues will be close to the unconfined values we set

$$N = n + \epsilon_{nl}, \tag{19}$$

where ϵ_{nl} is small in comparison to unity, and is treated as a perturbation parameter. Note that the Coulomb energy degeneracy is broken for a box of finite radius, so that N is not then independent of the angular momentum l. By developing the solution as a power series in x, Michels et al. [1] found a first-order expression for ϵ_{1s},

$$\epsilon_{1\mathrm{s}} = \left[\sum_{i=1}^{\infty} \frac{X^i}{i(i+1)!}\right]^{-1}. \tag{20}$$

This formula was extended to $\epsilon_{2\mathrm{s}}$ and $\epsilon_{2\mathrm{p}}$ by De Groot and Ten Seldam [3] and later to arbitrary n and l by Dingle [23]; see also Wilcox [24]. However, such formulae are not convenient in practice because they are valid only when $X \gg 1$ so that the series converge slowly. Rather more convenient expressions, as power series in $1/X$, have been derived by Dingle [23], who considered the asymptotic expansion of $F(a, b, x)$ for large x and obtained

$$\epsilon_{nl} = \frac{X^{2n} \mathrm{e}^{-X}}{(n-l-1)!(n+l)!} \frac{1 - (n+l)(n-l-1)/X + \cdots}{1 + (n-l)(n+l+1)/X + \cdots}, \tag{21}$$

whereas Laughlin et al. [25] derived the analogous first-order expression

$$\begin{aligned}\epsilon_{nl} = {} & \frac{(-1)^{n-l} X^{n+l+1} \mathrm{e}^{-X} L_{n+l}^{2l+1}(X)}{[(n+l)!]^2} \left[1 - \frac{(n-l)(n+l+1)}{X}\right. \\ & \left. + \frac{(n-l)(n+l+1)\left((n+l)(n-l-1) - 2\right)}{2X^2} + \cdots\right],\end{aligned} \tag{22}$$

where, to zero-order in ϵ_{nl},

$$X = \frac{2ZR}{n}. \tag{23}$$

Specifically, this gives for the 1s, 2s and 2p states

$$\begin{aligned}\epsilon_{1s} &= \left(4Z^2R^2 - 4ZR - 2 - \frac{4}{ZR} + O(R^{-2})\right)\mathrm{e}^{-2ZR},\\ \epsilon_{2s} &= \frac{1}{2}(ZR-2)(Z^3R^3 - 6Z^2R^2 - 24 + O(R^{-1}))\mathrm{e}^{-ZR},\\ \epsilon_{2p} &= \frac{1}{6}(Z^4R^4 - 4Z^3R^3 - 4Z^2R^2 - 24ZR + O(1))\mathrm{e}^{-ZR}.\end{aligned} \tag{24}$$

These expressions give highly accurate energy results with the accuracy increasing, of course, as R increases. The minimum confining radius which allows a specified maximum error depends on the state under consideration since the higher, more diffuse, states require larger and larger confining radii as the principal quantum number n increases. For example, for a relative error of less than 10^{-6} in the energy, box sizes greater than (approximately) 10, 26 and 25 are required for, respectively, the 1s, 2s and 2p states.

As noted above, the non-relativistic unconfined atom Coulomb degeneracy is broken for a confined atom with, for a given n, states of higher angular momentum l having lower energy than states of lower l for any finite R, ie

$$E_{nl} > E_{n,l+1}. \tag{25}$$

This result was established rigorously, for all R, by Scherbinin et al. [26] and by Laughlin et al. [25]. It can be deduced by use of the Sturm comparison theorem for differential equations [27] and the result [28]

$$\frac{\mathrm{d}E_{nl}}{\mathrm{d}R} = -\frac{1}{2}\left[\left(\frac{\mathrm{d}\psi_{nl}}{\mathrm{d}r}\right)_{r=R}\right]^2. \tag{26}$$

Equation (25) also holds perturbatively, as it is clear from (21) or (22) that

$$\epsilon_{n,l+1} \cong \frac{n-l-1}{n+l+1}\epsilon_{nl} < \epsilon_{nl}, \tag{27}$$

from which (25) immediately follows. This expression was also obtained by Wilcox [24] and it follows that, for large R, the (positive) energy shift of the 2s energy from the unconfined atom value is approximately 3 times greater than the 2p energy shift. Equivalently, the shifts in 3s, 3p and 3d energies are approximately in the ratio 10:5:1 [24].

Asymptotic energies $\Xi_{nl} = -Z^2/2(n+\epsilon_{nl})^2$ derived using (24) are compared with precise numerical values E_{nl} in Table 1 for a range of values of R for the 1s, 2s and 2p states of hydrogen. The E_{nl} were obtained by

Table 1 Comparison of precise numerical, E_{nl}, and asymptotic, Ξ_{nl}, energies for a hydrogen atom confined in a spherical box of radius R. For each value of R the first row refers to E_{nl}, the second to Ξ_{nl} (see Equations (7), (11), (19) and (24))

R	1s	2s	2p
4	−0.4833		
	−0.4852		
6	−0.4992773		
	−0.4992799		
8	−0.499975100		
	−0.499975074		
10	−0.49999926328	−0.1128	−0.1189
	−0.49999926293	−0.1169	−0.1201
15	−0.499999999922	−0.1244990	−0.12477133
	−0.499999999922	−0.1245041	−0.12477175
20	−0.500000000000	−0.124987114	−0.124994607
	−0.500000000000	−0.124987071	−0.124994593
25		−0.12499976371	−0.124999906047
		−0.12499976341	−0.124999905960
30		−0.124999996469	−0.124999998640
		−0.124999996467	−0.124999998640
35		−0.124999999954	−0.124999999983
		−0.124999999954	−0.124999999983
40		−0.124999999999	−0.125000000000
		−0.124999999999	−0.125000000000

solving (15) for α_{nl} using an iterative procedure and they are expected to agree with the exact values to the number of decimal places quoted; they differ from the data obtained in other accurate calculations [20,21,29] by at most 10^{-11}. It will be observed that, for all values of R quoted, $\Xi_{2s} > \Xi_{2p}$, and also that, for larger R, the shifts from the unconfined atom value, viz -0.125, are approximately in the ratio noted above.

The accuracy of the asymptotic energies Ξ_{nl} clearly improves rapidly as R increases. In general, for a given R, the errors $e_{nl} = |E_{nl} - \Xi_{nl}|$ satisfy $e_{1s} < e_{2p} < e_{2s}$, as expected. For higher, more diffuse, nl states an equivalent accuracy will be obtained only for increasingly larger values of the box radius R (note the presence of the factors R^{2n} and $\exp(-2ZR/n)$ in ϵ_{nl}, Equations (21) and (22)).

2.3. Oscillator strengths, dipole polarisabilities and nuclear shielding factors

The method developed by Laughlin et al. [25] allows the wave functions of a confined hydrogen-like atom to be written, to first-order, as

$$\psi_{nl}(x) = N_{nl} x^{l+1} \left[-1 + \epsilon_{nl} g_{nl}(x)\right] \mathrm{e}^{-x/2}. \tag{28}$$

Expressions for the functions g_{1s} and g_{2p} have been given, in terms of the exponential integral [30], by Laughlin [31], allowing properties to be developed to first order. In particular, the 1s–2p oscillator strength f_{1s-2p} and ground state polarisability α_1 have been calculated [31]; for example,

$$\begin{aligned}\alpha_1 \cong \frac{9}{2Z^4}\Bigg[1 &- \frac{1}{108}\left(\frac{1}{4}X^6 + \frac{3}{2}X^5 + \frac{3}{2}X^4 + 6X^3\right)\mathrm{e}^{-X} \\ &- \frac{\epsilon_{1s}}{108}\left(\frac{1}{4}X^4 + X^3 - 84X - 1002 + \cdots\right) \\ &- \frac{1}{2}\epsilon_{1s}^2\left(\frac{1}{X^4} + \frac{8}{X^5} + \frac{56}{X^6} + \cdots\right)\mathrm{e}^{X} + O(\mathrm{e}^{-2X})\Bigg],\end{aligned} \tag{29}$$

where, here,

$$X = 2ZR. \tag{30}$$

The first line of (29) is the zero-order polarisability $\alpha_1^{(0)}$ and, of course, it approaches the well-known free-atom value $9/2Z^4$ as the box radius becomes infinitely large. The next two terms in (29) constitute the first-order correction, $\alpha_1^{(1)}$, to $\alpha_1^{(0)}$. At first sight it might be thought that the ϵ_{1s}^2 factor makes the final term in (29) a second-order quantity but, due to the presence of e^X, this is not so. It is a result of the fact that the ϵ_{1s}^2 term in the normalisation of ψ_{1s} is *first order* (see [31]).

In an analogous manner we obtain the expression

$$\begin{aligned}\beta_1 \cong \frac{1}{Z}\Bigg[1 &- \frac{1}{12}(X^4 + 2X^3 - 14X^2 - 12X)\mathrm{e}^{-X} \\ &- \frac{1}{12}(X^4 - 10X^3 - 10X^2 + 9X\log X + 12X\cdots)\mathrm{e}^{-X} \\ &+ O(\mathrm{e}^{-2X})\Bigg],\end{aligned} \tag{31}$$

for the nuclear shielding factor β_1.

Table 2 Oscillator strengths and dipole polarisabilities of a hydrogen atom confined in a spherical box of radius R (the superscripts (asy) and (num) denote, respectively, asymptotic analytical and precise numerical values)

	Oscillator strength			Dipole polarisability		
R	$f_{1s-2p}^{(asy)}$	$f_{1s-2p}^{(num)}$	$f_{1s-2p}^{(other)}$	$\alpha_1^{(asy)}$	$\alpha_1^{(num)}$	$\alpha_1^{(other)}$
4	0.308183	0.927251		2.336311	2.377982	2.377982[b], 2.35782[c], 2.369[d] 2.3669[e], 2.377982[f]
5	0.414740	0.848799	0.6855[a]	3.402946	3.422454	2.9051[a], 3.40294[c], 3.441[d] 3.4081[e]
6	0.485886	0.755211	0.6063[a]	4.052278	4.058140	3.4652[a], 4.058140[b], 4.04707[c] 4.0432[e], 4.058140[f]
7	0.529854	0.665249	0.5438[a]	4.346177	4.347638	3.8079[a], 4.34342[c], 4.3345[e]
8	0.539336	0.590250		4.453633	4.453965	4.453965[b], 4.4419[e], 4.453965[f]
9	0.518882	0.533101		4.487342	4.487413	4.4757[e]
10	0.489301	0.492040	0.4382[a]	4.496800	4.496814	4.2343[a], 4.496814[b], 4.4851[e] 4.496814[f], 3.977[g]
15	0.421579	0.421525		4.499998	4.499998	
20	0.416383	0.416382	0.4191[a]	4.500000	4.500000	4.4628[a]
∞	0.416197	0.416197	0.4160[a]	4.500000	4.500000	4.4997[a], 4.5[b,c,d,e]

[a] Ref. [12].
[b] Ref. [33].
[c] Ref. [34].
[d] Ref. [35].
[e] Ref. [36].
[f] Ref. [37].
[g] Ref. [29].

Table 2 gives a comparison of the asymptotic values of f_{1s-2p} and α_1 with those of other authors for the hydrogen atom and for specimen values of $R \geq 4$. The asymptotic result for α_1 is in error by just 2% at $R = 4$, and the accuracy increases rapidly as R increases. The first-order correction $\alpha_1^{(1)}$ makes an important contribution to this accuracy; in fact, when one examines the deviation of $\alpha_1^{(0)}$ from the unconfined atom value $9/2$, it is found to be in error by approaching 50% for some $R \geq 10$, whereas $\alpha_1^{(0)} + \alpha_1^{(1)}$ is in error by only 0.5% at $R = 10$ and this error decreases rapidly as R increases [31].

In Table 3 we compare zero-order (first line of (31)) and first-order β_1 values with precise numerical data and the results of other researchers [33, 37]. We see that the value correct to first order is in error by little more than 3% at $R = 4$ and that accuracy increases dramatically as R increases, so that it converges very rapidly towards the free atom value $\beta_1 = 1$. Further, as

Table 3 Nuclear shielding factors β_1 for a hydrogen atom confined in a spherical box of radius R

R	Asymptotic expansion		Numerical[a]	Other
	Zero order	First order		
4	0.885	0.927	0.894473121	0.894473121[b,c], 0.894473120[c]
6	0.98873	0.98756	0.987274170	0.987274173[b], 0.987274170[c]
8	0.9993441	0.9991339	0.999131573	0.999131578[b], 0.999131573[c]
10	0.999970775	0.999957628	0.999957586	0.999957590[b], 0.999957586[c]
12	0.999998896	0.999998303	0.999998302	
14	0.999999963	0.999999940	0.999999940	
16	0.999999999	0.999999998	0.999999998	
18	1.000000000	1.000000000	1.000000000	

[a] Present calculation.
[b] Ref. [33].
[c] Ref. [37].

for the dipole polarisability, we note the significant contribution made by the first-order correction to the nuclear shielding factor.

3. EXPANSIONS FOR SMALL BOX RADII

3.1. Wave functions and energies

To effect a perturbation theory expansion for small R we transform to the independent variable

$$x = \frac{\pi \gamma_{nl} r}{R} \tag{32}$$

so that the radial equation (4) becomes

$$\left[\frac{\mathrm{d}^2}{\mathrm{d}x^2} - \frac{l(l+1)}{x^2} + \frac{\lambda}{x}\right]\psi_{nl}(x) = \mathcal{E}_{nl}\psi_{nl}(x), \tag{33}$$

where

$$\lambda = \frac{2RZ}{\pi \gamma_{nl}} \tag{34}$$

and the scaled energy $\mathcal{E}_{nl}$ is defined by

$$\mathcal{E}_{nl} = -\frac{2R^2 E_{nl}}{\pi^2 {\gamma_{nl}}^2}. \tag{35}$$

For sufficiently small R, λ/x may be treated as a perturbation. Thus, expanding

$$\psi_{nl}(x) = \psi_{nl}^{(0)}(x) + \lambda\psi_{nl}^{(1)}(x) + \lambda^2\psi_{nl}^{(2)}(x) + \cdots \tag{36}$$

and

$$\mathcal{E}_{nl} = \mathcal{E}_{nl}^{(0)} + \lambda\mathcal{E}_{nl}^{(1)} + \lambda^2\mathcal{E}_{nl}^{(2)} + \cdots \tag{37}$$

gives

$$\left[\frac{\mathrm{d}^2}{\mathrm{d}x^2} - \frac{l(l+1)}{x^2}\right]\psi_{nl}^{(0)}(x) = \mathcal{E}_{nl}^{(0)}\psi_{nl}^{(0)}(x) \tag{38}$$

and

$$\begin{aligned}\left[\frac{\mathrm{d}^2}{\mathrm{d}x^2} - \frac{l(l+1)}{x^2} - \mathcal{E}_{nl}^{(0)}\right]\psi_{nl}^{(i)} &= \mathcal{E}_{nl}{}^{(i)}\psi_{nl}^{(0)} + \mathcal{E}_{nl}^{(i-1)}\psi_{nl}^{(1)} + \cdots \\ &\quad + \mathcal{E}_{nl}^{(2)}\psi_{nl}^{(i-2)} + \left(\mathcal{E}_{nl}^{(1)} - \frac{1}{x}\right)\psi_{nl}^{(i-1)},\end{aligned} \tag{39}$$

for $i = 1, 2, 3, \ldots$. The solutions to (38) may be expressed in terms of the spherical Bessel functions $j_l(x)$, viz

$$\psi_{nl}^{(0)}(x) = N_{nl}x j_l(x), \tag{40}$$

with associated eigenvalues

$$\mathcal{E}_{nl}^{(0)} = -1, \tag{41}$$

where N_{nl} is a normalisation factor.

The boundary conditions (5) impose the restriction

$$j_l(\pi\gamma_{nl}) = 0, \tag{42}$$

thus giving γ_{nl} and the zero-order eigenvalues

$$E_{nl}^{(0)} = \frac{\pi^2\gamma_{nl}{}^2}{2R^2}. \tag{43}$$

We note that for s (i.e. $l = 0$) states

$$\gamma_{n0} = n, \quad n = 1, 2, 3, \ldots, \tag{44}$$

while for $l \geq 1$ the values of γ_{nl} have to be calculated numerically.

In first order

$$\mathcal{E}_{nl}^{(1)} = \int_0^{\pi \gamma_{nl}} \frac{\psi_{nl}^{(0)^2}(x)}{x} \mathrm{d}x = N_{nl}^2 \int_0^{\pi \gamma_{nl}} x j_l^2(x) \mathrm{d}x, \tag{45}$$

so that

$$E_{nl}^{(1)} = -\frac{Z\pi \gamma_{nl}}{R} \frac{\int_0^{\pi \gamma_{nl}} x j_l^2(x) \mathrm{d}x}{\int_0^{\pi \gamma_{nl}} x^2 j_l^2(x) \mathrm{d}x} \tag{46}$$

and, hence,

$$E_{n0}^{(1)} = -\frac{Z}{R} \int_0^{2n\pi} \frac{1 - \cos x}{x} \mathrm{d}x. \tag{47}$$

In general (Hirschfelder et al. [32]),

$$\begin{aligned} \mathcal{E}_{nl}^{(2i)} = & \int_0^{\pi \gamma_{nl}} \frac{\psi_{nl}^{(i-1)}(x) \psi_{nl}^{(i)}(x)}{x} \mathrm{d}x \\ & - \sum_{j=2}^{i} \mathcal{E}_{nl}^{(j)} \sum_{k=0}^{j-1} \int_0^{\pi \gamma_{nl}} \psi_{nl}^{(i+k-j)}(x) \psi_{nl}^{(i-k)}(x) \mathrm{d}x \end{aligned} \tag{48}$$

and

$$\begin{aligned} \mathcal{E}_{nl}^{(2i+1)} = & \int_0^{\pi \gamma_{nl}} \frac{\psi_{nl}^{(i)^2}(x)}{x} \mathrm{d}x \\ & - \sum_{j=2}^{i} \mathcal{E}_{nl}^{(j)} \sum_{k=0}^{j-1} \int_0^{\pi \gamma_{nl}} \psi_{nl}^{(i+1+k-j)}(x) \psi_{nl}^{(i-k)}(x) \mathrm{d}x \end{aligned} \tag{49}$$

for $i = 1, 2, 3, \ldots$, where we have imposed the conditions that

$$\int_0^{\pi \gamma_{nl}} \psi_{nl}^{(0)}(x) \psi_{nl}^{(i)}(x) \mathrm{d}x = 0, \quad i = 1, 2, 3, \ldots. \tag{50}$$

On converting back to the actual energy E_{nl} we have

$$E_{nl} = R^{-2} \sum_{i=0} E_{nl}^{(i)} (ZR)^i, \tag{51}$$

where

$$E_{nl}^{(i)} = -2 \left(\frac{2}{\pi \gamma_{nl}} \right)^{i-2} \mathcal{E}_{nl}^{(i)}, \quad i = 0, 1, 2, \ldots . \tag{52}$$

Defining $E_{nl}(Z, R)$ as the right-hand-side of (51) provides the identity

$$E_{nl}(Z, R) = Z^2 E_{nl}(1, ZR), \tag{53}$$

allowing us to restrict attention to the hydrogen-atom problem ($Z = 1$). The energy relation (53) holds in general, including for free hydrogen-like ions, i.e. in the limit $R \to \infty$.

3.2. Oscillator strengths

The oscillator strength, $f_{nl-n'l'}$, for absorption from a lower level nl to a higher level $n'l'$ is given by

$$g_n f_{nl-n'l'} = \frac{2}{3} l_> (E_{n'l'} - E_{nl}) \left| D_{nl-n'l'} \right|^2, \tag{54}$$

where

$$D_{nl-n'l'} = \int_0^R \psi_{nl}(r) r \psi_{n'l'}(r) \mathrm{d}r, \tag{55}$$

$g_n = 2l + 1$ and $l_>$ is the greater of l and l'. We now develop a perturbation theory expansion for $f_{nl-n'l'}$ in powers of λ. First, consider $D_{nl-n'l'}$:

$$D_{nl-n'l'} = \frac{R\sqrt{\rho}}{\pi \gamma_{nl}} \int_0^{\pi \gamma_{nl}} \left(\psi_{nl}^{(0)}(x) + \lambda \psi_{nl}^{(1)}(x) + \cdots \right) \\ \times x \left(\psi_{n'l'}^{(0)}(\rho x) + \lambda' \psi_{n'l'}^{(1)}(\rho x) + \cdots \right) \mathrm{d}x, \tag{56}$$

where

$$\rho = \frac{\gamma_{n'l'}}{\gamma_{nl}}, \tag{57}$$

so that

$$D_{nl-n'l'} = \frac{R\sqrt{\rho}}{\pi\gamma_{nl}}\sum_{i=0}\lambda^i\sum_{j=0}^{i}\rho^{-j}D_{i-j,j}, \tag{58}$$

with

$$D_{ij} = \int_0^{\pi\gamma_{nl}}\psi_{nl}^{(i)}(x)x\psi_{n'l'}^{(j)}(\rho x)\mathrm{d}x. \tag{59}$$

Similarly, from (51),

$$E_{n'l'} - E_{nl} = \frac{\Delta E_0}{R^2} + \frac{\Delta E_1}{R} + \Delta E_2 + \cdots. \tag{60}$$

Combining (58) and (60) with (54) gives

$$f_{nl-n'l'} = \sum_{i=0} f_i R^i, \tag{61}$$

where

$$f_i = \frac{2l_>\gamma_{n'l'}}{3\pi^2 g_n\gamma_{nl}{}^3}\sum_{j=0}^{i}\left(\frac{2Z}{\pi\gamma_{nl}}\right)^{i-j}\Delta E_j\sum_{k=0}^{i-j}C_kC_{i-j-k}, \tag{62}$$

with

$$C_k = \sum_{j=0}^{k}\rho^{-j}D_{k-j,j}. \tag{63}$$

Observe that, as a function of Z and R, the oscillator strength $f(Z, R)$ satisfies $f(Z, R) = f(1, ZR)$, so that again we need consider only $Z = 1$.

3.3. Dipole polarisabilities and nuclear shielding factors

The dipole polarisability α_1 of a hydrogen-like atom confined in its 1s ground state is

$$\alpha_1 = 2\sum_{n\geq 2}\frac{\left|\langle\Psi_{1s}(\mathbf{r})|r\cos\theta|\Psi_{np}(\mathbf{r})\rangle\right|^2}{E_{np} - E_{1s}}, \tag{64}$$

where the summation over n extends over all excited np states of the confined atom. Alternatively, we can express α_1 as

$$\alpha_1 = 2\langle \Psi_{1s}(\mathbf{r}) | r\cos\theta | \chi(\mathbf{r}) \rangle, \tag{65}$$

where χ satisfies

$$\left[-\frac{1}{2}\nabla^2 - \frac{Z}{r} - E_{1s}\right]\chi(\mathbf{r}) = r\cos\theta\,\Psi_{1s}(\mathbf{r}). \tag{66}$$

The substitution $\chi = r^{-1}P(r)\cos\theta$ gives the second-order inhomogeneous differential equation

$$\left[\frac{d^2}{dr^2} - \frac{2}{r^2} + \frac{2Z}{r} + 2E_{1s}\right]P = -2r\psi_{1s} \tag{67}$$

which must be solved subject to the boundary conditions

$$P(0) = P(R) = 0. \tag{68}$$

Employing the change of variable (32) gives

$$\left[\frac{d^2}{dx^2} - \frac{2}{x^2} + \frac{\lambda}{x} - \mathcal{E}_{1s}\right]P(x) = -2\left(\frac{R}{\pi}\right)^{5/2} x\psi_{1s}(x), \tag{69}$$

so that

$$\alpha_1 = \sum_{i=0} a_i R^{i+4}, \tag{70}$$

where

$$a_i = \frac{2}{3\pi^4}\left(\frac{2Z}{\pi}\right)^i \sum_{j=0}^{i} Q^{(1)}_{j,i-j}, \tag{71}$$

with

$$Q^{(p)}_{jk} = \int_0^\pi \psi^{(j)}_{1s}(x)x^p P^{(k)}(x)\mathrm{d}x \tag{72}$$

and

$$P(x) = \left(\frac{R}{\pi}\right)^{5/2} \left[P^{(0)}(x) + \lambda P^{(1)}(x) + \lambda^2 P^{(2)}(x) + \cdots\right]. \quad (73)$$

The equations for $P^{(k)}(x)$, $k = 0, 1, 2, \ldots$, are

$$\left[\frac{\mathrm{d}^2}{\mathrm{d}x^2} - \frac{2}{x^2} - \mathcal{E}_{1s}^{(0)}\right] P^{(0)} = -2x\psi_{1s}^{(0)}, \quad (74)$$

and

$$\left[\frac{\mathrm{d}^2}{\mathrm{d}x^2} - \frac{2}{x^2} - \mathcal{E}_{1s}^{(0)}\right] P^{(k)} = \left(\mathcal{E}_{1s}^{(1)} - \frac{1}{x}\right) P^{(k-1)} + \mathcal{E}_{1s}^{(2)} P^{(k-2)} + \cdots$$
$$+ \mathcal{E}_{1s}^{(k)} P^{(0)} - 2x\psi_{1s}^{(k)}, \quad k \geq 1. \quad (75)$$

Again we see that, as a function of Z and R, the dipole polarisability satisfies $\alpha_1(Z, R) = Z^{-4}\alpha_1(1, ZR)$, as expected.

The nuclear shielding factor β_1 can be evaluated in a similar fashion, viz

$$\beta_1 = 2\left\langle \Psi_{1s}(\mathbf{r}) \left| \frac{1}{r^2} \cos\theta \right| \chi(\mathbf{r}) \right\rangle, \quad (76)$$

which can be expressed as

$$\beta_1 = \sum_{i=0} b_i R^{i+1}, \quad (77)$$

with

$$b_i = \frac{2}{3\pi} \left(\frac{2Z}{\pi}\right)^i \sum_{j=0}^{i} Q_{j,i-j}^{(-2)}. \quad (78)$$

As a function of Z and R, the nuclear shielding factor satisfies $\beta_1(Z, R) = \beta_1(1, ZR)/Z$.

4. VARIATIONAL AND NUMERICAL SOLUTIONS

In addition to perturbation theory approaches, variational and purely numerical techniques can be used to calculate wave functions and energy levels and, hence, oscillator strengths, polarisabilities and other properties. In a variational approach the wave function is expanded as

$$\psi_{nl}(r) = \sum_{i=1} c_i \phi_i(r), \tag{79}$$

where $\{\phi_i\}$ is, in principle, a complete set of functions satisfying the boundary conditions

$$\phi_i(0) = \phi_i(R) = 0, \quad i = 1, 2, 3, \ldots . \tag{80}$$

Natural choices for the problem under consideration here are

$$\phi_i(r) = N_i r^l \sin \frac{i\pi r}{R} \tag{81}$$

and

$$\phi_i(r) = N_i r^{l+1} j_l(\pi \gamma_{il} r/R). \tag{82}$$

These sets are, of course, identical for $l = 0$ and, additionally, $\{\phi_i\}$ is orthonormal for this value of the angular momentum l (N_i is a normalisation factor). The set generated by (82) was used by Aguilera-Navarro et al. [38] in a variational calculation on a harmonic oscillator in a spherical box. It is more difficult to compute and less convenient to use than the set generated by (81), so the latter functions were used to calculate, for comparison, energies and oscillator strengths. Accurate values of these properties were also calculated by solving the differential equations (4) numerically, and electric dipole polarisabilities and nuclear shielding factors were calculated by numerical solution of Equation (67) followed by evaluation of, respectively, (65) and (76).

The perturbation theory Equations (39) may be conveniently solved numerically. We employ a finite difference method (Fox [39]). The Numerov method is not sufficiently accurate, apart from the s, ie $l = 0$, states where the $l(l + 1)/r^2$ term is absent; its order of accuracy for $l \geq 1$ is quadratic (Killingbeck [40]). It was found that a sixth-order method worked well in practice, giving highly accurate results. There was little difference between a fourth- and a sixth-order method in general, and progressing to an eighth-order one gave negligible, if any, improvement. For convenience, we employed a constant step-length across the whole range, giving a banded coefficient matrix. It is worth mentioning, however, that the Numerov method gives a tri-diagonal coefficient matrix (provided a constant step-length is used).

5. PERTURBATION THEORY RESULTS FOR SMALL BOX RADII

5.1. Energies

The perturbation theory equations were solved, as described in the previous section, up to order 15, allowing energies to be calculated to 31st order. The initial estimates for the zeros of the spherical Bessel functions are provided by a formula given by Abramowitz and Stegun [30], and these were then refined by a Newton–Raphson procedure. Some energy results are presented in Tables 4–6 in comparison with our precise numerical estimates (Section 4). We note that they converge rapidly for smaller R but, as expected, accuracy deteriorates as R becomes larger. The perturbation expansion results should be more accurate than the numerical estimates for the smallest value of R tabulated; for $R = 0.5$, the extremely accurate data computed by Aquino et al. [21] favour the perturbation results when there is a difference of 1 in the final decimal place quoted. Otherwise, in all cases where direct comparisons are available, our energies calculated by direct numerical solution of (4) agree with the exact values tabulated by Burrows and Cohen [20] and by Aquino et al. [21].

It may also be observed from Tables 4–6 that there are curve crossings in the plots of the E_{nl} versus R; for example, $E_{2\mathrm{s}} = E_{3\mathrm{d}}$ at $R = 2$, $E_{3\mathrm{p}} = E_{4\mathrm{f}}$ at $R = 6$, etc. These observations are contrary to the conjecture of Wilcox that there are no level crossings [24]. In fact, it has been proved by Scherbinin et al. [26,41] that the levels E_{nl} and $E_{n+1,l+2}$ are degenerate at $R = (l + 1)(l + 2)$, in agreement with our tabulated results. There are also additional level crossings, for example, the $E_{3\mathrm{s}}$ and $E_{4\mathrm{f}}$ curves cross at $R = 10.19735$.

The perturbation theory coefficients are presented in Table 7 for the 1s, 2p, 3d, 4f and 5g states. In general, they decrease rapidly with increasing order. It is, of course, of interest to know something about the convergence of the perturbation series. One way to increase the rate of convergence is to use a rational approximation to the perturbation theory power series solution. We have done this by use of Padé approximations. It was found that, in general, the diagonal $[N/N]$ and off-diagonal $[N/N + 1]$ approximants provided the best results and we have chosen the $[N/N + 1]$ approximant which was usually very slightly better than the $[N/N]$ approximant. An indication of the improved convergence provided by the Padé approximants is given in Table 8 where we present the maximum box radius which allows a relative accuracy of at least 10^{-6} in the energy. We observe that the Padé approximant approach generally allows a box radius at least 20% larger to achieve this level of accuracy. As expected, the higher the state the larger is the box radius which permits a given level of accuracy, though we also note that the $n = l+2$ states converge slightly more slowly than the $n = l+1$ states and other higher states of the same l. The reason for this behaviour is not understood, though it is sometimes corrected by the Padé approximations.

Table 4 Comparison of small-box perturbation and precise numerical energies for the ns ($n = 1, 2, 3, 4, 5$) levels of a hydrogen atom at the centre of a spherical cavity of radius R. For each value of R the first row refers to the perturbation energy whereas the second refers to the precise numerical energy. The final row of the table specifies the estimated radius of convergence, R_{rc}, of the perturbation theory expansion (see text)

R	1s	2s	3s	4s	5s
0.1	468.99303866	1942.72035456	4406.12165184	7857.62918498	12296.73165931
	468.99303870	1942.72035446	4406.12165174	7857.62918501	12296.73165922
0.5	14.74797003	72.67203919	170.58516419	308.19724748	485.41111873
	14.74797003	72.67203919	170.58516418	308.19724748	485.41111874
1.0	2.37399087	16.57025609	40.86312460	75.13049306	119.32706250
	2.37399087	16.57025609	40.86312460	75.13049306	119.32706250
2.0	−0.12500000	3.32750916	9.31415044	17.81609350	28.81350572
	−0.12500000	3.32750916	9.31415044	17.81609350	28.81350572
4.0	−0.48326530	0.42023563	1.87270207	3.96647655	6.69068964
	−0.48326530	0.42023563	1.87270207	3.96647655	6.69068964
6.0	−0.49932132	0.01276914	0.63173744	1.54636467	2.74538312
	−0.49927729	0.01272510	0.63173744	1.54636467	2.74538312
8.0	1.08289966	−1.66761315	0.24649166	0.74988946	1.41680996
	−0.49997510	−0.08473872	0.24649198	0.74988946	1.41680995
10.0			0.08864052	0.40515480	0.82638888
			0.09142232	0.40515436	0.82638888
12.0			−0.98878856	0.23123531	0.51920279
			0.01975295	0.23108321	0.51920248
14.0				0.15270939	0.34225521
				0.13409590	0.34222897
R_{rc}	6.04	6.03	9.15	12.3	15.4

Table 5 Same as Table 4 but for the np ($n = 2, 3, 4, 5$) levels of a hydrogen atom confined in a spherical box of radius R

R	2p	3p	4p	5p
0.1	991.00758944	2960.46230228	5918.18288891	9863.60475943
	991.00758945	2960.46230236	5918.18288910	9863.60475987
0.5	36.65887588	114.64355252	232.42796006	389.85089714
	36.65887588	114.64355252	232.42796007	389.85089715
1.0	8.22313832	27.47399530	56.75803389	95.99185333
	8.22313832	27.47399530	56.75803389	95.99185334
2.0	1.57601879	6.26900279	13.51058416	23.25908280
	1.57601879	6.26900279	13.51058416	23.25908280
4.0	0.14352708	1.26152121	3.03390630	5.44204180
	0.14352708	1.26152121	3.03390630	5.44204180
6.0	−0.05555556	0.42150945	1.19331568	2.25124834
	−0.05555556	0.42150945	1.19331568	2.25124834
8.0	−0.10445007	0.15736820	0.58292476	1.17128935
	−0.10445007	0.15736820	0.58292476	1.17128935
10.0	−0.11885954	0.04919076	0.31605624	0.68837033
	−0.11885954	0.04919076	0.31605624	0.68837033
12.0	−0.12325490	−0.00161068	0.17971400	0.43529186
	−0.12325506	−0.00161053	0.17971400	0.43529186
15.0	−0.12466557	−0.03514999	0.07688662	0.23745278
	−0.12477133	−0.03504423	0.07688662	0.23745278
20.0	0.18261672	−0.35921073	0.00807467	0.09516973
	−0.12499461	−0.05161142	0.00808668	0.09516973
25.0			−0.02329315	0.03675757
			−0.01650346	0.03675552
R_{rc}	15.4	15.4	21.8	28.2

An important property of a power series is its radius of convergence, determined by the singularity closest to the origin. This, of course, is not known in the present cases but we may attempt to estimate it as follows (Fernández [17]). The perturbation expansion (37) is a real power series in λ, so we choose the singular points of $\mathcal{E}_{nl}(\lambda)$ to be the complex conjugate pair λ_0 and λ_0^* (now allowing λ to be a complex variable) and consider the function

$$e(\lambda) = (\lambda^2 - 2\lambda_R\lambda + |\lambda_0|^2)^\alpha, \tag{83}$$

where $\lambda_0 = \lambda_R + i\lambda_I$. As $e(\lambda)$ satisfies the differential equation

$$(\lambda^2 - 2\lambda_R\lambda + |\lambda_0|^2)e'(\lambda) = 2\alpha(\lambda - \lambda_R)e(\lambda) \tag{84}$$

Table 6 Same as Table 4 but for the 3d, 4d, 5d, 4f, 5f and 5g levels of a hydrogen atom confined in a spherical box of radius R

R	3d	4d	5d	4f	5f	5g
0.1	1644.52992240	4115.58263203	7569.54251963	2426.39554900	5407.22202410	3333.30400341
	1644.52992273	4115.58263308	7569.54252146	2426.39555006	5407.22202638	3333.30391976
0.5	63.16018447	161.35700660	299.06471541	94.62659776	213.31044634	131.02461828
	63.16018448	161.35700663	299.06471547	94.62659782	213.31044645	131.02461827
1.0	14.96746409	39.31531986	73.60191934	22.89582548	52.39510296	32.03408911
	14.96746409	39.31531986	73.60191934	22.89582548	52.39510296	32.03408911
2.0	3.32750916	9.31415044	17.81609350	5.34209438	12.63104777	7.64667498
	3.32750916	9.31415044	17.81609350	5.34209438	12.63104777	7.64667498
4.0	0.62135578	2.06842517	4.15956991	1.14306295	2.92239311	1.72990025
	0.62135578	2.06842517	4.15956991	1.14306295	2.92239311	1.72990025
6.0	0.18034102	0.80201266	1.71649514	0.42150945	1.19331568	0.68753054
	0.18034102	0.80201266	1.71649514	0.42150945	1.19331568	0.68753054
8.0	0.04605825	0.38419924	0.89042559	0.18782457	0.61134967	0.34067969
	0.04605825	0.38419924	0.89042559	0.18782457	0.61134967	0.34067969
10.0	−0.00709278	0.20244288	0.52134810	0.08824159	0.35258417	0.18834187
	−0.00709278	0.20244288	0.52134810	0.08824159	0.35258417	0.18834187
15.0	−0.04664258	0.04062201	0.17697947	0.00271613	0.11302776	0.05024539
	−0.04664258	0.04062201	0.17697947	0.00271613	0.11302776	0.05024539
20.0	−0.05396757	−0.00553503	0.06824429	−0.02000000	0.03844567	0.00900191
	−0.05396756	−0.00553503	0.06824429	−0.02000000	0.03844567	0.00900191

(continued on next page)

Table 6 (continued)

R	3d	4d	5d	4f	5f	5g
25.0	−0.05532298	−0.02185374	0.02346293	−0.02743534	0.00818253	−0.00690812
	−0.05532145	−0.02185527	0.02346293	−0.02743534	0.00818253	−0.00690812
30.0	−0.05556089	−0.02793621	0.00226745	−0.03001920	−0.00591442	−0.01388889
	−0.05552808	−0.02796897	0.00226740	−0.03001920	−0.00591443	−0.01388889
40.0		−1.84859910	−0.01411336	−0.03116295	−0.01656709	−0.01870988
		−0.03094952	−0.01421302	−0.03115686	−0.01657318	−0.01870988
50.0					−0.01448878	−0.01977524
					−0.01929962	−0.01977527
R_{rc}	27.9	27.9	37.6	43.3	43.3	61.5

Table 7 Perturbation theory coefficients $E_{nl}^{(i)}$ for expansion of the energies in powers of R for the 1s, 2p, 3d, 4f and 5g levels of a hydrogen-like atom confined in a spherical box of radius R (see Equation (51) of text). The numbers in parentheses indicate the power of 10

i	1s	2p	3d	4f	5g
0	0.4934802201(1)	0.1009536428(2)	0.1660873096(2)	0.2441559682(2)	0.3347715596(2)
1	−0.2437653393(1)	−0.1850827538(1)	−0.1633563104(1)	−0.1516045663(1)	−0.1440949475(1)
2	−0.1093911226(0)	−0.2047417885(−1)	−0.7524779236(−2)	−0.3671178666(−2)	−0.2095981929(−2)
3	−0.1243810890(−1)	−0.8844162776(−3)	−0.1745627982(−3)	−0.5361664289(−4)	−0.2114839441(−4)
4	−0.1244523497(−2)	−0.3835437228(−4)	−0.4323234597(−5)	−0.8665179357(−6)	−0.2418022375(−6)
5	−0.8586931721(−4)	−0.1438217661(−5)	−0.1001800455(−6)	−0.1365543078(−7)	−0.2766024882(−8)
6	0.2260261125(−6)	−0.3713850664(−7)	−0.1943612115(−8)	−0.1938634755(−9)	−0.2963220064(−10)
7	0.1223900293(−5)	0.1860162473(−9)	−0.2296632855(−10)	−0.2172489972(−11)	−0.2745214958(−12)
8	0.2151719049(−6)	0.1028487918(−9)	0.3340357155(−12)	−0.9574020751(−14)	−0.1767151808(−14)
9	0.1875431634(−7)	0.7919408479(−11)	0.3411317751(−13)	0.4016666605(−15)	0.3451011081(−17)
10	−0.3915503619(−9)	0.3782051096(−12)	0.1394575654(−14)	0.1650862806(−16)	0.3856948799(−18)
11	−0.4398701730(−9)	0.9215595360(−14)	0.3963774735(−16)	0.4048913882(−18)	0.8672720790(−20)
12	−0.7950342758(−10)	−0.3495570771(−15)	0.7308293445(−18)	0.7455985697(−20)	0.1396508624(−21)
13	−0.6899154040(−11)	−0.5995288000(−16)	−0.1308921963(−21)	0.9692531947(−22)	0.1774260968(−23)
14	0.2828514866(−12)	−0.4240455436(−17)	−0.7262005052(−21)	0.3480167478(−24)	0.1651949930(−25)
15	0.2078600892(−12)	−0.1857816432(−18)	−0.3831856427(−22)	−0.2814293085(−25)	0.5405365898(−28)
16	0.3690993687(−13)	−0.2707243439(−20)	−0.1301940212(−23)	−0.1150432279(−26)	−0.2089811459(−29)
17	0.3061845967(−14)	0.3845030990(−21)	−0.3014358908(−25)	−0.2912363391(−28)	−0.6560452467(−31)
18	−0.1933466909(−15)	0.4487979611(−22)	−0.2582077285(−27)	−0.5478457756(−30)	−0.1253829967(−32)

(*continued on next page*)

Table 7 (continued)

i	1s	2p	3d	4f	5g
19	−0.1111993269(−15)	0.2826791647(−23)	0.1811255234(−28)	−0.6985262618(−32)	−0.1840260743(−34)
20	−0.1916352491(−16)	0.1055331157(−24)	0.1268586995(−29)	−0.9526419871(−35)	−0.2056126939(−36)
21	−0.1491351325(−17)	−0.1722017019(−27)	0.4994924739(−31)	0.2808216066(−35)	−0.1306863183(−38)
22	0.1322092555(−18)	−0.3952142762(−27)	0.1361635316(−32)	0.1076450585(−36)	0.1181486154(−40)
23	0.6388422836(−19)	−0.3671814545(−28)	0.2011498903(−34)	0.2684233918(−38)	0.6120207368(−42)
24	0.1063044456(−19)	−0.2041009937(−29)	−0.3989194712(−36)	0.4963870791(−40)	0.1363014363(−43)
25	0.7647807372(−21)	−0.5948323247(−31)	−0.4472693560(−37)	0.5949361167(−42)	0.2237421943(−45)
26	−0.9113328661(−22)	0.1850628048(−32)	−0.2047085628(−38)	−0.1067962675(−44)	0.2844039786(−47)
27	−0.3843212341(−22)	0.3971503202(−33)	−0.6391085499(−40)	−0.3196959476(−45)	0.2435225510(−49)
28	−0.6158011608(−23)	0.3115144659(−34)	−0.1256012612(−41)	−0.1145899032(−46)	−0.9724153527(−53)
29	−0.4033623692(−24)	0.1505608165(−35)	0.2415546938(−44)	−0.2779724375(−48)	−0.5996659714(−53)
30	0.6333123667(−25)	0.2789656896(−37)	0.1576444237(−44)	−0.4983106851(−50)	−0.1601316856(−54)
31	0.2387118290(−25)	−0.2983304837(−38)	0.8614423915(−46)	−0.5458804112(−52)	−0.2922224778(−56)

Table 8 Maximum box radii R_{PT} and R_{Pade} for, respectively, direct perturbation theory expansion and Padé approximant, which guarantee a relative accuracy of at least 10^{-6} in the energy

State	R_{PT}	R_{Pade}
1s	4.82	6.33
2s	4.67	6.09
3s	7.98	10.7
4s	9.98	15.5
5s	12.2	22.5
10s	25.0	54.9
2p	11.9	15.0
3p	11.1	14.4
4p	16.5	21.0
5p	22.1	30.9
10p	47.8	80.7
3d	22.1	26.5
4d	20.7	32.2
5d	28.2	36.3
10d	67.7	119.7
4f	33.7	39.9
5f	32.6	43.3
10f	84.6	119.5
5g	49.6	58.8
6g	48.2	55.3
10g	98.0	134.0

we easily obtain the recurrence relation

$$|\lambda_0|^2(i+1)e^{(i+1)} - 2\lambda_{\mathrm{R}}(i-\alpha)e^{(i)} - 2\alpha e^{(i-1)} \\ = -(i-1)e^{(i-1)}, \quad i = 1, 2, 3, \ldots, \tag{85}$$

for expansion of $e(\lambda)$ in a Taylor series

$$e(\lambda) = \sum_{i=0}^{\infty} e^{(i)}\lambda^i. \tag{86}$$

If $\mathcal{E}_{nl}(\lambda)$ behaves as $e(\lambda)$ then we can use (85) to estimate $|\lambda_0|$ and α. One approach is to consider the 3 inhomogeneous equations obtained from (85) with 3 successive values of i and see if the solutions for $|\lambda_0|^2$, λ_{R} and α converge as i increases. However, this method is not convenient to use as the equations are nonlinear. An alternative procedure is to define $\beta = \alpha\lambda_{\mathrm{R}}$ and solve 4 successive linear equations for $|\lambda_0|^2$, λ_{R}, α and β. This procedure

Table 9 Estimated radius of convergence of $E_{1s}(\lambda)$ as the number of terms, i, in the perturbation theory expansion is increased (see text)

i	$\|\lambda_0\|^2$	λ_R	α	β	$\alpha\lambda_R$
20	36.22	4.209	0.601	2.95	2.53
25	36 33	4.211	0.573	2.71	2.41
30	36.38	4.211	0.559	2.60	2.35
35	36.41	4.212	0.547	2.52	2.30
40	36.43	4.212	0.537	2.44	2.26
45	36.44	4.212	0.533	2.40	2.24
50	36.45	4.212	0.530	2.37	2.23

worked well in practice as is evidenced by Table 9 where we present results for the ground state energy E_{1s} for a series of values of i. Note that, to allow a test of the convergence of the method, we have extended the energy calculation up to order 51. β and $\alpha\lambda_R$ should agree if proper convergence has been obtained; the fact that they do not coincide indicates there is still some way to go. There are, of course, other singular points of $\mathcal{E}_{1s}(\lambda)$, further from the origin, which affect convergence; this method reveals only the one smallest in magnitude.

The estimated radius of convergence of each series is indicated in the final rows of Tables 4–6 and they would appear to agree well with the behaviour of the actual numerical energy values obtained from summation of the perturbation series. From Table 9 it seems that α is (slowly) approaching 0.5 as i increases. If we choose $\alpha = 0.5$ and solve 2 linear equations for $|\lambda_0|^2$ and λ_R we find very little difference from the results in Table 9; for example, for $i = 50$ we obtain $|\lambda_0|^2 = 36.49$ with $\alpha = 0.5$ as opposed to $|\lambda_0|^2 = 36.45$ with $\alpha = 0.53$, so it would appear that $E_{1s}(\lambda)$ has a square root branch point close to $|\lambda| = 6.04$. Other series $E_{nl}(\lambda)$ also behave in a similar manner, with square root branch points, which would appear to be a general feature of perturbation theory expansions for linear operators [42].

5.2. Oscillator strengths

With the perturbed wave functions and energies in hand, it is a straightforward matter to calculate oscillator strengths via Equation (61). The coefficients f_i, $(i = 0, 1, 2, \ldots, 15)$, in the expansion of f_{1s-2p} are listed in Table 10. In this case, however, we are limited to the order of the perturbation, viz. 15 in the present calculation (this limit can easily be changed by increasing the dimensions of the arrays in the computer programs used).

Results are presented in Table 11, together with comparison values obtained by direct numerical procedures, as described earlier. Again, it is observed that the accuracy of the present results deteriorates as R increases,

Table 10 Perturbation theory coefficients f_i, a_i and b_i for expansion of, respectively, oscillator strength f_{1s-2p} Equation (61), dipole polarisability α_1 Equation (70) and nuclear shielding factor β_1 Equation (77) in powers of R for a hydrogen atom confined in a spherical box of radius R. The numbers in parentheses indicate the power of 10

i	f_i	a_i	b_i
0	0.9666504480(0)	0.3634022344(−1)	0.3333333333(0)
1	0.2078397498(−1)	−0.7507322626(−2)	−0.1899254128(−1)
2	−0.1817122683(−2)	−0.8599289733(−4)	−0.1992688816(−2)
3	−0.8924964821(−3)	0.3758372586(−4)	−0.1451011784(−3)
4	−0.1548486093(−3)	0.6932310100(−5)	0.4409941353(−5)
5	−0.1298933083(−4)	0.6265117258(−6)	0.3650186359(−5)
6	0.8373507149(−6)	−0.9997068170(−8)	0.6562902609(−6)
7	0.4962936217(−6)	−0.1481604027(−7)	0.5647115455(−7)
8	0.8866174673(−7)	−0.2832410217(−8)	−0.3104706959(−8)
9	0.7314462105(−8)	−0.2605888187(−9)	−0.2002003142(−8)
10	−0.5891728766(−9)	0.1022520397(−10)	−0.3593002725(−9)
11	−0.3144472027(−9)	0.8379179830(−11)	−0.2947833364(−10)
12	−0.5510811639(−10)	0.1571518664(−11)	0.2421747569(−11)
13	−0.4301394159(−11)	0.1383673727(−12)	0.1271032427(−11)
14	0.4466706728(−12)	−0.8473893221(−14)	0.2204550518(−12)
15	0.2105203164(−12)	−0.5323078926(−14)	0.1674136559(−13)

though the perturbation series gives data accurate to 6 significant figures up to $R = 3$.

One point of interest is the behaviour of the 1s − 2p oscillator strength, f_{1s-2p}. It has been noted previously that f_{1s-2p} initially increases as R increases, reaching a maximum of approximately 0.9912 close to $R = 1.8$, and subsequently decreasing monotonically to its free-atom value of 0.416197 [31]. First, we calculate the $R = 0$ value,

$$f_0 = \frac{4(\gamma^2 - 1)}{3\pi^2 \sin^2 \gamma\pi} \left| \frac{\pi}{\gamma^2 - 1} \cos \gamma\pi - \frac{3\gamma^2 - 1}{\gamma(\gamma^2 - 1)^2} \sin \gamma\pi \right|^2, \tag{87}$$

where $\gamma = 1.430296653$ ($\gamma\pi = 4.493409458$ is the first zero of j_1), which gives $f_0 = 0.9666504480$ (Table 10). Next, calculating and evaluating the perturbation series for f_{1s-2p} results in a maximum of $f_{1s-2p} = 0.99122575$ at $R = 1.8751$.

5.3. Dipole polarisabilities and nuclear shielding factors

The dipole polarisability of a hydrogen-like atom confined at the centre of a spherical box of radius R is given by Equation (70). The coefficients a_i, $(i = 1, 2, \ldots, 15)$, of the expansion are listed in Table 10 and the values for

Table 11 Comparison of perturbation and precise numerical 1s–2p oscillator strengths for a hydrogen atom at the centre of a spherical box of radius R

R	Perturbation theory	Numerical
0.2	0.97072717	0.97072717
0.4	0.97461209	0.97461209
0.6	0.97825287	0.97825287
0.8	0.98159037	0.98159037
1.0	0.98455839	0.98455839
1.2	0.98708360	0.98708360
1.4	0.98908568	0.98908568
1.6	0.99047783	0.99047783
1.8	0.99116771	0.99116771
2.0	0.99105877	0.99105876
2.5	0.98664479	0.98664470
3.0	0.97515323	0.97515174
3.5	0.95547673	0.95546082
4.0	0.92725144	0.92713538
4.5	0.89113305	0.89051528
5.0	0.84879929	0.84632272
5.5	0.80263220	0.79520587
6.0	0.75521138	0.74024166

this polarisability are presented in Table 12, together with data obtained by other authors. We observe that our numerical values agree precisely (to the 8 figures quoted) with the exact results of Burrows and Cohen [37] and with the highly accurate results of Montgomery [33], and that the perturbation theory values are accurate to 6 significant figures up to $R = 3$. Dutt et al. [34] used the Buckingham approximation to the polarisability [43] and their values differ from our numerical values by less than 1%.

The coefficients b_i, ($i = 1, 2, 3, \ldots, 15$), in the expansion of the nuclear shielding factor β_1 in powers of R, Equation (77), are also listed in Table 10, and Table 13 presents a comparison of our perturbation theory and numerical values with other calculated results [33,37]. We note that perturbation theory provides at least 6 figure accuracy up to $R = 3$, and we also observe that our numerical and other calculated data are in harmony for all values of R for which comparisons are available. β_1 behaves as $R/3$ as $R \to 0$ [6].

6. INTERMEDIATE BOX SIZES

Short- and long-range problems are frequently much easier to deal with than intermediate-range ones, so it is not generally possible to develop perturbation theory expansions in the latter range of R values. Fernández

Table 12 Comparison of perturbation and precise numerical electric dipole polarisabilities for a hydrogen atom confined at the centre of a spherical box of radius R. The numbers in parentheses indicate the power of 10

R	Perturbation theory	Numerical[a]	Other
0.2	0.55737010(−4)	0.55737010(−4)	
0.4	0.85314879(−3)	0.85314879(−3)	0.00085[b]
0.6	0.41230862(−2)	0.41230862(−2)	0.00411[b]
0.8	0.12411540(−1)	0.12411540(−1)	0.01237[b]
1.0	0.28792023(−1)	0.28792023(−1)	0.02868[b], 0.28792023(−1)[c,d]
1.2	0.56585203(−1)	0.56585202(−1)	0.05632[b]
1.4	0.99091299(−1)	0.99091299(−1)	0.09857[b]
1.6	0.15934166(0)	0.15934166(0)	0.15841[b]
1.8	0.23987672(0)	0.23987672(0)	0.23833[b]
2.0	0.34255811(0)	0.34255811(0)	0.34016[b], 0.34255811[c,d], 0.3427[e]
2.5	0.70067527(0)	0.70067518(0)	0.69490[b]
3.0	0.11917095(1)	0.11917061(1)	1.02432[b]
3.5	0.17723329(1)	0.17722626(1)	1.75615[b]
4.0	0.23788855(1)	0.23779823(1)	2.35782[b], 2.3779823[c,d]
4.5	0.29518878(1)	0.29438223(1)	
5.0	0.34753879(1)	0.34224542(1)	3.40294[b]
5.5	0.40561843(1)	0.37929337(1)	
6.0	0.50499443(1)	0.40581405(1)	4.04707[b], 4.0581405[c,d]

[a] Present calculation.
[b] Ref. [34].
[c] Ref. [33].
[d] Ref. [37].
[e] Ref [6].

Table 13 Comparison of perturbation and precise numerical nuclear shielding factors for a hydrogen atom confined at the centre of a spherical box of radius R

R	Perturbation theory	Numerical[a]	Other
0.125	0.041365981	0.041365981	0.041365981[b,c]
0.25	0.082114602	0.082114602	0.082114602[b,c]
0.5	0.161660577	0.161660577	0.161660577[b,c]
1	0.312211769	0.312211769	0.312211769[b,c], 0.312211770[c]
2	0.572902117	0.572902116	0.572902116[b,c], 0.572902117[c]
3	0.768804934	0.768804022	
4	0.894592798	0.894473121	0.894473121[b,c], 0.894473120[c]
5	0.965866521	0.960017142	
6	0.111621475	0.987274170	0.987274173[b], 0.987274170[c]

[a] Present calculation.
[b] Ref. [33].
[c] Ref. [37].

and Castro [44] used a hypervirial-perturbational technique to give second-order energies for 1s, 2s and 2p states for a series of intermediate R values. An alternative way to proceed is now presented. As pointed out earlier, exact solutions are available at some discrete R values and these can be employed to obtain perturbation expansions for energies and polarisabilities. For example, by considering the lobes of the known unconfined atom solutions ϕ_{2s}, ϕ_{3s} and ϕ_{3p}, Laughlin et al. [25] derived the following first-order approximations to the energies of the 1s, 2s and 2p states

$$E_{1s} \cong -\frac{1}{2R^2}\left[1 + \frac{1-5e^{-2}}{1-7e^{-2}}(RZ-2)\right], \tag{88}$$

for R close to $2/Z$,

$$E_{1s} \cong -\frac{r_1^2}{9R^2}\left[\frac{1}{2} + \left(\frac{ZR}{r_1} - 1\right)\frac{1-(34-18\sqrt{3})e^{-2r_1/3}}{1-(52-28\sqrt{3})e^{-2r_1/3}}\right], \tag{89}$$

where $r_1 = 3(3-\sqrt{3})/2$, valid for R close to r_1/Z,

$$E_{2s} \cong -\frac{r_2^2}{9R^2}\left[\frac{1}{2} + \left(\frac{ZR}{r_2} - 1\right)\frac{1-(34+18\sqrt{3})e^{-2r_2/3}}{1-(52+28\sqrt{3})e^{-2r_2/3}}\right], \tag{90}$$

where $r_2 = 3(3+\sqrt{3})/2$, valid for R close to r_2/Z, and

$$E_{2p} \cong -\frac{2}{R^2}\left[1 + \frac{3-103e^{-4}}{9-437e^{-4}}(RZ-6)\right], \tag{91}$$

valid in the neighbourhood of $R = 6/Z$.

Each of the above expressions for E_{1s}, E_{2s} and E_{2p} gives the exact energy result when the impenetrable wall is situated at the appropriate node of an unconfined atom wave function, and the error increases as the wall moves away from the node.

These first-order expressions can, of course, be improved by including higher-order corrections and this we now do. We consider only the nodeless nl state, $n = l+1, l = 0, 1, 2, \ldots$, based on the inner lobe of the $(n+1, l)$ unconfined solution; generalisation to other states is straightforward.

The unconfined $(n = l+2, l)$ radial wave function has the (unnormalised) form

$$\phi_{l+2,l}(r) = r^{l+1}\,[n(l+1) - Zr]\,e^{-Zr/n} \tag{92}$$

and this is an exact solution of the confined problem, with eigenvalue $-Z^2/2(l+2)^2$, when $R = (l+1)(l+2)/Z$. Introducing the independent variable

$$x = \frac{r}{R} \tag{93}$$

and writing the zero-order solution as

$$\psi_{nl}^{(0)}(x) = x^{l+1}(1-x)\mathrm{e}^{-(l+1)x}, \tag{94}$$

we find that it satisfies the differential equation

$$\left[\frac{\mathrm{d}^2}{\mathrm{d}x^2} - \frac{l(l+1)}{x^2} + 2\frac{(l+1)(l+2)}{x}\right]\psi_{nl}^{(0)}(x) = (l+1)^2\psi_{nl}^{(0)}(x). \tag{95}$$

Since the equation we need to solve is

$$\left[\frac{\mathrm{d}^2}{\mathrm{d}x^2} - \frac{l(l+1)}{x^2} + 2\frac{ZR}{x}\right]\psi_{nl}(x) = \mathcal{E}_{nl}\psi_{nl}(x), \qquad \mathcal{E}_{nl} = -2R^2E_{nl}, \tag{96}$$

it follows that the required perturbation is λ_{nl}/x, with

$$\lambda_{nl} = 2[RZ - (l+1)(l+2)]. \tag{97}$$

The various orders of correction can be evaluated as described in Section 3, the only difference now being the presence of an extra $1/x$ term in the zero-order operator, cf. Equation (38), but this causes no difficulty in the numerical solutions. Explicitly we have

$$E_{1\mathrm{s}} = -\frac{1}{2R^2}\left[1 + 2(ZR-2)\mathcal{E}_{1\mathrm{s}}^{(1)} + 2^2(ZR-2)^2\mathcal{E}_{1\mathrm{s}}^{(2)} + \cdots\right] \tag{98}$$

and

$$E_{2\mathrm{p}} = -\frac{1}{2R^2}\left[4 + 2(ZR-6)\mathcal{E}_{2\mathrm{p}}^{(1)} + 2^2(ZR-6)^2\mathcal{E}_{2\mathrm{p}}^{(2)} + \cdots\right]. \tag{99}$$

Table 14 presents comparisons of energies thus calculated (to 31st order) with those of the short-range perturbation theory of Section 3.1 and with precise numerical values. There are two main points to note. First, as expected, the present procedure extends the range of validity of perturbation expansions to larger box radii R. Second, since the $1/R$ expansions of

Section 3.1 are accurate up to (and beyond) $R = (l+1)(l+2)/Z$ it is not necessary, in the present context, to extend the calculations to the inner lobes of higher nl states, with $n > l+2$, as these produce R values smaller than $(l+1)(l+2)/Z$.

The polarisability and nuclear shielding factor can also be evaluated in a manner analogous to that described previously in Section 3.3, so that we have

$$\alpha_1 = R^4 \sum_{i=0} a_i'(ZR-2)^i. \tag{100}$$

$$\beta_1 = R \sum_{i=0} b_i'(ZR-2)^i. \tag{101}$$

Polarisabilities and nuclear shielding factors, thus evaluated, are presented in Table 15, and again we can draw the same two conclusions as for energies in the preceeding paragraph. The exact analytical value of α_1 at $R = 2/Z$ was first evaluated by Fowler [6] and corrected by Laughlin [31]. The corresponding β_1 is given by

$$\beta_1(R = 2/Z) = \frac{12 - 88e^{-2}}{3Z(1 - 7e^{-2})}. \tag{102}$$

Fowler [6] gives an approximate value of 0.5729/Z for β_1 at $R = 2/Z$, which agrees with the above result (0.5729021162/Z) to the four decimal places quoted.

7. CONCLUDING REMARKS

Perturbation theory expansions for wave functions, energies and properties of a hydrogen-like atom confined at the centre of an impenetrable spherical cavity have been reviewed and extended. Three different expansions need to be employed to cover the whole range of cavity radii from very small to very large. For the states considered here, the short- and intermediate-range expansions provide energies accurate to at least 6 significant figures for cavity radii $R \leq 0.8R_{\mathrm{rc}}^{(i)}$, where $R_{\mathrm{rc}}^{(i)}$ is the radius of convergence of the intermediate-range series. The simple first-order analytical expressions for energy provided by the large-cavity expansions also give results accurate to at least 6 significant figures for $R \geq 0.8R_{\mathrm{rc}}^{(i)}$. Although very accurate values for properties are predicted by all three perturbation theory expansions within their ranges of validity, the accuracy is lower than for energies in the overlap region, with errors approaching 1% for the dipole polarisability α_1 and the nuclear shielding factor β_1 in the region between intermediate and large R. In the case of the 1s–2p oscillator strength, the situation is not so

Table 14 Comparison of perturbation and precise numerical energies for a hydrogen atom confined at the centre of a spherical box of radius R. The superscripts (s) and (i) on E_{PT} indicate, respectively, the small- and intermediate-box perturbation theory results of Sections 3 and 5. The final row of the Table specifies the estimated radius of convergence, R_{rc}, of the perturbation theory expansion

R	1s			2p		
	$E_{\mathrm{PT}}^{(s)}$	$E_{\mathrm{PT}}^{(i)}$	Numerical	$E_{\mathrm{PT}}^{(s)}$	$E_{\mathrm{PT}}^{(i)}$	Numerical
2	−0.12500000	−0.12500000	−0.12500000			
4	−0.48326530	−0.48326530	−0.48326530			
6	−0.49932132	−0.49927666	−0.49927729	−0.05555556	−0.05555556	−0.05555556
7		−0.49962354	−0.49986258	−0.08747902	−0.08747902	−0.08747902
8		−0.50311421	−0.49997510	−0.10445007	−0.10445007	−0.10445007
10				−0.11885954	−0.11885954	−0.11885954
12				−0.12325490	−0.12325506	−0.12325506
15				−0.12466557	−0.12477135	−0.12477133
20				0.18261672	−0.10821777	−0.12499461
R_{rc}	6.04	7.72		15.4	16.8	

Table 15 Comparison of perturbation and precise numerical dipole polarisabilities α_1 and nuclear shielding factors β_1 for a hydrogen atom confined at the centre of a spherical box of radius R. The superscripts (s) and (i) on α_1 and β_1 indicate, respectively, the small- and intermediate-box perturbation theory results of Sections 3 and 5

R	$\alpha_1^{(s)}$	$\alpha_1^{(i)}$	$\alpha_1^{(num)}$	$\beta_1^{(s)}$	$\beta_1^{(i)}$	$\beta_1^{(num)}$
2.0	0.342558	0.342558	0.342558	0.572902	0.572902	0.572902
2.5	0.700675	0.700675	0.700675	0.679595	0.679595	0.679595
3.0	1.191710	1.191706	1.191706	0.768805	0.768804	0.768804
3.5	1.772333	1.772263	1.772263	0.840196	0.840185	0.840185
4.0	2.378886	2.377983	2.377983	0.894593	0.894475	0.894473
4.5	2.951888	2.943853	2.943822	0.934464	0.933531	0.933521
5.0	3.475388	3.423358	3.422454	0.965867	0.960170	0.960017
5.5	4.056184	3.808972	3.792934	1.006792	0.979487	0.976985
6.0	5.049944	4.251138	4.058141	1.116215	1.019625	0.987274

favourable and clearly it would be desirable to improve the range of validity of the intermediate- and/or the large-cavity radii perturbation theories.

The model examined here has been used in the past to study the effects of pressure on hydrogen. Although not discussed in the present article, the pressure exerted by a hydrogen-like atom when enclosed in a spherical cavity can easily be determined, as it is given by

$$P = -\frac{1}{4\pi R^2}\frac{\mathrm{d}E}{\mathrm{d}R}, \tag{103}$$

and E is now known analytically to high accuracy in terms of the cavity radius R.

REFERENCES

[1] A. Michels, J. de Boer, A. Bijl, Physica 4 (1937) 981.
[2] A. Sommerfeld, H. Welker, Ann. Phys. 32 (1938) 56.
[3] S.R.. De Groot, C.A. Ten Seldam, Physica 12 (1946) 669.
[4] E. Ley-Koo, S. Rubinstein, J. Chem. Phys. 71 (1979) 351.
[5] P.O. Fröman, S. Yngve, N. Fröman, J. Math. Phys. 28 (1987) 1813.
[6] P.W. Fowler, Mol. Phys. 53 (1984) 865.
[7] W. Jaskólski, Phys. Rep. 271 (1996) 1.
[8] T. Sako, G.H.F. Diercksen, J. Phys. B: At. Mol. Opt. Phys. 36 (2003) 1433.
[9] T. Sako, G.H.F. Diercksen, J. Phys. B: At. Mol. Opt. Phys. 36 (2003) 1681.
[10] V.K. Dolmatov, A.S. Baltenkov, J.-P. Connerade, S.T. Manson, Radiat. Phys. Chem. 70 (2004) 417.
[11] J.-P. Connerade, V.K. Dolmatov, P.A. Lakshmi, S.T. Manson, J. Phys. B: At. Mol. Opt. Phys. 32 (1999) L239.
[12] B. Saha, P.K. Mukherjee, G.H.F. Diercksen, Astron. Astrophys. 396 (2002) 337.
[13] J. Gorecki, W. Byers Brown, J. Phys. B: At. Mol. Opt. Phys. 22 (1989) 2659.
[14] E.P. Wigner, Phys. Rev. 94 (1954) 77.

[15] V.C. Aguilera-Navarro, W.M. Kloet, A.U. Zimerman, Rev. Bras. Fis. 1 (1971) 55.
[16] B.F. Gray, I. Gonda, J. Chem. Phys. 62 (1975) 2007.
[17] F.M. Fernández, Introduction to Perturbation Theory in Quantum Mechanics, CRC Press, Florida, 2001.
[18] V.I. Pupyshev, J. Phys. B: At. Mol. Opt. Phys. 33 (2000) 961.
[19] S. Goldman, C. Joslin, J. Phys. Chem. 96 (1992) 6021.
[20] B.L. Burrows, M. Cohen, Int. J. Quantum Chem. 106 (2005) 478.
[21] N. Aquino, G. Campoy, H.E. Montgomery, Int. J. Quantum Chem. 107 (2007) 1548.
[22] V.I. Pupyshev, A.V. Scherbinin, J. Phys. B: At. Mol. Opt. Phys. 32 (1999) 4627.
[23] R.B. Dingle, Proc. Camb. Phil. Soc. 49 (1953) 103.
[24] W. Wilcox, Am. J. Phys. 57 (1988) 526.
[25] C. Laughlin, B.L. Burrows, M. Cohen, J. Phys. B: At. Mol. Opt. Phys. 35 (2002) 701.
[26] A.V. Scherbinin, V.I. Pupyshev, A. Ermilov, Physics of Clusters, World Scientific, Singapore, 1997.
[27] D.L. Kreider, R.G. Kuller, D.R. Ostberg, F.W. Perkins, An Introduction to Linear Analysis, Addison-Wesley Publishing Company, Reading, 1966.
[28] I. Gonda, B.F. Gray, J. Chem. Soc. Faraday Trans. II 71 (1975) 2016.
[29] N. Aquino, Int. J. Quantum Chem. 54 (1995) 107.
[30] M. Abramowitz, I.A. Stegun, Handbook of Mathematical Functions, Dover Publications, New York, 1965.
[31] C. Laughlin, J. Phys. B: At. Mol. Opt. Phys. 37 (2004) 4085.
[32] J.O. Hirschfelder, W. Byers Brown, S.T. Epstein, Int. J. Quantum Chem. 1 (1964) 235.
[33] H.E. Montgomery, private communication, 2004.
[34] R. Dutt, A. Mukherjee, Y.P. Varshni, Phys. Lett. A 280 (2001) 318.
[35] S.H. Patil, J. Phys. B: At. Mol. Opt. Phys. 35 (2002) 255.
[36] K.D. Sen, J. Garza, R. Vargas, N. Aquino, Phys. Lett. A 295 (2002) 299.
[37] B.L. Burrows, M. Cohen, Phys. Rev. A 72 (2005) 032508.
[38] V.C. Aguilera-Navarro, J.F. Gomes, A.H. Zimerman, E. Ley koo, J. Phys. A: Math. Gen. 16 (1983) 2943.
[39] L. Fox, The Numerical Solution of Two-Point Boundary Problems in Ordinary Differential Equations, Clarendon Press, Oxford, 1957.
[40] J.P. Killingbeck, Microcomputer Algorithms: Action from Algebra, Adam Hilger, Bristol, 1991.
[41] A.V. Scherbinin, V.I. Pupyshev, A. Ermilov, Izv. Akad. Nauk. Ser. Fiz (Russ. Phys. Bull.) 61 (1997) 1799.
[42] F.M. Fernández, G.A. Arteca, E.A. Castro, J. Phys. A: Math. Gen. 20 (1987) 6303.
[43] A.D. Buckingham, Adv. Chem. Phys. 12 (1967) 107.
[44] F.M. Fernández, E.A. Castro, J. Math. Phys. 23 (1982) 1103.

CHAPTER 7

Comparative Study Between the Hartree-Fock and Kohn-Sham Models for the Lowest Singlet and Triplet States of the Confined Helium Atom

Jorge Garza [a] and **Rubicelia Vargas** [a]

Contents

1. INTRODUCTION

An atom confined within a sphere, of radius R_c, with rigid walls has been used as a model to give an insight into the behavior of electrons confined under high pressures [1–3]. For many-electron atoms, this model has been applied by using the Hartree-Fock (HF) [4,5] and the Kohn-Sham (KS) model [6], some applications of these methods can be found in Refs. [7–12].

[a] Departamento de Química, División de Ciencias Básicas e Ingeniería, Universidad Autónoma Metropolitana-Iztapalapa, San Rafael Atlixco 186, Col. Vicentina. Iztapalapa. C. P. 09340. México D. F., México

E-mail address: jgo@xanum.uam.mx (J. Garza).

Advances in Quantum Chemistry, Vol. 57
ISSN: 0065-3276, DOI: 10.1016/S0065-3276(09)00611-X

We think that a comparison between the HF and KS methods is pertinent; in particular to obtain the KS performance when just the exchange contribution is considered.

For a long time, two-electron systems have been employed to study the performance of many exchange-correlation functionals of the KS model [13–17], within the Density Functional Theory framework [18]. Recently, the lowest singlet and triplet states of the confined helium atom have been reported by two methodologies. For the first one, the wave function is expanded on Hylleraas functions (H-WF), giving a benchmark for these two states [19]. For the second one, the Kohn-Sham method was used and its equations were solved by using the local density approximation (LDA) [20], coupled with the self-interaction correction (SIC) of Perdew and Zunger [21] and the optimized effective potential (OEP) [22,23] in the Krieger-Li-Iafrate (KLI) approximation [24–26], according to the method given in Refs. [27,28]. For the triplet state, appreciable discrepancies were reported between the KS model and the H-WF approach when the electrons were confined in small regions. However, the KS approach used in Ref. [19] included the exchange and correlation contributions simultaneously and it is not clear which of these contributions is responsible for such a discrepancy with respect to the H-WF method. Thus, in this work, we present a detailed study of this system considering just the exchange contribution, in order to see if the exchange contribution is reasonably described by the LDA and LDA–SIC approaches.

2. HARTREE-FOCK AND KOHN-SHAM MODELS FOR TWO-ELECTRON SYSTEMS

2.1. Exchange energy

We start with the HF equations, where the total energy for a two-electron system is [29]

$$E_{\mathrm{HF}} = \langle 1|h|1\rangle + \langle 2|h|2\rangle + \frac{1}{2}(\langle 12\|12\rangle + \langle 21\|21\rangle) \tag{1}$$

with

$$h = -\frac{1}{2}\nabla^2 + v(\mathbf{r}), \tag{2}$$

$$\langle ij\|kl\rangle = \langle ij|kl\rangle - \langle ij|lk\rangle, \tag{3}$$

and

$$\langle ij|kl\rangle = \iint d\mathbf{x}_1 d\mathbf{x}_2 \frac{\chi_i^*(\mathbf{x}_1)\chi_j^*(\mathbf{x}_2)\chi_k(\mathbf{x}_1)\chi_l(\mathbf{x}_2)}{r_{12}}. \tag{4}$$

In these equations, υ represents the external potential, χ_i a spin-orbital and $\mathbf{x}$ the four coordinates, the spatial ($\mathbf{r}$) and the spin (ω). For a closed-shell system $\chi_1(\mathbf{x}) = \psi_1(\mathbf{r})\alpha(\omega)$ and $\chi_2(\mathbf{x}) = \psi_1(\mathbf{r})\beta(\omega)$, such that the total energy has the form

$$E_{\mathrm{HF}} = 2(1|h|1) + \iint d\mathbf{r_1}d\mathbf{r_2}\frac{\rho_1(\mathbf{r_1})\rho_1(\mathbf{r_2})}{r_{12}}, \tag{5}$$

with $\rho_1 = |\psi_1|^2$.

For an open-shell system, $\chi_1(\mathbf{x}) = \psi_1(\mathbf{r})\alpha(\omega)$ and $\chi_2(\mathbf{x}) = \psi_2(\mathbf{r})\alpha(\omega)$. In this case the energy expression has an additional contribution

$$\begin{aligned} E_{\mathrm{HF}} = {} & (1|h|1) + (2|h|2) + \iint d\mathbf{r_1}d\mathbf{r_2}\frac{\rho_1(\mathbf{r_1})\rho_2(\mathbf{r_2})}{r_{12}} \\ & - \iint d\mathbf{r_1}d\mathbf{r_2}\frac{\psi_1^*(\mathbf{r_1})\psi_2(\mathbf{r_1})\psi_2^*(\mathbf{r_2})\psi_1(\mathbf{r_2})}{r_{12}}. \end{aligned} \tag{6}$$

Evidently, Equations (5) and (6) differ from each other by the exchange contribution. It is important to remark that for a two-electron system there is no spin contamination. Thus, neither closed-shell nor open-shell presents this problem.

Within the KS model, the energy of the system has the form [20]

$$E_{\mathrm{KS}} = \langle 1|h|1\rangle + \langle 2|h|2\rangle + J[\rho] + E_{\mathrm{XC}}[\rho_\alpha, \rho_\beta], \tag{7}$$

where

$$J[\rho] = \frac{1}{2}\iint d\mathbf{r}_1 d\mathbf{r}_2\frac{\rho(\mathbf{r}_1)\rho(\mathbf{r}_2)}{r_{12}}, \tag{8}$$

and E_{XC} represents the exchange-correlation functional. In principle this is an exact approach if E_{XC} is known exactly, however, it is unknown and for that reason there are several approximations to this functional. In this work, we consider just the exchange contribution within the LDA

$$E_x^{\mathrm{LDA}}[\rho_\alpha, \rho_\beta] = -C_x\sum_\sigma \int d\mathbf{r}\rho_\sigma^{4/3}(\mathbf{r}), \tag{9}$$

where $\sigma = \alpha, \beta$ and $C_x = \frac{3}{2}(\frac{3}{4\pi})^{1/3}$. For this approximation, we apply the SIC of Perdew and Zunger. In such an approach, the exchange energy for the two-electron system has the expression [21],

$$E_x^{\mathrm{SIC}}[\rho_\alpha, \rho_\beta] = E_x^{\mathrm{approx}}[\rho_\alpha, \rho_\beta] - \sum_{a=1}^{2}(J[\rho_a] + E_x^{\mathrm{approx}}[\rho_a, 0]), \tag{10}$$

in our case approx = LDA. For a closed-shell system the total energy becomes

$$E_{\mathrm{KS}}^{\mathrm{SIC}} = 2(1|h|1) + J[\rho] + E_x^{\mathrm{approx}}[\rho_1, \rho_1] - 2J[\rho_1] - 2E_x^{\mathrm{approx}}[\rho_1, 0]. \tag{11}$$

In this case, $\rho = 2\rho_1$ and for the exchange functional used in this work, $E_x^{\mathrm{approx}}[\rho_1, \rho_1] = 2E_x^{\mathrm{approx}}[\rho_1, 0]$, therefore

$$E_{\mathrm{KS}}^{\mathrm{SIC}} = 2(1|h|1) + \iint \mathrm{d}\mathbf{r}_1 \mathrm{d}\mathbf{r}_2 \frac{\rho_1(\mathbf{r}_1)\rho_1(\mathbf{r}_2)}{r_{12}}. \tag{12}$$

For an open-shell system, the Coulomb term is not effectively reduced, so the total energy is given by

$$\begin{aligned} E_{\mathrm{KS}}^{\mathrm{SIC}} &= (1|h|1) + (2|h|2) + \iint \mathrm{d}\mathbf{r}_1 \mathrm{d}\mathbf{r}_2 \frac{\rho_1(\mathbf{r}_1)\rho_2(\mathbf{r}_2)}{r_{12}} \\ &\quad + E_x^{\mathrm{approx}}[\rho_1 + \rho_2, 0] - E_x^{\mathrm{approx}}[\rho_1, 0] - E_x^{\mathrm{approx}}[\rho_2, 0]. \end{aligned} \tag{13}$$

Comparing Equation (5) with (12), and (6) with (13), it is clear that while the SIC functional due to Perdew and Zunger [21] gives the HF energy for a two-electron closed-shell system, for an open-shell system this is no the case. We see that the difference found in the open-shell system, between E_{HF} and $E_{\mathrm{KS}}^{\mathrm{SIC}}$, is related to the exchange contribution. Assuming that the SIC functional removes the self-interaction term in the KS equations, it follows that

$$\begin{aligned} -\iint \mathrm{d}\mathbf{r}_1 \mathrm{d}\mathbf{r}_2 \frac{\psi_1^*(\mathbf{r}_1)\psi_2(\mathbf{r}_1)\psi_2^*(\mathbf{r}_2)\psi_1(\mathbf{r}_2)}{r_{12}} &= E_x^{\mathrm{approx}}[\rho_1 + \rho_2, 0] + E_x^{\mathrm{approx}}[\rho_1, 0] \\ &\quad + E_x^{\mathrm{approx}}[\rho_2, 0]. \end{aligned} \tag{14}$$

At this point a question arises, how must be E_x^{approx} be chosen to satisfy Equation (14)? Interestingly, Equation (14) may be used as a starting point to design new exchange functionals.

2.2. Exchange potential

The HF equations for the considered systems are, for the closed-shell system [29]

$$\left(h(\mathbf{r}_1) + \int d\mathbf{r}_2 \frac{\rho_1(\mathbf{r}_2)}{r_{12}} \right) \psi_1(\mathbf{r}_1) = \varepsilon_1 \psi_1(\mathbf{r}_1), \tag{15}$$

and for the open-shell system we have two equations, for the first electron

$$h(\mathbf{r}_1)\psi_1^\alpha(\mathbf{r}_1) + \int d\mathbf{r}_2 \frac{\psi_2^\alpha(\mathbf{r}_2)\psi_2^\alpha(\mathbf{r}_2)}{r_{12}} \psi_1^\alpha(\mathbf{r}_1) - \int d\mathbf{r}_2 \frac{\psi_2^\alpha(\mathbf{r}_2)\psi_1^\alpha(\mathbf{r}_2)}{r_{12}} \psi_2^\alpha(\mathbf{r}_1) = \varepsilon_1^\alpha \psi_1^\alpha(\mathbf{r}_1) \tag{16}$$

and for the second electron

$$h(\mathbf{r}_1)\psi_2^\alpha(\mathbf{r}_1) + \int d\mathbf{r}_2 \frac{\psi_1^\alpha(\mathbf{r}_2)\psi_1^\alpha(\mathbf{r}_2)}{r_{12}} \psi_2^\alpha(\mathbf{r}_1) - \int d\mathbf{r}_2 \frac{\psi_1^\alpha(\mathbf{r}_2)\psi_2^\alpha(\mathbf{r}_2)}{r_{12}} \psi_1^\alpha(\mathbf{r}_1) = \varepsilon_2^\alpha \psi_2^\alpha(\mathbf{r}_1). \tag{17}$$

For the closed-shell system, the KS equations within the OEP-KLI approach are [27,28]

$$\left(h(\mathbf{r}_1) + \int d\mathbf{r}_2 \frac{\rho_{1s}(\mathbf{r}_2)}{r_{12}} \right) \psi_{1s}(\mathbf{r}_1) = \varepsilon_{1s} \psi_{1s}(\mathbf{r}_1), \tag{18}$$

and for the open-shell system

$$\left(h(\mathbf{r}_1) + \int d\mathbf{r}_2 \frac{\rho(\mathbf{r}_2)}{r_{12}} + V_{x\alpha}^{\mathrm{OEP-KLI}}(\mathbf{r}_1) \right) \psi_i^\alpha(\mathbf{r}_1) = \varepsilon_i^\alpha \psi_i^\alpha(\mathbf{r}_1), \quad i = 1\mathrm{s}, 2\mathrm{s}. \tag{19}$$

The OEP-KLI exchange potential is obtained from

$$V_{x\alpha}^{\mathrm{OEP-KLI}}(\mathbf{r}) = \upsilon_{x\alpha}^{\mathrm{approx}}(\mathbf{r}) + \frac{\rho_{1\alpha}(\mathbf{r}) u_x^{1\alpha}(\mathbf{r})}{\rho_\alpha(\mathbf{r})} + \frac{\rho_{2\alpha}(\mathbf{r}) u_x^{2\alpha}(\mathbf{r})}{\rho_\alpha(\mathbf{r})} + \frac{\rho_{1\alpha}(\mathbf{r}) C_1^\alpha}{\rho_\sigma(\mathbf{r})}, \tag{20}$$

with

$$\upsilon_x^{\mathrm{approx}}(\mathbf{r}) = \delta E_x^{\mathrm{approx}} / \delta\rho(\mathbf{r}), \tag{21}$$

$$u_x^{i\alpha}(\mathbf{r}_1) = -\int d\mathbf{r}_2 \frac{\rho_{i\alpha}(\mathbf{r}_2)}{r_{12}} - \frac{\delta E_x^{\mathrm{LDA}}[\rho_{i\sigma}, 0]}{\delta\rho_{i\alpha}(\mathbf{r}_1)}, \quad i = 1\mathrm{s}, 2\mathrm{s}, \tag{22}$$

$$C_1^\alpha = \frac{\bar{V}^\alpha_{x,1s} - \bar{u}_x^{1s\alpha}}{1 - M_{11}^\alpha}, \tag{23}$$

$$\bar{V}^\alpha_{x,1s} = \int d\mathbf{r}\rho_{1\alpha}(\mathbf{r})\left(\frac{\rho_{1\alpha}(\mathbf{r})u_x^{1\alpha}(\mathbf{r})}{\rho_\alpha(\mathbf{r})} + \frac{\rho_{2\alpha}(\mathbf{r})u_x^{2\alpha}(\mathbf{r})}{\rho_\alpha(\mathbf{r})}\right), \tag{24}$$

$$\bar{u}_x^{1s\alpha} = \int d\mathbf{r}\rho_{1\alpha}(\mathbf{r})u_x^{1s\alpha}(\mathbf{r}), \tag{25}$$

$$M_{11}^\alpha = \int d\mathbf{r}\frac{\rho_{1\alpha}(\mathbf{r})\rho_{2\alpha}(\mathbf{r})}{\rho_\alpha(\mathbf{r})}. \tag{26}$$

Comparing the HF and KS–SIC–OEP–KLI equations, we see that for a two-electron closed-shell system the equations are the same; this is not true for the open-shell system.

By checking the results from Table II of Ref. [19], we see that for the closed-shell system there are discrepancies between the HF and KS-SIC-OEP-KLI methods. Such differences can be attributed to the basis set used for the HF method, or to the methodology used in the KS approach. The aim of this work is twofold. First, we recomputed the Table II from Ref. [19] by using an optimized basis set for the HF calculation. Second, we compute the lowest triplet state of the confined helium atom by the HF method and its results are compared with those obtained by the KS approach. In this way a reliable comparison between HF and KS is made for the lowest singlet and triplet states of the confined helium atom.

3. METHODOLOGY

We solve the HF equations by using the same approach as that used by Ludeña [4]. In this approach, the HF orbitals are expanded on Modified Slater Type Orbitals (MSTO). We wrote a code from scratch, in Fortran 77, to solve the HF equations for free and confined atoms. We did this so that for each confinement radius the MSTO exponents were optimized. For the singlet state we use five MSTO's, three 1s and two 2s, and for the triplet state the basis set was built with seven MSTO's, three 1s, two 2s, one 3s and one 4s. For the solution of the KS equations, we use a numerical code that has been proved in several applications. Details concerning this code and its applications can be found in Refs. [6,9–12].

4. RESULTS AND DISCUSSION

4.1. Closed-shell confined helium atom

In Table 1, we report the total energies obtained with the HF and KS methods for the lowest singlet state of the confined helium atom, for several

Table 1 Total energy (TE) and exchange energy (XE), for the confined helium atom in its lowest singlet state. All quantities are in hartrees

R_c	LDA		LDA–SIC		HF		
	TE	XE	TE	XE	TE	XE	TE[a]
1.0	1.35361	−1.73038	1.06120	−2.03175	1.06120	−2.03175	1.06122
1.2	−0.41119	−1.48907	−0.66463	−1.75235	−0.66462	−1.75235	−0.66461
1.4	−1.34814	−1.32261	−1.57418	−1.55950	−1.57417	−1.55950	−1.57417
1.5	−1.64898	−1.25824	−1.86422	−1.48488	−1.86422	−1.48488	−1.86422
1.6	−1.87831	−1.20334	−2.08422	−1.42125	−2.08423	−1.42124	−2.08422
1.8	−2.19199	−1.11576	−2.38268	−1.31973	−2.38268	−1.31973	−2.38267
2.0	−2.38363	−1.05047	−2.56258	−1.24412	−2.56258	−1.24412	−2.56253
2.4	−2.57980	−0.96425	−2.74236	−1.14477	−2.74237	−1.14477	−2.74221
2.5	−2.60706	−0.94924	−2.76661	−1.12763	−2.76661	−1.12763	−2.76644
3.0	−2.68210	−0.89898	−2.83105	−1.07119	−2.83105	−1.07118	−2.83083
4.0	−2.71813	−0.86257	−2.85859	−1.03342	−2.85859	−1.03342	−2.85852
5.0	−2.72290	−0.85465	−2.86139	−1.02682	−2.86139	−1.02682	−2.86134
6.0	−2.72354	−0.85311	−2.86165	−1.02589	−2.86165	−1.02589	−2.86151

[a] Ref. [4].

confinement radii. Also, in the same table, we include the exchange energies obtained with the three methods considered in this work. An additional column is included in Table 1, to compare with previous results reported for the HF method [4]. Comparing our HF results with those obtained previously, we found small differences between them, although there are confinement radii where our HF results predict deeper energies than those obtained without exponent optimization. The dependence of the exponents on the confinement radius is depicted in Figure 1. We see from this figure that the confinement radius has a strong impact on the exponents. Thus, exponent optimization is required for each confinement radius.

If we accept the wave function expanded with 40 Hylleraas functions, reported in Ref. [19], as the reference to describe the ground state of the confined helium atom, and our HF wave function as the reference for the exchange contribution, we can compare these numbers to obtain the correlation energy involved in this state of the confined helium atom. In Table 2 we present the correlation energy estimated by the difference $E_{\text{H-WF}}-E_{\text{HF}}$, where $E_{\text{H-WF}}$ denotes the total energy obtained with a wave function expanded by 40 Hylleraas functions and E_{HF} denotes the total energy obtained by the HF method with MSTO's and optimized exponents. In the same table we include the correlation energy reported by other methods. The small variations shown by the correlation energy, when the confinement radius is changed, are impressive, confirming such an observation from early reports [30]. Results from previous predictions differ appreciably from our approach, since in some cases the correlation energy

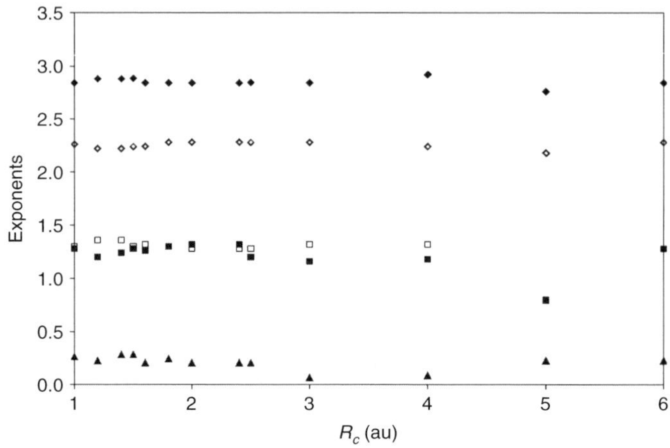

Figure 1 Exponents of the Modified Slater Type Orbitals as a function of the confinement radius, R_c. The solid points correspond to the 1s orbitals and the open points to the 2s orbitals.

Table 2 Correlation energy (CE) estimated as the difference between the wave function expanded with 40 Hylleraas functions (H-WF) and the HF wave function obtained with optimized exponents, for the lowest singlet state of confined helium atom. All quantities are in hartrees

R_c	H-WF[a]	HF	CE	CE[b]	CE[c]
2.0	−2.60403	−2.56258	−0.0415	−0.0401	−0.0383
2.2	−2.71453	−2.67328	−0.0413	−0.0402	−0.0402
2.4	−2.78356	−2.74237	−0.0412		
2.6	−2.82712	−2.78590	−0.0412		
2.8	−2.85480	−2.81350	−0.0413		
3.0	−2.87246	−2.83105	−0.0414		
3.5	−2.89354	−2.85187	−0.0417		
4.0	−2.90042	−2.85859	−0.0418	−0.0407	−0.0419
4.5	−2.90264	−2.86072	−0.0419		
5.0	−2.90337	−2.86139	−0.0420	−0.0408	−0.0485
5.5	−2.90361	−2.86159	−0.0420		
6.0	−2.90368	−2.86165	−0.0420	−0.0409	
6.5	−2.90370	−2.86167	−0.0420		
7.0	−2.90370	−2.86168	−0.0420	−0.0409	
8.0	−2.90371	−2.86168	−0.0420	−0.0409	
10.0	−2.90372	−2.86168	−0.0420		

[a] Ref. [19].
[b] Ref. [30].
[c] Ref. [31].

is overestimated, as it can be seen from Table 2. The approach presented in this work for predicting the correlation energy must be used carefully, since we are assuming complete basis sets in the HF and in the Hylleraas methodology. Thus, these results can be used as a benchmark for other correlated methods.

Comparing the HF and LDA–SIC total energies from Table 1, we found just 4 confinement radii where there are small differences, none more than 1×10^{-5} hartrees. Such differences may be attributed to the basis set size, for the HF method, or to the mesh, for the KS method. These differences could be removed by increasing the size of the basis set or the number of points in the mesh, but we considered that such an effort is not necessary to show that both methods predict the same energies, which is in agreement with the equations presented above.

From Table 1 we see the deficiencies of the LDA method, since it underestimates both total and exchange energies. Such discrepancies are increased for small confinement radii. As we discussed in the introduction, the main difference between LDA and LDA–SIC is the self-interaction error, thus this spurious contribution is more important when the electrons are localized within small regions, as it has been pointed out before [21].

4.2. Open-shell confined helium atom

In Table 3 we present the total energies obtained by the KS and HF methods for the lowest triplet state of the confined helium atom. If we accept our HF results as reliable values, then it is clear that the LDA–SIC values differ appreciably from these. It is worth noting that LDA–SIC overestimates the total energy, with regard to the HF method, for large confinement radii. Such an overestimate makes no sense; in Ref. [19] the total energy reported for the lowest triplet state of the confined helium atom, at $R_c = 10$ au, with the Hylleraas expansion (with 40 functions) predicts a total energy of −2.17262 hartrees, which is above of the exchange-only LDA–SIC prediction at this R_c. For small confinement radii, the behavior is reversed and an underestimation is obtained. Evidently, the LDA–SIC method fails when two electrons are unpaired. The most important result obtained from this table is the exchange energy predicted by the LDA–SIC method compared with the HF method; for any confinement radius the LDA–SIC method underestimates this quantity. In Table 4 we present the energy components obtained by HF and LDA–SIC methods, for the lowest triplet state of the confined helium atom. From this table it is clear that the LDA–SIC method underestimates the exchange energy and, for many confinement radii, each component is also underestimated with regard to the HF method, although this underestimation is different for each one. In Table 5 we report the difference, LDA–SIC minus HF, for each energy component. From this table it is clear that the smallest differences are found in the one-electron contribution (core contribution). The two-electron contributions present the largest differences. In particular, for $R_c = 1.2$ au these contributions differ appreciably from the HF results, although for

Table 3 Total energy (TE) and exchange energy (XE), for the confined helium atom in its lowest triplet state. All quantities are in hartrees.

R_C	LDA		LDA–SIC		HF	
	TE	XE	TE	XE	TE	XE
1.2	8.56702	−1.81683	8.41587	−1.96959	8.40193	−2.17924
1.4	5.13103	−1.58778	5.00063	−1.71953	4.98113	−1.74755
1.6	3.00804	−1.41817	2.89276	−1.53450	2.87651	−1.55950
1.8	1.62429	−1.28824	1.52017	−1.39312	1.49650	−1.42802
2.0	0.68450	−1.18615	0.58854	−1.28256	0.56647	−1.31526
2.2	0.02505	−1.10432	−0.06510	−1.19466	−0.08845	−1.22950
2.4	−0.45012	−1.03769	−0.53637	−1.12392	−0.56008	−1.15993
2.6	−0.80012	−0.98275	−0.88402	−1.06650	−0.90729	−1.10280
2.8	−1.06276	−0.93697	−1.14555	−1.01956	−1.16774	−1.05540
3.0	−1.26301	−0.89848	−1.34563	−0.98096	−1.36626	−1.01574
3.5	−1.58957	−0.82566	−1.67429	−0.91084	−1.68995	−0.94130
4.0	−1.77283	−0.77569	−1.86091	−0.86535	−1.87167	−0.89078
4.5	−1.88130	−0.74026	−1.97246	−0.83437	−1.97936	−0.85531
5.0	−1.94822	−0.71443	−2.04173	−0.81226	−2.04590	−0.82960
5.5	−1.99086	−0.69514	−2.08601	−0.79587	−2.08837	−0.81046
6.0	−2.01871	−0.68042	−2.11499	−0.78335	−2.11615	−0.79587
6.5	−2.03724	−0.66902	−2.13430	−0.77358	−2.13467	−0.78456
7.0	−2.04976	−0.66008	−2.14733	−0.76586	−2.14718	−0.77566
8.0	−2.06419	−0.64741	−2.16231	−0.75475	−2.16156	−0.76300
10.0	−2.07409	−0.63450	−2.17274	−0.74315	−2.17156	−0.75005

reasonable confinements the Coulomb interaction does not exhibit large changes when the confinement radius is changed. Thus, the exchange energy predicted by the LDA–SIC method shows important deviations from the HF results when the confinement radius is reduced.

Evidently the LDA differs from the HF results for any radii. However, according to the Table 3 this method always underestimates the total and exchange energies, respectively. In a sense, this is a good result since we can design a correlation functional to include this deficiency. Such an approach cannot be used with the LDA–SIC method because, depending on the confinement radius, the total energy is overestimated or underestimated.

As in the closed-shell case, we estimate the correlation energy involved in the lowest triplet state of the confined helium atom. In Table 6 we report the total energy obtained from Table IV of Ref. [19], also the HF results from this work, and results obtained from correlated methods. We see that the results obtained with the HF method, coupled with exponent optimization, compete with the other correlated methods, since for some confinement radii our results are below those reported by the other authors. Comparing our Hartree-Fock results with those obtained with 40-terms Hylleraas wave

Table 4 Exchange (Exch), Coulomb (Coul) and core (Core) energies, for the lowest triplet state of the confined helium atom. All quantities are in hartrees.

R_C	HF			LDA–SIC		
	Exch	Coul	Core	Exch	Coul	Core
1.2	−2.17924	3.46950	7.11167	−1.96959	3.27889	7.10656
1.4	−1.74755	2.86093	3.86775	−1.71953	2.85835	3.86180
1.6	−1.55950	2.55570	1.88031	−1.53450	2.54573	1.88153
1.8	−1.42802	2.29936	0.62515	−1.39312	2.30514	0.60815
2.0	−1.31526	2.12730	−0.24556	−1.28256	2.11504	−0.24394
2.2	−1.22950	1.97505	−0.83399	−1.19466	1.96171	−0.83215
2.4	−1.15993	1.85035	−1.25049	−1.12392	1.83600	−1.24845
2.6	−1.10280	1.74684	−1.55133	−1.06650	1.73156	−1.54908
2.8	−1.05540	1.65992	−1.77226	−1.01956	1.64382	−1.76982
3.0	−1.01574	1.58622	−1.93674	−0.98096	1.56945	−1.93412
3.5	−0.94130	1.44448	−2.19313	−0.91084	1.42674	−2.19019
4.0	−0.89078	1.34439	−2.32528	−0.86535	1.32669	−2.32224
4.5	−0.85531	1.27116	−2.39521	−0.83437	1.25418	−2.39228
5.0	−0.82960	1.21594	−2.43225	−0.81226	1.20011	−2.42957
5.5	−0.81046	1.17334	−2.45125	−0.79587	1.15876	−2.44891
6.0	−0.79587	1.13987	−2.46016	−0.78335	1.12650	−2.45815
6.5	−0.78456	1.11327	−2.46338	−0.77358	1.10096	−2.46167
7.0	−0.77566	1.09192	−2.46344	−0.76586	1.08054	−2.46201
8.0	−0.76300	1.06094	−2.45950	−0.75475	1.05093	−2.45849
10.0	−0.75005	1.02867	−2.45018	−0.74315	1.01998	−2.44957

function, it is clear that the correlation energy is quite small. Even when the correlation energy is increased for small confinement radii this quantity is still small compared to the total energy. Thus, if we apply the HF method with an adequate basis set then we can give a good description of the lowest triplet state of the confined helium atom.

4.3. Electron density differences between Hartree-Fock, LDA and LDA–SIC

We have discussed the differences observed between HF, LDA and LDA–SIC methods when the total energy is considered. In this section, we present the differences obtained on local quantities, such as the orbital densities and the total electron density for the lowest triplet state. We do not consider the lowest singlet state of the confined helium atom in this analysis, because the differences found above were related to the triplet state. In Table 7 we report the difference

$$\Delta_i = \int \mathbf{dr} |\rho_i^{\mathrm{HF}}(\mathbf{r}) - \rho_i^{\mathrm{KS}}(\mathbf{r})|, \tag{27}$$

Table 5 Differences of the total energy (TE), exchange energy (Exch), Coulomb energy (Coul) and core energy (Core) obtained between LDA–SIC and HF methods. All quantities are in hartrees.

R_C	TE	Exch	Coul	Core
1.2	0.01394	0.20965	−0.19061	−0.00511
1.4	0.01950	0.02802	−0.00258	−0.00595
1.6	0.01625	0.02500	−0.00997	0.00122
1.8	0.02370	0.03348	0.00212	−0.01190
2.0	0.02207	0.03270	−0.01226	0.00163
2.2	0.02335	0.03485	−0.01333	0.00184
2.4	0.02371	0.03601	−0.01435	0.00205
2.6	0.02327	0.03630	−0.01528	0.00225
2.8	0.02219	0.03584	−0.01610	0.00244
3.0	0.02062	0.03478	−0.01677	0.00262
3.5	0.01566	0.03046	−0.01774	0.00294
4.0	0.01076	0.02543	−0.01771	0.00304
4.5	0.00689	0.02094	−0.01698	0.00293
5.0	0.00417	0.01734	−0.01584	0.00267
5.5	0.00236	0.01459	−0.01458	0.00235
6.0	0.00116	0.01252	−0.01337	0.00201
6.5	0.00037	0.01097	−0.01231	0.00171
7.0	−0.00015	0.00980	−0.01138	0.00144
8.0	−0.00075	0.00825	−0.01001	0.00101
10.0	−0.00119	0.00689	−0.00870	0.00061

where i represents the orbital 1s, 2s or the total electron density. In this way we have a measure of the distance between the KS (exchange only) and HF densities.

The discrepancies observed for the LDA method with regard to the HF method, for the total energy, are also seen with the orbital densities. We see from Table 7 that the LDA method gives the orbital densities with the biggest distances from the HF orbital densities. The LDA–SIC method gives small distances, particularly for large confinement radii. However, such distances are not zero. In particular, the 1s orbital gives bigger distances than the 2s orbital. Thus, the exchange-only LDA and LDA–SIC methods give orbital densities with an appreciable distance from those orbitals obtained by the HF method. We note here that, for small confinement radii, the LDA or LDA–SIC total density gives a smaller distance, with regard to the HF total density, than that obtained with each orbital density. This result suggests that whereas one LDA or LDA–SIC orbital density is overestimated, with respect to the HF method, the other one is underestimated, such that the total density is compensated and it is close to that obtained from the HF method.

Table 6 Total energy (TE) and correlation energy (CE) estimated as the difference between the wave function expanded in 40 Hylleraas terms (H-WF) and the HF wave function obtained with optimized exponents, for the lowest triplet state of the confined helium atom. All quantities are in hartrees

R_c	TE_{HF}	$TE_{H\text{-}WF}$[a]	TE[b]	TE[c]	CE
2.0	0.56647	0.56026	0.9809	0.5862	−0.00621
2.2	−0.08845	−0.09421			−0.00576
2.4	−0.56008	−0.56542			−0.00534
2.6	−0.90729	−0.91225			−0.00496
2.8	−1.16774	−1.17233			−0.00459
3.0	−1.36626	−1.37051	−1.1193	−1.3679	−0.00425
3.5	−1.68995	−1.69347			−0.00352
4.0	−1.87167	−1.87461	−1.7277	−1.8734	−0.00294
4.5	−1.97936	−1.98185			−0.00249
5.0	−2.04590	−2.04804	−1.9615	−2.0473	−0.00214
5.5	−2.08837	−2.09024			−0.00187
6.0	−2.11615	−2.11782	−2.0658	−2.1171	−0.00167
6.5	−2.13467	−2.13617			−0.00150
7.0	−2.14718	−2.14856	−2.1166	−2.1477	−0.00138
8.0	−2.16156	−2.16278	−2.1429	−2.1617	−0.00122
10.0	−2.17156	−2.17262	−2.1647	−2.1714	−0.00106

[a] Ref. [19].
[b] Ref. [32].
[c] Ref. [33].

Table 7 Hartree-Fock and Kohn-Sham density differences for the lowest triplet state of the confined helium atom. The differences were obtained by $\Delta_i = \int d\mathbf{r}|\rho_i^{HF}(\mathbf{r}) - \rho_i^{KS}(\mathbf{r})|$, with i = 1s, 2s and total, for several confinement radii. All quantities are in hartrees

R_c	LDA–SIC			LDA		
	Δ_{1s}	Δ_{2s}	Δ_{tot}	Δ_{1s}	Δ_{2s}	Δ_{tot}
2.0	0.1019	0.0893	0.0212	0.1146	0.1001	0.0210
3.0	0.1143	0.0906	0.0313	0.1233	0.0981	0.0316
4.0	0.0846	0.0629	0.0343	0.0973	0.0689	0.0394
6.0	0.0433	0.0315	0.0302	0.0668	0.0408	0.0487
8.0	0.0305	0.0201	0.0258	0.0584	0.0333	0.0530
10.0	0.0264	0.0172	0.0246	0.0557	0.0324	0.0559

5. CONCLUSIONS

In this work we show the importance of exponent optimization on the Roothaan approach to the solution of the Hartree-Fock equations for the lowest singlet and triplet state of the confined helium atom. By using this procedure, we found for the lowest triplet state the Hartree-Fock method as an appropriate method to describe it. By contrasting the exchange-only

SIC-OEP-KLI approach with the Hartree-Fock method, we found the SIC-OEP-KLI method as inadequately describing the unpaired pair of electrons, since this method overestimates the total energy of the system for large confinement radii, while it underestimates the one-electron and two-electron contributions to the energy. For small confinement radii, the SIC-OEP-KLI approach underestimates the total energy; in particular for very small confinement radii the two-electron contributions show large discrepancies from those results obtained with the Hartree-Fock method.

ACKNOWLEDGMENTS

This work was partially financed by CONACYT, from the projects 60614 (J. G.) and 60116 (R. V.).

REFERENCES

[1] A. Michels, J. de Boer, A. Bijl, Physica 4 (1937) 981.
[2] S.R. de Groot, C.A. Ten Seldam, Physica 12 (1946) 669.
[3] A. Sommerfeld, H. Welker, Ann. Phys. 32 (1938) 56.
[4] E. Ludeña, J. Chem. Phys. 69 (1978) 1770.
[5] J.-P. Connerade, V.K. Dolmatov, J. Phys. B 31 (1998) 3557.
[6] J. Garza, R. Vargas, A. Vela, Phys. Rev. E. 58 (1998) 3949.
[7] J.-P. Connerade, V.K. Dolmatov, P.A. Lakshmi, J. Phys. B 33 (2000) 251.
[8] V.K. Dolmatov, A.S. Baltenkov, J.-P. Connerade, S.T. Manson, Radiat. Phys. Chem. 70 (2004) 417.
[9] J. Garza, R. Vargas, A. Vela, K.D. Sen, J. Mol. Struct. (THEOCHEM) 501 (2000) 183.
[10] K.D. Sen, J. Garza, R. Vargas, A. Vela, Chem. Phys. Lett. 325 (2000) 29.
[11] K.D. Sen, J. Garza, R. Vargas, A. Vela, Proc. Indian Nat. Sci. Acad. 70A (2004) 675.
[12] J. Garza, R. Vargas, N. Aquino, K.D. Sen, J. Chem. Sci. 117 (2005) 379.
[13] P.M. Laufer, J.B. Krieger, Phys. Rev. A 33 (1986) 1480.
[14] C. Filippi, C.J. Umrigar, M. Taut, J. Chem. Phys. 100 (1994) 1290.
[15] S. Kais, D.R. Herschbach, N.C. Handy, C.W. Murray, G.J. Laming, J. Chem. Phys. 99 (1993) 417.
[16] J. Jung, J.E. Alvarellos, J. Chem. Phys. 118 (2003) 10825.
[17] J. Jung, P. García-González, J.E. Alvarellos, R.W. Godby, Phys. Rev. A 69 (2004) 052501.
[18] R.G. Parr, W.Y. Yang, Density-Functional Theory of Atoms and Molecules, Oxford University Press, New York, 1989.
[19] N. Aquino, J. Garza, A. Flores-Riveros, J.F. Rivas-Silva, K.D. Sen, J. Chem. Phys. 124 (2006) 054311.
[20] W. Kohn, L.J. Sham, Phys. Rev. 140 (1965) A1133.
[21] J.P. Perdew, A. Zunger, Phys. Rev. B 23 (1981) 5048.
[22] R.T. Sharp, G.K. Horton, Phys. Rev. 90 (1953) 317.
[23] J.D. Talman, W.F. Shadwick, Phys. Rev. A 14 (1976) 36.
[24] J.B. Krieger, Y. Li, G.J. Iafrate, Phys. Rev. A 45 (1992) 101.
[25] J.B. Krieger, Y. Li, G.J. Iafrate, Phys. Rev. A 46 (1992) 5453.
[26] J.B. Krieger, Y. Li, G.J. Iafrate, Phys. Rev. A 47 (1993) 165.
[27] J. Garza, J.A. Nichols, D.A. Dixon, J. Chem. Phys. 112 (2000) 1150.
[28] J. Garza, J.A. Nichols, D.A. Dixon, J. Chem. Phys. 112 (2000) 7880.
[29] A. Szabo, N.S. Ostlund, Modern Quantum Chemistry, 1st ed. revised, McGraw-Hill, New York, 1989.
[30] E.V. Ludeña, M. Gregori, J. Chem. Phys. 71 (1979) 2235.
[31] B.M. Gimarc, J. Chem. Phys. 47 (1967) 5110.
[32] S.H. Patil, J.P. Varshney, Can. J. Phys. 82 (2004) 647.
[33] A. Banerjee, C. Kamal, A. Chowdhury, Phys. Lett. A 350 (2006) 121.

CHAPTER 8

Thomas–Fermi–Dirac–Weizsäcker Density Functional Formalism Applied to the Study of Many-electron Atom Confinement by Open and Closed Boundaries

Salvador A. Cruz[a]

Contents

1. INTRODUCTION

Spatial limitation effects on the properties of a quantum system have motivated a rapidly increasing research interest due to the remarkable differences observed in the physical and chemical properties between a confined system and its free counterpart. The problem is not only of academic importance but has an underlying technological relevance for the development of new materials with unconventional properties [1–3].

[a] Departamento de Física, Universidad Autónoma Metropolitana - Iztapalapa, Apdo. Postal 55 534, 09340, México, D.F., México

E-mail address: cruz@xanum.uam.mx.

Advances in Quantum Chemistry, Vol. 57
ISSN: 0065-3276, DOI: 10.1016/S0065-3276(09)00612-1

The wealth of problems related to the study of the properties of quantum confined systems is not new, as stated by Ludeña in one of his pioneering papers, dealing with Hartree–Fock (HF) calculations of the ground state wavefunctions of compressed atoms [4]. However, further ongoing developments in different materials research areas, such as nanotube and fullerene synthesis acting as atom/molecule traps [5–10], as well as zeolitic channels [11,12] and semiconductor quantum dots [1,3, 13,14], have brought a burst of interest in this challenging topic. In this connection, let me briefly mention that different models and strategies have been proposed to analyze confined quantum systems over the years. The interested reader is directed to the excellent review papers by Fröman et al. [15], Jaskolski [16] and Dolmatov et al. [7].

Among the different models used to study the properties of confined atoms and molecules, the boxed-in model, whereby an atom/molecule is caged within a box of varying geometry with either Dirichlet or Neumann boundary conditions, has proven to be adequate. This class of model also includes the case of open boundaries represented by geometric planes of different shapes. For hydrogenic systems the majority of these studies have dealt with infinitely hard closed and open boundaries with different geometries [17–32]. The advantage of this class of models – in the hydrogenic case – is that exact solutions to the Schrödinger wave equation may be found [17–21,24,25,27,33–36], which have served as important reference calculations to calibrate simpler, approximate, treatments [16,22,26,29–32,37–41]. Also, within this approach, two-electron atoms and molecules have been studied using several approximate techniques for the case of closed [42–55] and open [56–59] boundaries with different symmetries.

The case of boxed-in many-electron atoms – other than helium – has been addressed mainly for spherical cavities with hard and soft walls using either ab-initio HF procedures [4,60], Dirac–Fock [61], density functional Kohn–Sham theory [62,63] and the Thomas–Fermi–Dirac–Weizsäcker (TFDλW) statistical atomic model [64,65]. The latter method being the first one used to systematically analyze the effect of barrier height on the many-electron atom ground-state energy evolution for varying cavity size. Interestingly, the study of a many-electron atom confined by a penetrable spherical box allows one to predict the conditions for electron escape from the confining region [65], leading the system to the true ionized confined condition in contrast with the case of infinitely hard confinement.

In this contribution, the adequacy of the TFDλW method for the study of many-electron atom confinement within different confinement conditions – by closed and open boundaries – is explored. We begin by making a brief review of the main strategy followed in the TFDλW method to account for the study of many-electron atoms confined by hard and soft spherical boxes. Here, important quantitative corrections to our previous studies are introduced, leading to better agreement with other reference calculations, and new results for other systems are presented. The method is then applied – for the first time – to many-electron atoms confined by hard prolate spheroidal boxes and for the case of an atom confined by an infinitely hard plane.

The novel nature of the results of this work makes it difficult to compare with equivalent ones in the literature, hence detailed numerical values are presented for relevant sample systems for future reference. Furthermore, may the author advance here that the main emphasis of this contribution is to provide new quantitative information on the predictions of the TFDλW method for the different confinement geometries treated. This may also be useful reference material for equivalent studies using other methods.

The paper is organized as follows. In Section 2, the main aspects of the TFDλW method applied to spherical – hard and soft – cages are presented followed by a discussion on the origin of the quantitative corrections to previous work. Also, in this Section, the treatment of hard-wall spheroidal confinement is presented and applied to the case of C and Ne. Section 3 deals with confinement by an open boundary – a plane in this case – showing sample calculations for C and Ne. Finally, in Section 4 the conclusions of this work are presented and future directions are envisaged. Atomic units are used throughout unless otherwise stated.

The author wishes here to devote this paper to the memory of Professor Jose Luis Marín Flores, who passed away on November 16, 2007. J.L. Marin was an enthusiastic friend, disciple, coauthor and researcher in the topic of confined quantum systems. His contributions in this field – particularly in the use of the variational method – have been a cornerstone for the development of the present work.

2. MANY-ELECTRON ATOM CONFINEMENT BY CLOSED BOUNDARIES

2.1. Hard spherical box

In this Section we begin by briefly reviewing some fundamental aspects concerning the application of the TFDλW method to many-electron atoms enclosed by an impenetrable spherical cavity. The interested reader is directed to a previous work [64] for a more detailed analysis. Here we shall present some new complementary results for further assessment of the model.

The TFDλW functional for an atom enclosed within a spherical cavity of radius R is constructed as:

$$E[\rho, R] = T[\rho, R] + V_{\mathrm{en}}[\rho, R] + V_{\mathrm{ee}}[\rho, R], \tag{1}$$

where ρ is the total electron density and $T[\rho, R]$ is the kinetic energy, given here as the well known first two terms of its gradient expansion [66–69]

$$T[\rho, R] = T_0[\rho, R] + \lambda T_W[\rho, R] \tag{2}$$

with

$$T_0[\rho, R] = \frac{3}{10}(3\pi^2)^{2/3} \int_\Gamma \rho^{5/3} \mathrm{d}\tau \tag{3}$$

the free-electron gas kinetic energy, and

$$T_W[\rho, R] = \frac{1}{8} \int_\Gamma \frac{\nabla\rho \cdot \nabla\rho}{\rho} \mathrm{d}\tau \tag{4}$$

the inhomogeneity correction introduced by Weizsäcker [70]. Here, Γ indicates the domain of integration within the cavity volume. The factor λ before the Weizsäcker term in Equation (2) has been derived formally as $\lambda = 1/9$ after the analysis of the second order correction to the gradient expansion of the Thomas–Fermi kinetic energy [71]. In spite of this, other values for λ have been used in an ad hoc manner – hence the acronym TFDλW for the method – so that close correspondence with HF energies is obtained [64,65,69]. This choice – although empirical – adds some flexibility for the set up of the correct free-atom HF energy reference value when the cavity radius is large enough. Accordingly, we shall adopt this criterion in this work.

The electron–nuclear attraction energy $V_{\mathrm{en}}[\rho, R]$ and electron–electron repulsion energy $V_{\mathrm{ee}}[\rho, R]$ in Equation (1) are given, respectively, as

$$V_{\mathrm{en}}[\rho, R] = -Z \int_\Gamma \rho r^{-1} \mathrm{d}\tau, \tag{5}$$

with Z the nuclear charge and

$$V_{\mathrm{ee}}[\rho, R] = \frac{1}{2} \int_\Gamma \rho\varphi \mathrm{d}\tau + X[\rho, R], \tag{6}$$

where the first term is the classical Coulomb repulsion, with φ the electrostatic potential due to charge density ρ and the second term is the Dirac exchange-only energy given as:

$$X[\rho, R] = -\frac{3}{4}(3/\pi)^{1/3} \int_\Gamma \rho^{4/3} \mathrm{d}\tau. \tag{7}$$

Following the same reasoning as in Ref. [64], the energy functional given by Equation (1) may be constructed through an appropriate ansatz variational representation for the electron density defined as:

$$\rho(r) = \sum_{i=1}^{\mathrm{occ}} \rho_i(r), \tag{8}$$

where $\rho_i(r)$ is the ansatz electron density of orbital "i" derived from angularly averaged Slater-type nodeless functions, i.e.:

$$\rho_i(r) = |\varphi_i(r)|^2, \tag{9}$$

where

$$\phi_i(r) = N_i r^{n_i - 1} \mathrm{e}^{-\zeta_i r}(1 - r/R)^{\nu} \tag{10}$$

with n_i the principal quantum number, ς_i an associated variational parameter and N_i a normalizing factor such that for an orbital population ω_i:

$$\int_{\Gamma} \rho_i(r)\mathrm{d}\tau = \omega_i. \tag{11}$$

The cut-off factor $(1 - r/R)^{\nu}$ in Equation (10) is chosen so that the Dirichlet boundary condition is satisfied for each orbital (and density), i.e. $\phi_i(r = R) = 0$. We note at this stage that the exponent value $\nu = 1/2$ in the cut-off term was used in our previous work [64]. Here we shall take $\nu = 1$ as a complementary analysis noting that, in this case, both the density and its slope become zero at the boundary.

Accordingly, the energy functional given by Equation (1) may be cast in terms of the density $\rho(\zeta_1, \ldots \zeta_n; r, R)$, where ζ_i are variational parameters such that the total ground-state energy energy is obtained by requiring that:

$$\frac{\partial E}{\partial \zeta_i} = 0, \quad (i = 1, \ldots, n) \tag{12}$$

and subject to the constriction

$$\int_{\Gamma} \rho \mathrm{d}\tau = N, \tag{13}$$

where N is the total number of electrons.

As mentioned before, the aforementioned scheme was applied for the first time [64] to free and confined atoms of the first three rows in the Periodic Table, using $\nu = 1/2$ and $\lambda = 1/8$, with reasonable success when compared with corresponding available calculations from the literature based on more sophisticated methods. In this comparison, some systematic quantitative deviations were observed for small cage radii and deemed as a limitation of the TFDλW method. However, in the course of the present investigation an important quantitative correction – for small cage radii – concerning the evaluation of the classical electron–electron interaction given by the first term in Equation (6) has been done. Before proceeding any further, let us take the opportunity here to briefly explain the origin and consequences of this correction.

In our previous treatment of the classical electron–electron repulsion integral $\frac{1}{2}\int \rho\varphi \mathrm{d}\tau$, we resorted to the solution of Poisson's equation to obtain the electrostatic potential $\varphi(r)$ within the confinement region, which demanded the proper evaluation of two integration constants [64]. While the first integration constant was well defined by analyzing the behavior of the potential close to the origin, the second integration constant required a judicious choice, due to its arbitrary character particularly at $r = R$ and such that it becomes zero faster than $1/R$ as $R \to \infty$ and keeps the correct behavior for small values of R. After analyzing the behavior of the electrostatic potential close to the origin, the second integration constant was found to be adequately represented by a simple exponential decaying closure relation meeting both, the asymptotic, and close to the origin limits. While this procedure allowed an analytical and simplified treatment of the electron–electron repulsion integral, we have now evaluated this integral exactly as:

$$V_{\mathrm{Coul}} = \frac{1}{2}\int\int_{\Gamma} \rho(r_1)\rho(r_2)r_{12}^{-1}\mathrm{d}\tau_1\mathrm{d}\tau_2, \tag{14}$$

where r_1 and r_2 denote the position of electron 1 and 2 relative to the nucleus and r_{12} their relative separation. Using the familiar multipolar expansion in spherical coordinates for r_{12}^{-1}, the above integral may be shown to become:

$$V_{\mathrm{Coul}} = 8\pi^2\left[\int_0^R \rho(r_2)r_2\left(\int_0^{r_2} \rho(r_1)r_1^2\mathrm{d}r_1 + r_2\int_{r_2}^R \rho(r_1)r_1\mathrm{d}r_1\right)\mathrm{d}r_2\right] \tag{15}$$

with $\rho(r)$ being the total electron density given by Equations (8)–(10) and R the cavity radius.

Figure 1 shows – for Ne as an example – the cavity size dependence of the ratio $(V_{\mathrm{Coul,exact}}/V_{\mathrm{Coul,Poisson}})$ between the classical electron–electron repulsion integrals calculated through Equation (15) $(V_{\mathrm{Coul,exact}})$ and the

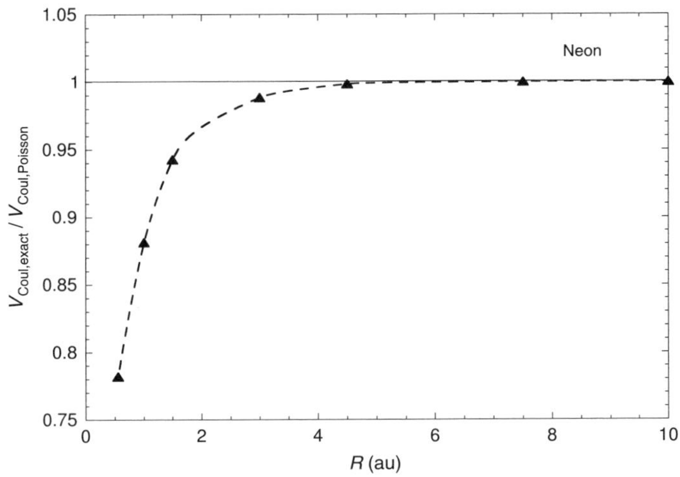

Figure 1 Relative difference – as a function of cavity radius for the case on Ne –between the exact calculation of the classical electron–electron repulsion integral using Equation (15) ($V_{\mathrm{Coul,exact}}$) and that obtained through the first term in Equation (6) using the Poisson solution ($V_{\mathrm{Coul,Poisson}}$) (see text). The dashed line is drawn to guide the eye.

first term in Equation (6) ($V_{\mathrm{Coul,Poisson}}$) with φ obtained after solution of the Poisson equation as described before. For comparison purposes we have used the same orbital parameters in each case. As the reader must be aware from this figure, excellent correspondence between both methods is achieved for a wide range of R values, except for the smaller ones. Although this calculation gives proof of the adequacy of the closure relation used in the solution of Poisson's equation [64,65], it also shows its breaking point. This behavior brings, as a consequence, important corrections in the cage-size energy evolution defined through the optimization process implied by Equation (12), mainly for small values of R. We note at this stage that both methods coincide for the free-atom case.

Table 1 shows the ground-state energy and optimized orbital parameters for C and Ne for selected values of R, calculated using Equation (15), as compared to corresponding values (shown in parenthesis) obtained in Ref. [64], using the Poisson method, and setting $\nu = 1/2$ and $\lambda = 1/8$ in both calculations. The HF values reported by Ludeña [4] are also shown for numerical comparison. In order to gather a better assessment on the consequences of the aforementioned corrections in the TFDλW method, a sample calculation for Ne has been carried out with $\nu = 1/2$ and the Weizsäcker factor $\lambda = 0.12331$ adjusted so that the free-atom energy corresponds to three decimal digits accuracy with that of Clementi et al. ($E = -128.547$ hartrees) [72]. Figure 2 shows the results of this calculation using Equation (15) (continuous line) and using the Poisson method (dotted

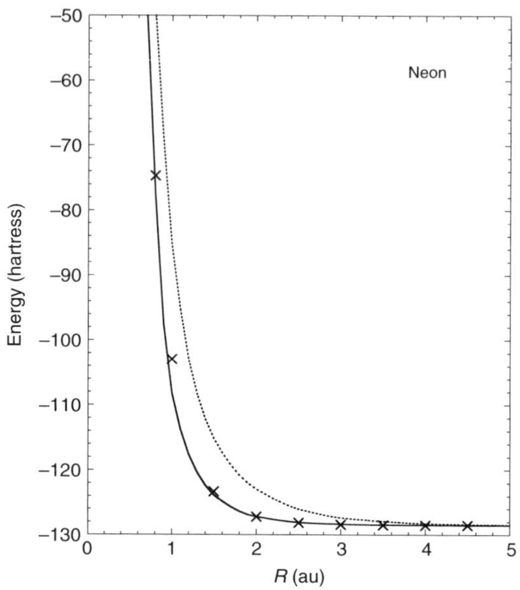

Figure 2 TFDλW predictions of the ground-state energy evolution of Ne as a function of confinement radius for the spherical impenetrable case using the exact calculation of the classical electron–electron repulsion integral (continuous line) and the Poisson solution (dotted line). Also shown for comparison are the HF values (crosses) calculated by Ludeña [4].

line) as compared with Ludeña's values (crosses). Clearly, an improvement of the TFDλW energy values is observed relative to the HF calculations. Hence, in spite of the simpler approximate Poisson solution, for a confined system the exact evaluation of Equation (15) for the classical electron–electron repulsion energy is the obvious choice to avoid inaccuracies throughout all calculations.

Having removed the already discussed existing quantitative deficiencies, the TFDλW method seems to be improved in its predictions yet still remains simple enough to explore the ground-state energy evolution of confined many-electron atoms. In the following analysis use will be made of the corrected method. Taking this into consideration, we now explore the case where the cut-off term in Equation (10) becomes $\nu = 1$. The reason for this possible choice is that in this case both the electron density and its derivative become zero at $r = R$. This representation might be more consistent with previously used ansatz wave functions used in variational calculations for confined hydrogen and helium [37,38,46].

We now proceed to present new calculations for O, Ne, Si and Ge with $\nu = 1$ and the respective λ factors adjusted so that the corresponding free-atom HF energies are equal to three-digit accuracy to those of Clementi et al. [72] [E(O, Si, Ge) = (−74.809, −288.854, −2075.359) hartrees].

Table 1 Ground state energy ($E_{TFD\lambda W}$) and orbital parameters (ς_i) obtained in this work for C and Ne for selected values of confinement radii (R) as compared to corresponding values obtained in Ref. [64] (shown in parenthesis) and with HF values (E_{HF}) reported by Ludeña [4]. $\lambda = 1/8$ and $\nu = 1/2$ were set in Equations (2) and (10), respectively. All values are given in atomic units

		Carbon			
R (au)	$-E_{TFD\lambda W}$	$-E_{HF}$[a]	ς_{1s}	ς_{2s}	ς_{2p}
5	37.673 (37.567)	37.668	5.47054 (5.47361)	2.59350 (2.25875)	1.06542 (1.29362)
3	37.516 (36.971)	37.257	5.39746 (5.41386)	2.43829 (1.95635)	1.03117 (1.95635)
2	36.738 (34.793)	35.059	5.31371 (5.38896)	1.67352 (2.20629)	1.67352 (2.20629)
		Neon			
4.5	128.359 (128.238)	128.535	9.56069 (9.56438)	5.36999 (5.36513)	1.90937 (1.95014)
3	128.209 (127.232)	128.415	9.50138 (9.52203)	5.30684 (5.26512)	1.79588 (2.02487)
2	127.073 (122.787)	127.231	9.42157 (9.48471)	5.19947 (4.85826)	1.66315 (2.35794)

[a] Ref. [4].

Table 2 displays the obtained ground-state energy values and optimal orbital parameters for a selected set of cavity radii. The λ factor for each system is shown within parenthesis. The relative effect of confinement on the energy evolution of the O, Si and Ge systems – as an example – may be better appreciated from Figure 3, where the ground-state *energy shift* is plotted against cavity radius. Indeed, this behavior may be related to pressure as discussed elsewhere [4,16,64]. However, this analysis will not be pursued since – as mentioned in the Introduction – our purpose here is to provide further quantitative information on the energy evolution based on the TFDλW method.

2.2. Soft spherical box

As a natural extension of the previously discussed model, in a recently published paper [65] we have presented a more realistic description of confinement by enclosing the many-electron system within a spherical cage with penetrable walls. A brief overview is presented here for completeness with some new results.

We now consider a many-electron atomic system confined by a spherical cage of radius R with a confining barrier potential V_c, such that

Table 2 Ground-state energy $E_{\mathrm{TFD}\lambda W}$ and optimized orbital parameters (ς_i) obtained in this work for O, Ne, Si and Ge for selected confinement radii (R). All calculations were performed using $\nu = 1$ in Equation (10) and the free-atom optimal λ factors are shown in parenthesis. All values are given in atomic units

	R (au)	∞	8	5	3	2
	$-E_{\mathrm{TFD}\lambda W}$	74.809	74.799	74.776	74.603	73.722
Oxygen	ς_{1s}	7.56192	7.43363	7.35372	7.20651	7.02015
($\lambda = 0.134105$)	ς_{2s}	4.06892	3.93733	3.85267	3.68946	3.45017
	ς_{2p}	1.67332	1.51835	1.38393	1.03654	0.49057
	$-E_{\mathrm{TFD}\lambda W}$	128.547	128.528	128.484	128.235	127.147
Neon	ς_{1s}	9.69016	9.56253	9.48358	9.33904	9.15524
($\lambda = 0.12331$)	ς_{2s}	5.48940	5.35966	5.27766	5.12610	4.92822
	ς_{2p}	2.06138	1.91429	1.79815	1.51715	1.03267
	$-E_{\mathrm{TFD}\lambda W}$	288.854	288.846	288.825	288.579	287.033
	ς_{1s}	13.85420	13.72804	13.65016	13.50978	13.33291
Silicon	ς_{2s}	8.12448	7.99708	7.91900	7.77272	7.57845
($\lambda = 0.12545$)	ς_{2p}	3.76596	3.62872	3.53325	3.34161	3.06677
	ς_{3s}	1.99956	1.87315	1.76800	1.30603	0.37948
	ς_{3p}	1.60657	1.39581	1.15734	0.73418	0.37756
	$-E_{\mathrm{TFD}\lambda W}$	2075.359	2075.351	2075.323	2074.914	2072.432
	ς_{1s}	32.58500	32.46228	32.38617	32.24962	32.07673
	ς_{2s}	19.44750	19.31750	19.24174	19.10687	18.94667
Germanium	ς_{2p}	11.98290	11.85417	11.77350	11.62323	11.41846
($\lambda = 0.141215$)	ς_{3s}	7.74503	7.63491	7.54821	7.38605	7.16861
	ς_{3p}	7.77043	7.63297	7.54811	7.38526	7.17072
	ς_{3d}	3.55262	3.40966	3.30229	3.05906	2.56660
	ς_{4s}	1.87935	1.68432	1.53178	1.09822	1.43779
	ς_{4p}	1.82696	1.65537	1.39075	1.09776	1.43666

$$V_c = \begin{cases} 0 & 0 \le r_i < R \\ V_0(i) = V_0 & R \le r_i < \infty \end{cases}, \quad i = 1, \ldots, N \tag{16}$$

where r_i denotes the radial position of electron "i" relative to the nucleus of atomic number Z, located at the origin, and $V_0(i) = V_0$, the height of the potential barrier felt by each electron outside the cage. The constant potential V_0 felt by each electron outside the cage is assumed to correspond to an effective confining potential due to the surrounding medium. Hence, the confining barrier potential is considered as a step-like function of finite height V_0. This assumption requires an appropriate description of the TFDλW energy functional for both the interior and exterior regions together with the corresponding *ansatz* orbital density representations subject to continuity boundary conditions at the wall.

Equation (1) now becomes:

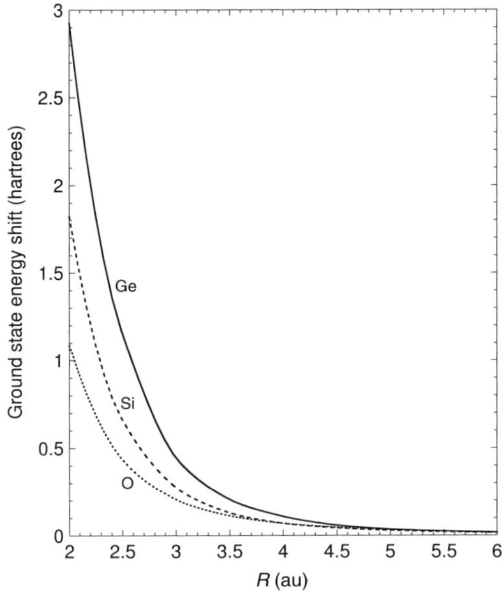

Figure 3 Ground-state energy shift dependence on cavity radius for spherical hard-wall confinement of O, Si and Ge calculated in this work with $\nu = 1$ in Equation (10).

$$E[R, V_0] = E_{\rm int}[\rho_{\rm int}, R] + E_{\rm ext}[\rho_{\rm ext}, R, V_0], \tag{17}$$

where $E_{\rm int}$ and $E_{\rm ext}$ correspond to the energy contribution from the interior and exterior regions, define, respectively, as:

$$E_{\rm int}[\rho_{\rm int}, R] = T_{\rm int}[\rho_{\rm int}, R] + V_{\rm en}[\rho_{\rm int}, R] + V_{\rm ee}[\rho_{\rm int}, R] \tag{18}$$

and

$$E_{\rm ext}[\rho_{\rm ext}, R, V_0] = T_{\rm ext}[\rho_{\rm ext}, R, V_0] + E_{\rm barrier}[\rho_{\rm ext}, R, V_0] \tag{19}$$

with $\rho_{\rm int}$ and $\rho_{\rm ext}$ different representations for the interior and exterior total electronic charge densities which will be defined further below.

$T_{\rm int}(T_{\rm ext})$ in Equations (18) and (19) are the sum of the first two contributions in the gradient expansion of the kinetic energy functional [Equations (2)–(4)] for the interior and exterior regions, respectively. $V_{\rm en}$ is the electron–nucleus interaction given by Equation (5) and $V_{\rm ee}$ is electron–electron interaction given by Equation (6) with its first term substituted by Equation (15) and defined only for the interior region due to the assumptions of the model. Finally, the second term in Equation (19) corresponds to the energy contribution due to the confining barrier:

$$E_{\text{barrier}} = V_0 \int_\Gamma \rho_{\text{ext}} \mathrm{d}\tau. \tag{20}$$

Note in this case the symbol Γ appearing in all the relevant equations denotes the domain of definition for either the interior or exterior region.

In contrast with the impenetrable case discussed in the previous section, here we require appropriate ansatz orbital densities to define a total interior $\rho_{\text{int}}(r)$ and exterior $\rho_{\text{ext}}(r)$ atomic density for the corresponding regions, namely:

$$\rho_{\text{int}}(r) = \sum_{i=1}^{\text{occ}} \rho_{\text{int},i}(r), \quad (0 \leq r \leq R) \tag{21}$$

$$\rho_{\text{ext}}(r) = \sum_{1=1}^{\text{occ}} \rho_{\text{ext},i}(r), \quad (R \leq r < \infty) \tag{22}$$

with $\rho_{\text{int},i}(r)$ and $\rho_{\text{ext},i}(r)$ the orbital densities for the interior and exterior region, respectively, and the sum being carried over all occupied orbitals.

According to Ref. [65], an appropriate guiding criterion to define a physically plausible exterior orbital density consists in using the functional form for the wave function derived from exact solutions to the Schrödinger equation for the hydrogen atom confined by a soft spherical box [34], i.e.:

$$\rho_{\text{ext},i}(r) = |\varphi_{\text{ext},i}|^2 \tag{23}$$

with

$$\varphi_{\text{ext},i}(r) = N_{\text{ext},i} r^{-l-1} \mathrm{e}^{-\beta_i r}, \tag{24}$$

where l_i is the angular momentum quantum number of orbital "i", β_i a corresponding variational parameter and $N_{\text{ext},i}$ a normalizing factor defined by the condition:

$$\int_0^R \rho_{\text{int},i}(r)\mathrm{d}\tau + \int_R^\infty \rho_{\text{ext},i}(r)\mathrm{d}\tau = \omega_i, \tag{25}$$

where ω_i is the electron population of orbital "i" and $\rho_{\text{int},i}(r)$ the ansatz orbital density for the interior region, which is proposed in terms of Slater-type nodeless functions in a similar fashion as in the impenetrable case [Equation (10)] as:

$$\rho_{\text{int},i}(r) = |\varphi_{\text{int},i}|^2, \tag{26}$$

where

$$\varphi_{\text{int},i}(r) = N_{\text{int},i} r^{n_i-1} \mathrm{e}^{-\alpha_i r}(1-\gamma r/R)^{\nu}. \tag{27}$$

As in Equation (10), n_i corresponds to the principal quantum number, α_i is a variational parameter and $N_{\text{int},i}$ the normalizing factor satisfying Equation (25) together with $N_{\text{ext},i}$. Furthermore, the factor $(1-\gamma r/R)^{\nu}$ appearing in Equation (27) now becomes a matching term to couple the interior and exterior orbital densities at the boundary. Note that this factor depends on a third variational parameter $\gamma (0 \le \gamma \le 1)$ which is assumed the same for all orbital densities of a given atom. In this study – as in Ref. [65] – we shall use $\nu = 1/2$ in Equation (27) for simplicity, although other choices can be made as was explored in the previous section for the impenetrable case.

Finally, the interior and exterior densities and their derivative must satisfy continuity boundary conditions, which may be imposed on each of the corresponding orbital components as:

$$\rho_{\text{int},i}|_{r=R} = \rho_{\text{ext},i}|_{r=R} \tag{28}$$

$$\left.\frac{\partial \rho_{\text{int},i}}{\partial r}\right|_{r=R} = \left.\frac{\partial \rho_{\text{ext},i}}{\partial r}\right|_{r=R}. \tag{29}$$

These two equations together with Equation (25) constitute the necessary subsidiary conditions to properly define the characteristics of the interior and exterior densities. Furthermore, from Equations (28) and (29) and Eqs. (23), (24) and (26), (27), the following coupling relations among the variational parameters $(\alpha_i, \beta_i, \gamma)$ are found to be satisfied:

$$\beta_i = \alpha_i + \frac{\gamma}{2R(1-\gamma)} - \frac{n_i + l_i}{R} > 0 \tag{30}$$

and

$$(2n_i + 2l_i + 1)\gamma + 2\alpha_i R(1-\gamma) > 2(n_i + l_i). \tag{31}$$

Hence, considering these subsidiary conditions, the energy functional $E(\alpha_i, \beta_i, \gamma; R, V_0)$ given by Equation (17) is variationally optimized for a given value of R and V_0 relative to only two sets of variational parameters, i.e.:

$$\frac{\partial E}{\partial \alpha_i} = \frac{\partial E}{\partial \gamma} = 0; \quad (i = 1, \ldots, occ). \tag{32}$$

A detailed analysis of the energy evolution of He, Li, C and Ne confined within spherical cavities with different radii and barrier heights has been

Table 3 Energy values and orbital parameters for Ne confined by a penetrable spherical box of barrier height $V_0 = 0$ for selected values of R calculated using Equation (15) for the electron–electron interaction as compared with those obtained in Ref. [65] using the Poisson solution (values in parenthesis). All values are given in atomic units

R (au)	$-E_{TFD\lambda W}$	α_{1s}	α_{2s}	α_{2p}	γ
10	128.547	9.68855	5.48724	2.05989	0.032232
	(128.547)	(9.68857)	(5.48778)	(2.05990)	(0.032221)
4	128.547	9.68585	5.48559	2.05721	0.032384
	(128.545)	(9.68614)	(5.48493)	(2.05869)	(0.032246)
3	128.552	9.68456	5.48402	2.05437	0.032378
	(128.502)	(9.68734)	(5.48076)	(2.08188)	(0.031065)
2	128.559	9.69212	5.48242	2.10588	0.001394
	(127.787)	(9.65128)	(5.38557)	(2.20710)	(0.229603)

recently reported using the method outlined here [65]. In that study the electron–electron interaction was also calculated by resorting to the Poisson method discussed in the previous section. Here we first investigate possible differences with the previous calculation by using the exact expression for this interaction as given by Equation (15). To this end, the case of Ne ($\lambda = 0.12331$) confined by a spherical box with barrier height $V_0 = 0$ is considered as an example.

Table 3 shows the results of this calculation. Interestingly, the differences observed between both types of calculations are smaller for the lower barrier height in contrast with the hard-wall case, as may be gathered from Table 1. This effect may be due to the flexibility added for the electron density to permeate through the boundary with probable compensating effects making the Poisson method still reasonably accurate. In any event, the classical electron–electron repulsive energy must be accurately evaluated to avoid eventual inaccuracies inherent in the Poisson method. Furthermore, in the case of finite barrier heights the electron ionization energy may reach the barrier height as the cavity radius is reduced and electron escape from the cavity may occur [46,65]. The critical point at which this process may take place depends crucially on the energy evolution of the system as a function of R and the barrier height. The interested reader is directed to the latter references where the details of this important effect are discussed.

Following the aims of this paper, new sample calculations for the ground-state energy evolution and orbital parameters for O, Si and Ge are presented in Table 4 for selected cavity radii and a barrier height $V_0 = 1$ hartree. The energy shift relative to the free-atom condition is also shown for each case. Comparison with the values obtained for the impenetrable box [Table 2] indicates a less steep variation as the cage radius is reduced, which reflects

Table 4 Ground-state energy $E_{\text{TFD}\lambda W}$, energy shift ΔE and optimized orbital parameters (α_i, γ) obtained in this work for O, Ne, Si and Ge confined by a penetrable spherical box with barrier height $V_0 = 1$ hartree for selected values of R. All calculations were performed using $\upsilon = 1/2$ in Equation (27) with the free-atom optimal λ factors shown in parenthesis. All values are given in atomic units

	R (au)	7.5	5	3	2
	$-E_{\text{TFD}\lambda W}$	74.809	74.808	74.701	73.854
	ΔE	0.000	0.001	0.108	0.955
	α_{1s}	7.85774	7.55732	7.56282	7.40461
Oxygen	α_{2s}	4.06572	4.06368	4.03913	4.08083
($\lambda = 0.134105$)	α_{2p}	1.67013	1.67028	1.76789	1.85047
	γ	0.048033	0.047818	0.026181	0.018716
	$-E_{\text{TFD}\lambda W}$	288.852	288.848	288.687	287.835
	ΔE	0.002	0.006	0.167	1.019
	α_{1s}	13.8225	13.82690	13.68790	13.60780
Silicon	α_{2s}	8.09659	8.08731	7.94416	7.87410
($\lambda = 0.12545$)	α_{2p}	3.73962	3.73619	3.55966	3.45584
	α_{3s}	1.84207	1.80354	1.70006	1.19316
	α_{3p}	1.71430	1.75694	1.02284	1.09731
	γ	0.397769	0.318422	0.959219	0.933352
	R (au)	7.5	5	3.5	3.0
	$-E_{\text{TFD}\lambda W}$	2075.357	2075.340	2075.180	2075.030
	ΔE	0.002	0.019	0.179	0.329
	α_{1s}	32.57520	32.52650	32.44420	32.42870
	α_{2s}	19.47130	19.40910	19.36560	19.29340
Germanium	α_{2p}	11.95800	11.91520	11.82950	11.81100
($\lambda = 0.141215$)	α_{3s}	7.57442	7.63715	7.29748	7.54282
	α_{3p}	7.82162	7.72085	7.73735	7.60346
	α_{3d}	3.53970	3.48427	3.36641	3.32904
	α_{4s}	1.88996	1.83610	1.90886	1.57635
	α_{4p}	1.80704	1.78831	1.46737	1.15901
	γ	0.186947	0.591577	0.941858	0.956597

the more flexible behavior of the confinement model in simulating realistic systems.

2.3. Hard prolate spheroidal box

In spite of the simplicity implied by a spherical box enclosing a many-electron atom located at its center, only few reference calculations are available in the literature whereby more sophisticated and powerful approaches than the TFDλW method have been applied, mainly for the impenetrable case.

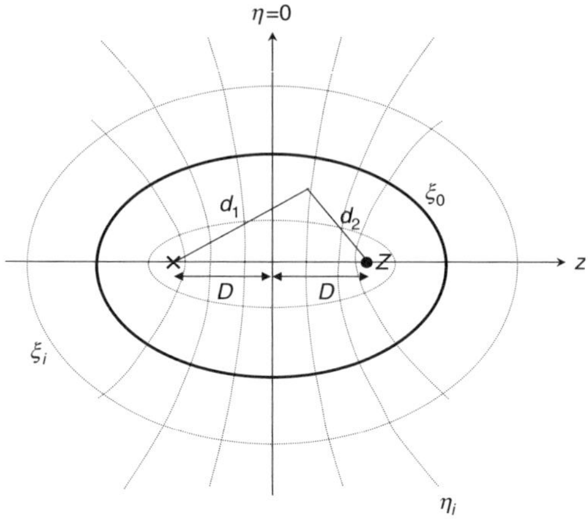

Figure 4 Schematic diagram showing an atom of nuclear charge Z located at one of the foci ($z = D$) of a prolate spheroidal box (bold line) characterized by $\xi = \xi_0$, which forms part of a family of confocal prolate spheroids (ξ_i), each one orthogonal to a family of confocal hyperboloids (η_i) (all shown with thin dotted lines). d_1 and d_2 denote an electron position relative to the nucleus and to the other focal point. The x–y plane corresponds to $\eta = 0$.

A completely new and challenging problem for the study of many-electron atom confinement concerns the case of a cavity with a symmetry different to the spherical one. As mentioned in the Introduction, one and two-electron atoms have been studied under different confinement cavity geometries showing new interesting properties in contrast with the spherically symmetric case. To the author's knowledge, no attempt has been done so far to model the ground-state energy behavior of a many-electron atom (other than helium) enclosed by a non-spherical cavity. In this Section we address this problem for the case of an impenetrable prolate spheroidal cavity by applying the TFDλW method.

Let us consider a many-electron atom of nuclear charge Z confined by a hard prolate spheroidal cavity. In this study the nuclear position will correspond to one of the foci as shown in Figure 4. In terms of prolate spheroidal coordinates, the nuclear position then corresponds to one of the foci for a family of confocal orthogonal prolate spheroids and hyperboloids defined, respectively, by the variables ξ and η as [73]:

$$\xi = \frac{d_1 + d_2}{2D}, \quad (1 \leq \xi < \infty), \tag{33}$$

$$\eta = \frac{d_1 - d_2}{2D}, \quad (-1 \leq \eta \leq 1), \tag{34}$$

where azimuthal symmetry is considered and d_1, d_2 denote the distances of a point in space from the foci separated a distance $2D$ along the z-direction.

Within this coordinate representation, the confining spheroidal boundary corresponds to $\xi = \xi_0$.

Setting the nucleus at the right hand focal position, the electron distance to the nucleus becomes $d_2 = D(\xi - \eta)$. Hence, following the ansatz given by Equation (10) based on the angularly averaged Slater-type nodeless functions the proposed orbital density becomes:

$$\rho_i(\xi, \eta) = |\varphi_i(\xi, \eta)|^2 \tag{35}$$

with

$$\varphi_i(\xi, \eta) = N_i D^{n_i - 1}(\xi - \eta)^{n_i - 1} \mathrm{e}^{-\alpha_i D(\xi - \eta)}(1 - \xi/\xi_0)^{\nu}, \tag{36}$$

where N_i is the normalizing factor and α_i the orbital exponent, such that:

$$\int_\Gamma \rho_i(\xi, \eta) \mathrm{d}\tau = \omega_i \tag{37}$$

with ω_i the orbital population and $\mathrm{d}\tau = 2\pi D^3(\xi^2 - \eta^2)\mathrm{d}\xi\mathrm{d}\eta$. As usual, Γ denotes the domain of integration within the confinement volume, i.e. $1 \le \xi \le \xi_0$ and $-1 \le \eta \le 1$.

As in the spherical case the cut-off function $(1 - \xi/\xi_0)^\nu$ guarantees compliance with the Dirichlet boundary condition at the surface, i.e. $\varphi_i(\xi = \xi_0, \eta) = 0$. Here we shall use $\nu = 1$ in Equation (36).

In this new geometry the total density is expressed as

$$\rho(\xi, \eta) = \sum_{i=1}^{\mathrm{occ}} \rho_i(\xi, \eta) \tag{38}$$

and the TFDλW functional becomes:

$$E[\rho, \xi_0] = T[\rho, \xi_0] + V_{\mathrm{en}}[\rho, \xi_0] + V_{\mathrm{ee}}[\rho, \xi_0], \tag{39}$$

with T representing the total kinetic energy including the free-electron gas term and the Weizsäcker correction:

$$T[\rho, \xi_0] = \frac{3}{10}(3\pi^2)^{2/3} \int_\Gamma \rho(\xi, \eta)^{5/3} \mathrm{d}\tau + \frac{\lambda}{8} \int_\Gamma \frac{\nabla\rho(\xi, \eta) \cdot \nabla\rho(\xi, \eta)}{\rho(\xi, \eta)} \mathrm{d}\tau. \tag{40}$$

V_{en} is the electron–nuclear attraction energy:

$$V_{\mathrm{en}}[\rho, \xi_0] = -\frac{Z}{D} \int_\Gamma \rho(\xi, \eta)(\xi - \eta)^{-1} \mathrm{d}\tau, \tag{41}$$

and V_{ee} the two-electron repulsion energy composed of the classical Coulomb term and the Dirac exchange energy:

$$V_{ee}[\rho,\xi_0]=\frac{1}{2}\int_\Gamma \rho(\xi,\eta)(r_{12})^{-1}\rho(\xi',\eta')\mathrm{d}\tau\mathrm{d}\tau' - \frac{3}{4}(3/\pi)^{1/3}\int_\Gamma \rho(\xi,\eta)^{4/3}\mathrm{d}\tau. \tag{42}$$

As discussed in the previous sections, the classical electron–electron interaction given by the first term in Equation (44) has to be calculated using the multipolar expansion in prolate spheroidal coordinates [74]:

$$\frac{1}{r_{12}}=\frac{1}{D}\sum_{l=0}^{\infty}(2l+1)\sum_{m=0}^{\infty}\varepsilon_m i^m\left[\frac{(l-m)!}{(l+m)!}\right]^2 P_l^m(\eta_1)P_l^m(\eta_2)Q_l^m(\xi_<)Q_l^m(\xi_>) \times\cos m(\varphi_1-\varphi_2), \tag{43}$$

with $\varepsilon_0=1$ and $\varepsilon_m=2$ $(m>0)$ and $P_l^m(z)$, $Q_l^m(z)$ the associated Legendre functions of the first and second kind, respectively. However, since we are interested in the ground-state, $m=0$, the latter expression greatly simplifies, leading to the following explicit relation for the first term in Equation (42) in terms of the total density [Equation (38)]:

$$\frac{1}{2}\int \rho(\xi,\eta)(r_{12})^{-1}\rho(\xi',\eta')\mathrm{d}\tau\mathrm{d}\tau' = 2\pi^2D^5\sum_l(2l+1) \times\left\{\int_1^{\xi_0} g_l(\xi)\mathrm{d}\xi\left[Q_l(\xi)\int_1^{\xi} g_l(\xi')P_l(\xi')\mathrm{d}\xi' + P_l(\xi)\int_\xi^{\xi_0} g_l(\xi')Q_l(\xi')\mathrm{d}\xi'\right]\right\}, \tag{44}$$

where

$$g_l(u)=\int_{-1}^{1}\rho(u,s)P_l(s)[u^2-s^2]\mathrm{d}s \tag{45}$$

for $u=\xi$ (ξ') and $s=\eta$ (η').

Following a parallel reasoning as in the spherical case, given an ellipsoidal cavity defined by ξ_0 and a focal position D for the nucleus, the ground-state energy is obtained after optimization of the total energy functional given by Equation (39) relative to the orbital exponents. All integrals given by Equations (40)–(45) were evaluated numerically with a 96-point Gauss–Legendre quadrature within the optimization process. In all cases, energy convergence to 10^{-4} au was obtained for $l_{max}=30$ in the l-expansion of Equation (44).

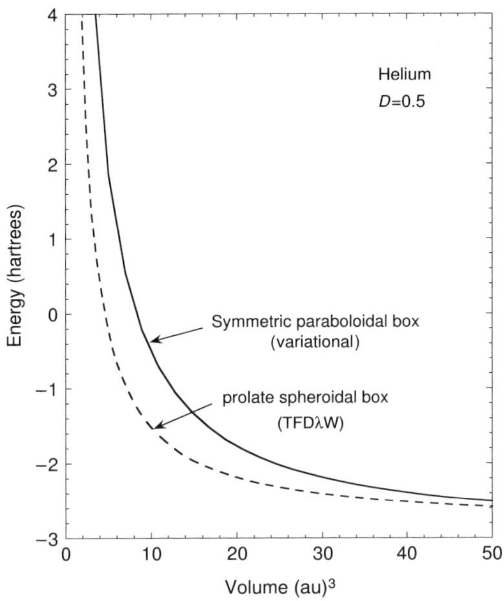

Figure 5 Ground-state energy evolution of He – located at the focal position $D = 0.5$ au – as a function of cage volume for hard-wall spheroidal confinement. Dashed curve: results of this work. Continuous line: values obtained by Ley-Koo et al. [49] for He confinement by symmetric paraboloidal boxes corresponding approximately to the same shape and volume as the prolate spheroidal box considered here [51].

To the author's knowledge, there are no similar calculations available in the literature to compare with for many-electron atoms, except for the case of helium-like systems reported by Corella et al. [51]. These authors calculate variationally the ground-state energy of He as a function of cage volume [$V = 4\pi D^3 \xi_0(\xi_0^2 - 1)/3$] for a set of focal distances ($D = 0.1$, 0.5, and 1.0 au). After comparing their results with calculations by Ley-Koo et al. [49] for the He atom enclosed by symmetric paraboloidal boxes – which yield approximately a similar enclosing geometry and volume – they conclude that the best numerical correspondence between both calculations is achieved for $D = 0.5$ au. Unfortunately, the former authors do not provide numerical values. We have performed a similar calculation – shown in Figure 5 – for He ($\lambda = 0.02544$) and $D = 0.5$ au (dashed line), comparing results with the tabulated values of Ley-Koo (continuous line). Although the TFDλW results of this case have the correct qualitative behavior, the energy evolution is generally underestimated as compared with the variational calculation. This seems to be a characteristic of the TFDλW method for the case of He, as was also observed in the case of spherical confinement [64,65]. For larger atomic systems, however, the quantitative assessment of the predictions by this method are deemed to improve, as was also discussed in the previous sections.

Table 5 Ground-state energy $E_{TFD\lambda W}$ and orbital parameters α_i of C for different confining spheroids defined by ξ_0 and located at selected focal distances D. $\nu = 1$ was set in Equation (36) for all calculations. All quantities are given in atomic units.

	Carbon ($\lambda = 0.12479$)				
ξ_0	5	4	3	2.	1.5
$D = 0.5$ au					
$-E_{TFD\lambda W}$	36.791	35.543	31.416	11.868	–35.679
α_{1s}	5.29067	5.21413	5.09225	4.84366	4.56525
α_{2s}	2.29969	2.22287	2.26582	2.63432	2.71304
α_{2p}	0.36420	0.24001	0.09854	0.07099	0.19648
$D = 1.0$ au					
$-E_{TFD\lambda W}$	37.556	37.399	36.808	33.215	22.558
α_{1s}	5.44058	5.39677	5.31516	5.09733	4.72130
α_{2s}	2.55776	2.50799	2.42683	2.39983	2.30822
α_{2p}	0.92257	0.82588	0.73269	0.56039	0.34773
$D = 2.0$ au					
$-E_{TFD\lambda W}$	37.661	37.6359	37.5519	36.944	34.564
α_{1s}	5.51154	5.48979	5.44675	5.32334	5.08603
α_{2s}	2.63978	2.61718	2.57184	2.46361	2.37077
α_{2p}	1.13914	1.10104	1.03210	0.89277	0.55506

Accordingly, we now continue our discussion and present sample calculations for larger systems.

Tables 5 and 6 show the numerical values of the ground-state energy evolution and orbital parameters for C and Ne as a function of ξ_0 for selected values of the focal position $D = 0.5$, 1 and 2 au. In both cases, the optimal free-atom λ factor has been used. These results are shown as a quantitative reference. However, a better interpretation may be gathered from Figures 6 and 7, where the corresponding energies are plotted against the ellipsoid semi-major axis ($D\xi_0$). We first note from these two figures a similar qualitative behavior for the energy evolution. For large values of $D\xi_0$ (≥ 4 au), the energy curves tend to coalesce, corresponding to a spherical cage of radius $R = D\xi_0$. Depending on the value for the focal position (D) the rapid change in the box eccentricity ($1/\xi_0$) makes the energy have a steep rise. As expected, the closer the atom is to the origin (a lower value for D), the less eccentric is the cage and hence the comparatively apparent energy shift response to lower values of the semi-major axis. However, this conclusion might be misleading by looking only at the energy behavior as a function of the size of the semi-major axis. Figure 8 shows, as an example, the energy response when the atom is located at a particular focal position and the aspect ratio – given by the quotient between minor and major axes (b/a) – changes. Interestingly,

Table 6 Ground-state energy $E_{TFD\lambda W}$ and orbital parameters α_i of Ne for different confining spheroids defined by ξ_0 and located at selected focal distances D. $\nu = 1$ was set in Equation (36) for all calculations All quantities are given in atomic units.

Neon ($\lambda = 0.12331$)					
ξ_0	5	4	3	2.	1.5
$D = 0.5$ au					
$-E_{TFD\lambda W}$	126.505	123.877	114.870	70.782	−36.410
α_{1s}	9.4189	9.34395	9.22272	8.92684	8.41209
α_{2s}	5.21516	5.15452	5.09995	5.38152	5.77402
α_{2p}	1.47038	1.35521	1.32410	1.40835	1.34568
$D = 1.0$ au					
$-E_{TFD\lambda W}$	128.178	127.822	126.547	118.287	92.996
α_{1s}	9.55893	9.51651	9.43653	9.22206	8.78797
α_{2s}	5.35589	5.31440	5.23735	5.09806	5.11183
α_{2p}	1.86283	1.79720	1.70304	1.59218	1.29493
$D = 2.0$ au					
$-E_{TFD\lambda W}$	128.469	128.405	128.203	126.786	120.388
α_{1s}	9.62607	9.60478	9.56258	9.44151	9.20793
α_{2s}	5.42442	5.40319	5.36163	5.24659	5.10576
α_{2p}	1.98249	1.95507	1.90095	1.77451	1.57140

the most sensible atom position for changes in b/a corresponds to the lowest value of D. This behavior may be understood in terms of the stronger rigidity imposed by the nuclear field on the electronic cloud when the nucleus is close to the central position of the box. The farther the nucleus is from the center, the larger the volume for the electronic density to readjust. Of course, in a realistic situation one would expect the nucleus to be positioned close to the central region of the cavity. This would imply a higher electronic sensitivity to changes in cavity shape.

3. MANY-ELECTRON ATOM CONFINEMENT BY OPEN BOUNDARIES

3.1. Hard-wall confinement

The family of confocal ellipsoids and hyberboloids represented by the prolate spheroidal coordinates allows us now to treat the case of a many-electron atom spatially limited by an open surface in half-space. A special case of the family of hyperboloids corresponds to an infinite plane defined by $\eta = 0$ according to Equations (35) and (36). We now treat the specific case of an atom whose nuclear position is located at the focus a distance D from the plane as shown in Figure 4.

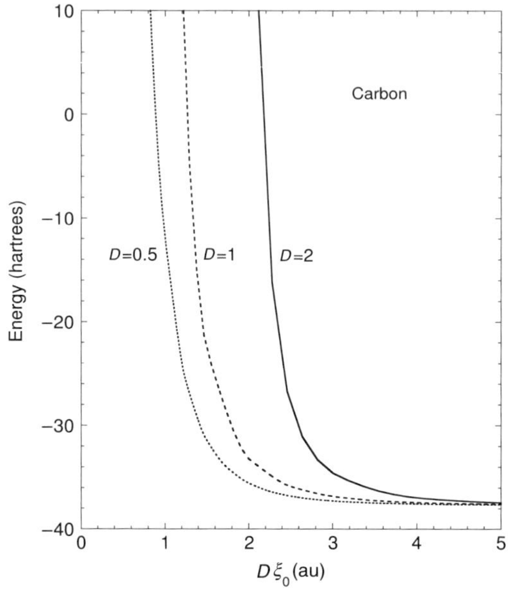

Figure 6 Semi-major axis $(D\xi_0)$ dependence of the ground-state energy of C enclosed by a hard spheroidal box and located at focal positions $D = 0.5$ au (dotted line), 1 au (dashed line) and 2 au (continuous line).

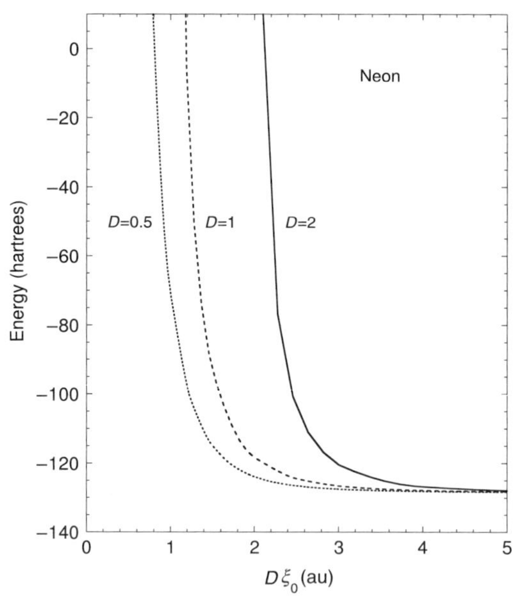

Figure 7 Semi-major axis $(D\xi_0)$ dependence of the ground-state energy of Ne enclosed by a hard spheroidal box and located at focal positions $D = 0.5$ au (dotted line), 1 au (dashed line) and 2 au (continuous line).

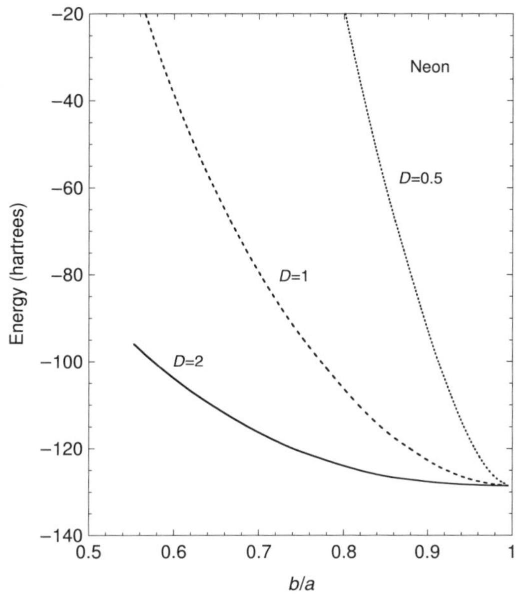

Figure 8 Comparative changes in the ground-state energy behavior of Ne located at different focal positions ($D = 0.5, 1$, and 2 au) of a confining spheroidal box as the aspect ratio (b/a) changes, where $b =$ semi-minor axis and $a =$ semi-major axis.

The coordinate transformation as well as the definition of the TFDλW energy functional – as indicated in Equation (39) with the substitution $\eta_0 \rightarrow \xi_0 = 0$ – have exactly the same structure as in the previous section. The only ingredient we need to add is the definition of a new ansatz orbital function to comply with the new boundary condition, i.e. according to Equation (36), we define the ansatz orbital function as:

$$\varphi_i(\xi, \eta) = N_i D^{n_i - 1} (\xi - \eta)^{n_i - 1} \mathrm{e}^{-\alpha_i D(\xi - \eta)} \eta^{\nu}, \tag{46}$$

where the factor η^{ν} now acts as a cut-off term such that $\varphi_i(\xi, \eta = 0) = 0$ and the range of definition Γ of the relevant variables becomes $1 \leq \xi < \infty$ and $0 \leq \eta \leq 1$.

Using Equation (46) into Equations (35) and (38) the total density is constructed and all integrations implied by Equations (40)–(45) are numerically evaluated within the corresponding domain with the same procedure and precision as in the spheroidal case during energy optimization. It is worth pointing out here that since D takes values in the interval $0 \leq D < \infty$, the change of variable $\xi = u/D$ is more adequate for numerical evaluation when $D > 1$ au. Also, as D increases, a greater number of terms must be included in the l-expansion in Equation (44) for proper convergence due to the higher

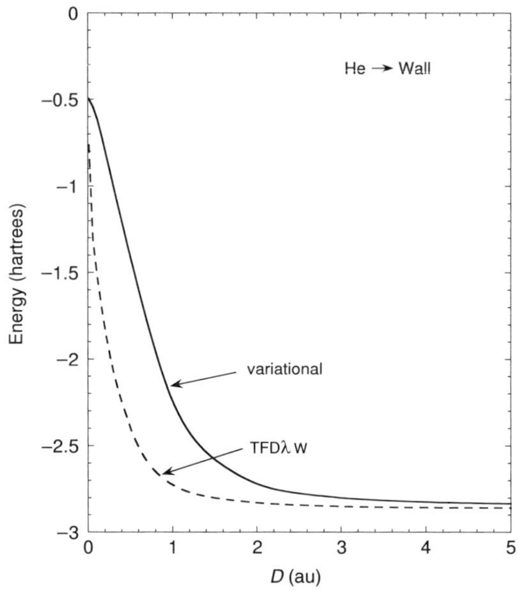

Figure 9 He ground-state energy dependence on distance to a hard-wall obtained in this work (dashed line) as compared with corresponding variational calculations [58] (solid line).

eccentricity posed by the coordinate system. Our calculations required typically $l_{\max} = 50$ for the largest value of D (=10) considered here for an energy convergence within 10^{-4} hartrees.

As in the case of spheroidal confinement, no similar calculations are available in the literature for the energy evolution of a many-electron atom close to a hard planar boundary, except for recent variational calculations performed for He using a similar cut-off function as in Equation (46) with $\nu = 1/2$ [58]. Figure 9 shows the He ground-state energy dependence on distance to the wall obtained through the TFDλW method (dashed line) as compared with the variational calculations of Ref. [58] (solid line). As expected, important quantitative deficiencies are observed for the TFDλW predictions for He, although showing reasonable qualitative agreement with the variational calculation. Notwithstanding the poor description of the TFDλW method for the He energy evolution close to a hard wall, further assessment is necessary for the case of larger atomic systems. Hence, as in the case of spheroidal confinement, we give quantitative information for the case of larger systems below, which may be useful for future comparison.

Table 7 shows the results of this calculation for the ground-state energy evolution of C and Ne using $\nu = 1/2$ in Equation (46), for a set of selected distances to the wall, together with corresponding values of the optimal orbital parameters. For completeness, in Figures 10 and 11, the energy evolution as a function of distance to the wall is displayed for each case (continuous line)

Table 7 Ground-state energy and orbital parameters of C and Ne as a function of distance D to a hard-wall obtained in this work with $\nu = 1/2$ in Equation (46). All values are given in atomic units

	D (au)	0.001	0.1	0.25	0.5	1	1.5	2	3	5	10
	$-E_{\mathrm{TFD}\lambda W}$	12.225	22.549	31.670	35.274	36.897	37.343	37.514	37.627	37.672	37.686
Carbon	α_{1s}	1.86046	3.66838	4.71773	5.10137	5.33349	5.41378	5.45372	5.49425	5.52714	5.55213
($\lambda = 0.12479$)	α_{2s}	1.06053	1.40737	1.85506	2.32028	2.56080	2.57730	2.59199	2.62593	2.65896	2.68358
	α_{2p}	0.28794	0.37225	0.46725	0.61924	0.61913	1.06670	1.12698	1.15939	1.18749	1.21432
	$-E_{\mathrm{TFD}\lambda W}$	43.248	92.378	113.686	121.790	126.431	127.720	128.150	128.405	128.508	128.544
Neon	α_{1s}	3.24435	5.99357	8.74513	9.19350	9.44763	9.52722	9.56675	9.60701	9.64030	9.66562
($\lambda = 0.12331$)	α_{2s}	2.43770	2.89863	4.64656	5.17694	5.27567	5.33304	5.36966	5.40853	5.44060	5.46491
	α_{2p}	0.68045	0.80273	1.17002	1.54199	1.85657	1.91732	1.94332	1.97730	2.01032	2.03604

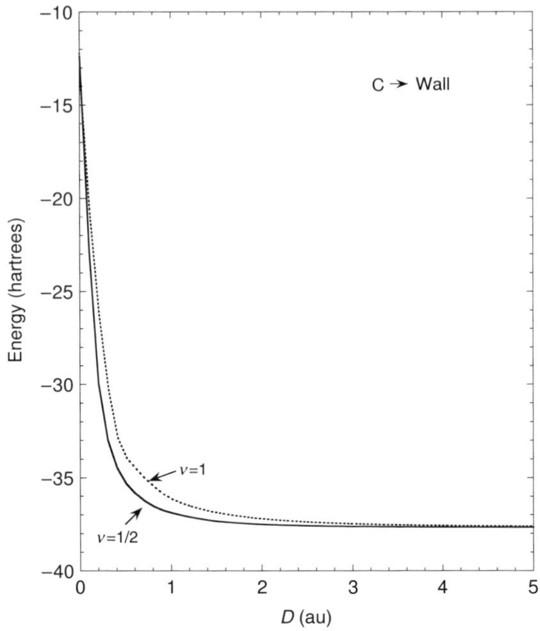

Figure 10 C ground-state energy dependence on distance to a hard-wall obtained in this work with $\nu = 1/2$ (continuous line) and $\nu = 1$ (dotted line) in Equation (46).

and compared with the corresponding curve constructed by using $\nu = 1$ in Equation (46) (dotted line). Important differences are noticed between the $\nu = 1$ and $\nu = 1/2$ representations for both systems, mainly in the region 0.25 au $\leq D \leq 2$ au, pointing to a more adequate variational representation for $\nu = 1/2$ within this scheme. On the other hand, a rapid energy ascent is observed for atom-wall distances $D \leq 2$ au, reaching a finite value for $D \approx 0$ (-12.225 hartrees for C and -43.248 hartrees for Ne for $\nu = 1/2$). This behavior – also observed in the He case – is concomitant with symmetry-breaking effects appearing in the atomic states due to the presence of the hard wall, as originally deemed for the case of hydrogenic systems [17–21], where the ground-state energy of the system is allowed to evolve from that of s-symmetry for large values of D to a $2p_z$-symmetry when $D = 0$. The corresponding analysis in the case of a many-electron atom is still an open question and the results presented here may be of use to surveying this question.

Before ending this section, this author would like to point out several issues which could be further analyzed, in the light of the results presented here, but have not been pursued. First, the distortion of the electronic cloud as the atom approaches the wall will induce an increasing net dipole moment, which may be estimated with the tabulated values. Second, it may be proven [58,59] that the ground-state energy-shift as a function of distance to

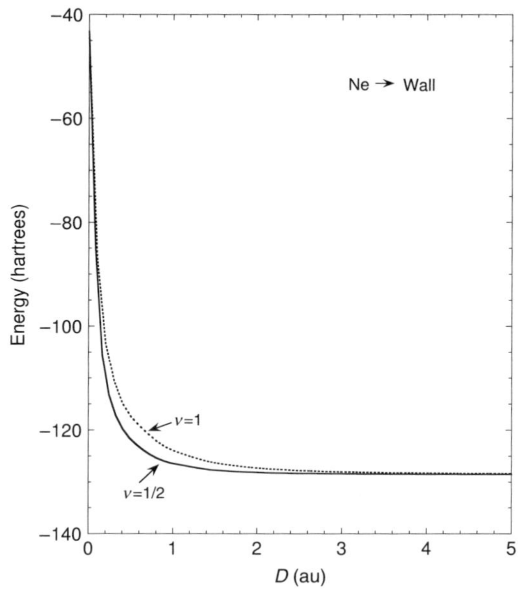

Figure 11 Ne ground-state energy dependence on distance to a hard-wall obtained in this work with $\nu = 1/2$ (continuous line) and $\nu = 1$ (dotted line) in Equation (46).

the wall is related to the elastic scattering potential and seems to constitute an upper limit to the continuous planar potential widely used in atom-surface scattering studies [75].

4. CONCLUSIONS

A critical assessment of the adequacy of the TFDλW method for studying the ground-state energy behavior of many-electron atom systems confined by spherical and spheroidal cavities and by a hard plane representing an open boundary has been presented. Rather than presenting a general overview of the method, our aim here has been to provide sufficient new quantitative information for the energy evolution of sample systems not treated before under non-spherical confinement. In spite of the found overall weakness of the TFDλW method to quantitatively account for the He energy evolution, its predictions for larger systems are deemed to improve substantially, as shown for the case of spherical confinement. Furthermore, the results of this work indicate the correct physical behavior of the systems considered when compared with corresponding available information for He. Other reference calculations are desirable to compare with the predictions of the method discussed here. We hope the results of this work will stimulate potential readers to explore the use of – similar or different – approaches to advance the knowledge on the confining geometry behavior of many-electron atoms (and molecules).

REFERENCES

[1] T. Sako, G.H.F. Dierksen, J. Phys. B 36 (2003) 1681.
[2] R.J. Hemley, Ann. Rev. Phys. Chem. 51 (2000) 763.
[3] M. Bayer, O. Stern, P. Hawrylak, S. Fafard, A. Forchel, Nature 405 (2000) 923.
[4] E.V. Ludeña, J. Chem. Phys. 69 (1978) 1770.
[5] J.J. Cioslowski, Am. Chem. Soc. 113 (1991) 4139.
[6] O. Shameema, C.N. Ramachandran, N. Sathyamurthy, J. Phys. Chem. A 110 (2006) 2.
[7] V.K. Dolmatov, A.S. Baltenkov, J.P. Connerade, S.T. Manson, Radiat. Phys. Chem. 70 (2004) 417.
[8] M.E. Madjet, H.S. Chakraborty, S.T. Manson, Phys. Rev. Lett. 99 (2007) 243003.
[9] A. Kaczmarek, R. Zaleśny, W. Bartkowiak, Chem. Phys. Lett. 449 (2007) 314 and references therein.
[10] P.V. Yurenev, A.V. Scherbinin, V.I. Pupyshev, Int. J. Quantum Chem. 106 (2006) 2201.
[11] Z.K. Tang, Y. Nozue, T. Goto, J. Phys. Soc. Japan 61 (1992) 2943.
[12] O.A. Yeshchenko, I.M. Dmitruk, S.V. Koryakov, I.P. Pundyk, Y.A. Barkanov, Solid State Comm. 133 (2005) 109.
[13] J.L. Marin, R. Riera, S.A. Cruz, J. Phys. C 10 (1998) 1349.
[14] P. Hawrylak, Phys. Rev. B 60 (1999) 5597.
[15] P.O. Fröman, S. Yngve, N. Fröman, J. Math. Phys. 28 (1987) 1813.
[16] W. Jaskolski, Physics Reports 1 (1996) 271.
[17] J.D. Levine, Phys. Rev. A 140 (1965) 586.
[18] S. Satpathy, Phys. Rev. B 28 (1983) 4585.
[19] Z. Liu, D.L. Lin, Phys. Rev. B 28 (1983) 4413.
[20] Y. Shan, T.F. Jiang, T.C. Lee, Phys. Rev. B 31 (1985) 5487.
[21] H.C. Tseng, Y. Shan, J. Phys. C 1 (1989) 2225.
[22] Z.J. Shen, X.Z. Yuan, B.C. Yang, Y. Shen, Phys. Rev. B 48 (1993) 1977.
[23] J.H. You, Z.G. Ye, M.L. Du, Phys. Rev. B 41 (1990) 8180.
[24] E. Ley-Koo, R.M.G. Garcia-Castelan, J. Phys. A 24 (1991) 1481.
[25] E. Ley-Koo, S. Mateos-Cortes, Int. J. Quantum Chem. 46 (1993) 609.
[26] S.A. Cruz, E. Ley-Koo, J.L. Marin, A. Taylor-Armitage, Int. J. Quantum Chem. 54 (1995) 3.
[27] D.S. Krämer, W.P. Schleich, V.P. Yakovlev, J. Phys. A 31 (1998) 4493.
[28] J.D. Serna-Salazar, J. Machecha-Gomez, Phys. Chem. Chem. Phys. 2 (2000) 4061.
[29] H.C. Lee, H.C. Tseng, Chinese J. Phys. 38 (2000) 49.
[30] S.H. Patil, Z. Naturforsch. A 59 (2004) 455.
[31] Y. Shan, J. Phys. B 23 (1990) L1.
[32] A.F. Kovalenko, M.F. Holovko, J. Phys. B 25 (1992) L 233.
[33] N. Aquino, G. Campoy, H.F. Montgomery, Int. J. Quantum Chem. 107 (2007) 1548.
[34] E. Ley-Koo, S. Rubinstein, J. Chem. Phys. 71 (1979) 351.
[35] E. Ley-Koo, S.A. Cruz, J. Chem. Phys. 74 (1981) 4603.
[36] B.L. Burrows, M. Cohen, Int. J. Quantum Chem. 106 (2006) 478.
[37] J.L. Marin, S.A. Cruz, J. Phys. B 24 (1991) 2899.
[38] J.L. Marin, S.A. Cruz, J. Phys. B 25 (1992) 4365.
[39] Y.P. Varshni, Z. Naturforsch. A 57 (2002) 915.
[40] S.H. Patil, Y.P. Varshni, Can. J. Phys. 82 (2004) 647.
[41] M.E. Changa, A.V. Scherbinin, V.I. Pupyshev, Int. J. Quantum Chem. 96 (2004) 167.
[42] B.M. Guimarc, J. Chem. Phys. 47 (1967) 5110.
[43] J. Gorecki, W. Byers Brown, J. Phys. B. 21 (1988) 403.
[44] N. Aquino, A. Flores-Riveros, J.F. Rivas-Silva, Phys. Lett. A 307 (2003) 326 and references therein.

[45] N. Aquino, J. Garza, A. Flores-Riveros, J.F. Rivas-Silva, K.D. Sen, J. Chem. Phys. 124 (2006) 054311.
[46] C. Diaz-Garcia, S.A. Cruz, Phys. Lett. A 353 (2006) 332.
[47] A. Banerjee, C. Kamal, A. Chowdhury, Phys. Lett. A 350 (2006) 121.
[48] D. Zorrilla Cuenca, J. Sánchez Márquez, M. Fernández Nuñez, R. Rodríguez Huertas, Int. J. Quantum Chem. 107 (2007) 879.
[49] E. Ley-Koo, A. Flores-Flores, Int. J. Quantum Chem. 66 (1998) 123.
[50] W.F. Xie, Chin. Phys. Lett. 23 (2006) 1742.
[51] A. Corella-Madueno, R.A. Rosas, J.L. Marin, R. Riera, Int. J. Quantum Chem. 77 (2000) 509.
[52] Y.P. Varshni, Eur. Phys. J. D 22 (2003) 229.
[53] R. Le Sar, D.R. Herschbach, J. Phys. Chem. 85 (1981) 2798.
[54] R. Le Sar, D.R. Herschbach, J. Phys. Chem 87 (1983) 520.
[55] T. Pang, Phys. Rev. A 49 (1994) 1709.
[56] C. Le Sech, G. Hadinger, M. Aubert-Frecon, Z. Phys. D 32 (1994) 219.
[57] E. Ley-Koo, K.P. Volke-Sepúlveda, Int. J. Quantum Chem. 65 (1997) 269.
[58] S.A. Cruz, E. Ley-Koo, R. Cabrera-Trujillo, AIP Conf. Proc. 963 (2007) 175.
[59] S.A. Cruz, E. Ley-Koo, R. Cabrera-Trujillo, Phys. Rev. A 78 (2008) 032905.
[60] J.P. Connerade, V.K. Dolmatov, P.A. Lakshmi, J. Phys. B 33 (2000) 251.
[61] J.P. Connerade, R.J. Semaoune, J. Phys. B 33 (2000) 3467.
[62] J. Garza, R. Vargas, A. Vela, K.D. Sen, J. Mol. Struct. (Theochem) 501/502 (2000) 183.
[63] J. Garza, R. Vargas, N. Aquino, K.D. Sen, J. Chem. Sci. 117 (2005) 379.
[64] S.A. Cruz, S.A.C. Díaz-García, G. Covarrubias, Int. J. Quantum Chem. 102 (2005) 897.
[65] C. Díaz- García, S.A. Cruz, Int. J. Quantum Chem. 108 (2008) 1572.
[66] W.P. Wang, R.G. Parr, Phys. Rev. A 16 (1977) 891.
[67] W.P. Wang, Phys. Rev. A 25 (1982) 2901.
[68] E. Hernández, J.L. Gázquez, Phys. Rev. A 25 (1982) 107.
[69] W. Yang, Phys. Rev. A 34 (1986) 4575.
[70] C.F. Weizsäcker, Z. Phys. 96 (1935) 431.
[71] D.A. Kirshnits, Zh. Eksp. Teor. Fiz. 32 (1957) 115 [Sov. Phys. JETP 5 (1957) 64].
[72] E. Clementi, C. Roetti, At. Nucl. Data Tables 14 (1974) 177.
[73] G. Arfken, Mathematical Methods for Physicists, 2nd ed., Academic Press, New York, 1970.
[74] P.M. Morse, H. Feshbach, Methods of Theoretical Physics, Part II, McGraw-Hill, New York, 1953.
[75] D.S. Gemmell, Rev. Mod. phys. 46 (1974) 129.

CHAPTER 9

Confined Atoms Treated as Open Quantum Systems

Richard F.W. Bader [a]

Contents

1. PHYSICALLY CONFINED SYSTEMS ARE OPEN SYSTEMS

If a confined system, atom or molecule, is restrained by a physical boundary rather than by a mathematically imposed boundary condition, it is an open system. This chapter is concerned with the former case, of a system bounded by the surfaces it shares with the atoms of the confining material, surfaces that enable the transfer of matter and momentum. This places the problem in the realm of the quantum mechanics of a 'proper open system', a system

[a] Department of Chemistry, McMaster University, Hamilton ON, L8S 4M1, Canada

Advances in Quantum Chemistry, Vol. 57
ISSN: 0065-3276, DOI: 10.1016/S0065-3276(09)00613-3

Ω bounded by a surface $S(\Omega; \mathbf{r})$ of zero-flux in the gradient vector field of the electron density $\rho(\mathbf{r})$, Equation (1) [1,2].

$$\nabla\rho(\mathbf{r}) \bullet \mathbf{n}(\mathbf{r}) = 0 \quad \forall\mathbf{r} \in S(\Omega, \mathbf{r}). \tag{1}$$

Treating a confined system as open brings into play the important role played by the interaction of the system of interest with its surroundings; to what extent are changes in the pressure–volume (pv) product of the open system mediated by changes in the atoms of the confining surfaces; are, for example, the changes accompanied by a significant transfer of electronic charge?

In addition to defining the bounding surface of an open system, the gradient vector field of the electron density also provides a definition of structure in terms of bond paths, lines of maximum density linking neighbouring atoms. [3] Thus the open system approach delineates the structure, not only of the open system, but of its interaction with its surroundings and of how these structures may change with changes in the pv product.

2. MECHANICS OF AN OPEN SYSTEM

Properties of a proper open system Ω are defined by Heisenberg's equation of motion obtained from the variation of the state vector within the system and on its boundaries [1,4], in the manner determined by Schwinger's principle of stationary action [5]. For a stationary state, the equation of motion for an observable $\hat{G}$ is given in Equation (2),

$$\begin{aligned}(N/2)&\left\{(\mathrm{i}/\hbar)\,\langle\Psi|\,[\hat{H},\hat{G}]\,|\Psi\rangle_{\Omega} + cc\right\} \\ &= (1/2)\left\{\oint \mathrm{d}S(\Omega;\mathbf{r})\mathbf{J}_G(\mathbf{r}) \bullet \mathbf{n}(\mathbf{r}) + cc\right\},\end{aligned} \tag{2}$$

where the current for property G is given by

$$\mathbf{J}_G(\mathbf{r}) = (\hbar/2m\mathrm{i})N \int \mathrm{d}\tau' \left\{\Psi^*\nabla_r(\hat{G}(\mathbf{r})\Psi) - \nabla_r\Psi^*(\hat{G}(\mathbf{r})\Psi)\right\} \tag{3}$$

and its property density by

$$\rho_G(\mathbf{r}) = N/2\left\{\int \mathrm{d}\tau'\Psi^*(\hat{G}(\mathbf{r})\Psi) + (\hat{G}(\mathbf{r})\Psi)^*\Psi\right\}. \tag{4}$$

The re-writing of the Heisenberg equation in the Schrödinger representation as in Equation (2) from the field theoretic language of Schwinger leads to the transformation of every property into an effective single-particle density and associated current defined in real space; the symbol $\int d\tau'$ implies a summation over spins and the integration over all coordinates save those denoted by $\mathbf{r}$, the coordinates of the electron integrated over the open system. The symbol $\langle\rangle_\Omega$ implies the same mode of integration, followed by integration of $\mathbf{r}$ over Ω. Aside from the summation over spins, this mode of integration is that introduced by Schrödinger in his fourth paper to define the 'electric density', the same paper in which he introduced the current density and the equation of continuity relating one to the other [6]. The resulting real space representation of density and current for every property is a most important result for it yields, through the Heisenberg equation, a real-space 'dressed' representation of *all* of a system's definable properties, including the total energy and Ehrenfest force that involve two-particle interactions. It is clear from the real-space representation of the property densities, that the properties of an open system are additive, the properties of the total system equalling the sum of the contributions from its component open systems [1].

The basic single-particle nature of the expectation values follows from the generator acting only on the coordinates of a single electron in the field theoretic expressions, Schwinger pointing out that "the essence of field theory is to provide a conceptually simpler and more fundamental description depending on the particle as the basic entity" [7].

2.1. Atomic expressions of the Ehrenfest and virial theorems

The Ehrenfest and virial theorems are obtained from the Heisenberg equation of motion for an open system, Equation (2), with $\hat{G}$ given by the electronic operators $-i\hbar\nabla$ and $\mathbf{r} \bullet \mathbf{p}$, respectively. The commutator term for the momentum operator yields $-\nabla_{\mathbf{r}}\hat{V}$, which determines the force exerted on the electron at $\mathbf{r}$ by the remaining electrons and by the nuclei, all in fixed positions. Taking the expectation value of this force in the manner denoted by $N \int d\tau'$ yields an expression for $\vec{F}(\mathbf{r})$, the force exerted on an electron at position $\mathbf{r}$ by the *average distribution* of the remaining electrons and by the rigid nuclear framework – a 'dressed density' – giving the force exerted on the electron density. The physics of an open system defines a corresponding 'dressed' density distribution for every measurable property, one whose integration over an atomic basin yields the atom's additive contribution to that property. A dressed density distribution for some particular property accounts for the corresponding interaction of the density at some point in space with the remainder of the molecule. The force density is an example of a physical quantity that clearly involves two-electron operators, and yet is expressible in terms of a real-space density. In a stationary state the force exerted on an atom is given by the surface integral of the momentum flux density expressed in terms of the stress tensor, Equation (5) [4,8]:

$$\vec{F}(\Omega) = \int_{\Omega} \mathrm{d}\mathbf{r} \int \mathrm{d}\tau' \left\{ \psi^*(-\nabla_r \hat{V})\psi \right\}$$
$$= \int_{\Omega} \mathrm{d}\mathbf{r}\, \vec{F}(\mathbf{r}) = -\oint \mathrm{d}S(\Omega; \mathbf{r_s})\, \overleftrightarrow{\sigma}(\mathbf{r}) \bullet \mathbf{n}(\mathbf{r}). \quad (5)$$

The stress tensor $\overleftrightarrow{\sigma}(\mathbf{r})$, which is the momentum current or 'momentum flux density' [9], is given by

$$\overleftrightarrow{\sigma}(\mathbf{r}) = (\hbar^2/4m)\left\{(\nabla\nabla + \nabla'\nabla') - (\nabla\nabla' + \nabla'\nabla)\right\}\Gamma^{(1)}(\mathbf{r}, \mathbf{r}')|_{\mathbf{r}=\mathbf{r}'}. \quad (6)$$

The stress tensor has the dimensions of an energy density (E/v) or equivalently, pressure (F/A), and is sometimes referred to as the pressure tensor. Wong [9] and Takabayashi [10] have used classical arguments based on an assumed homogeneity of matter to derive an expression relating the pressure to $-(1/3)\mathrm{Tr}\sigma(\mathbf{r})$. Unlike the assumed homogeneous nature of classical matter, the density of quantum matter and its mechanical properties, including $\sigma(\mathbf{r})$, exhibit large local variations, even within a single atom. Thus while the equilibrium pressure in a classical continuum is constant throughout the system, the same is not true for a quantum system. Equation (5) states that the force exerted on the density of an open system is equal and opposite to the force exerted on it by the surroundings as determined by the flux in its momentum.

The expectation value of the commutator for the virial operator $\hat{G}(\mathbf{r}) = \hat{\mathbf{r}} \bullet \hat{\mathbf{p}}$ yields $2T(\Omega) + \nu_b(\Omega)$, twice the atom's electronic kinetic energy, $T(\Omega)$, together with the virial of the Ehrenfest force exerted over the basin of the atom, $\nu_b(\Omega)$ [4]. In a stationary state these contributions are balanced by $\nu_s(\Omega)$, the virial of the Ehrenfest force acting over the surface of the atom. Expressing by $\nu(\Omega)$, the total virial for atom Ω, the virial theorem for a stationary state may be stated as [4]

$$-2T(\Omega) = \nu(\Omega) = \nu_b(\Omega) + \nu_s(\Omega). \quad (7)$$

The virials of the Ehrenfest force exerted over the basin and the surface of the atom, with the origin for the coordinate $\mathbf{r}$ placed at the nucleus of atom Ω, are given respectively in Equations (8) and (9)

$$\nu_b(\Omega) = -\int_{\Omega} \mathrm{d}\mathbf{r}\, \mathbf{r}_{\Omega} \bullet \nabla \bullet \overleftrightarrow{\sigma}(\mathbf{r}) = \int_{\Omega} \mathrm{d}\mathbf{r}\, \mathbf{r}_{\Omega} \bullet \vec{F}(\mathbf{r}) \quad (8)$$

$$\nu_s(\Omega) = \oint \mathrm{d}S(\Omega, \mathbf{r_s})\mathbf{r}_{\Omega} \bullet \overleftrightarrow{\sigma}(\mathbf{r}) \bullet \mathbf{n}(\mathbf{r}). \quad (9)$$

Each theorem obtained from Equation (5) can be stated in a local form, the local form of the viral theorem being

$$(\hbar^2/4m)\nabla^2\rho(\mathbf{r}) = 2G(\mathbf{r}) + \nu(\mathbf{r}). \tag{10}$$

$G(\mathbf{r})$ is the positive definite form of the kinetic energy density [11] and the virial field $\nu(\mathbf{r})$ may be expressed as

$$\nu(\mathbf{r}) = -\mathbf{r} \bullet \nabla \bullet \overset{\leftrightarrow}{\sigma}(\mathbf{r}) + \nabla \bullet (\mathbf{r} \bullet \overset{\leftrightarrow}{\sigma}(\mathbf{r})) = \mathrm{Tr}\ \overset{\leftrightarrow}{\sigma}(\mathbf{r}). \tag{11}$$

Integration of Equation (10) over an atom yields the atomic virial theorem, Equation (7). The virial field $\nu(\mathbf{r})$ is a dressed density distribution of particular importance. It describes the energy of interaction of an electron at some position $\mathbf{r}$ with all of the other particles in the system, averaged over the motions of the remaining electrons. When integrated over all space it yields the total potential energy of the molecule, including the nuclear energy of repulsion, and for a system in electrostatic equilibrium, with $\nu = V$, it equals twice the molecule's total energy. The virial field condenses all of the electron–electron, electron–nuclear and nuclear–nuclear interactions described by the many-particle wave function into an energy density that is distributed in real space.

The virial theorem plays a dominant role in the definition of pressure, in both classical and quantum mechanics. The following section demonstrates that the pv product for a proper open system is proportional to the surface virial, Equation (9), the virial of the Ehrenfest forces exerted by the surroundings on the open system [9,12].

3. VIRIAL THEOREM AND THE DEFINITION OF PRESSURE

The opening discussion will demonstrate that the definition of pressure is a problem that requires the physics of an open system, classical or quantum. This is an understandable result since the pressure acting on a system is the force exerted per unit area of the surface enclosing the system, the flux in the momentum density per unit area per unit time of the bounding surface. This understanding calls into question the use of the result obtained from the classical virial theorem for an ideal gas to define the pressure acting on a quantum system.

A standard result obtained from the classical virial theorem, Equation (12), shows that for a confined gas of uniform density – a closed isolated system – the virial of the 'wall forces' equals $3pv$.

$$\left(\sum_{i=1}^{N}(2T_i + \mathbf{r}_i \bullet \mathbf{F}_i)\right) = 0. \tag{12}$$

The term $\Sigma \mathbf{r}_i \bullet \mathbf{F}_i$ consists in general of external and internal contributions, the former being the virial of the constraining forces that result from the exchange of momentum of the gas molecules with the walls of the container, the latter being the virial of the intermolecular forces acting within the system. Calculating the exchange of momentum with the walls yields $-3pv$ and, for a system with no internal forces, Equation (12) yields the ideal gas equation $NkT = pv$ [13]. Unfortunately, this result has been carried over to the quantum mechanical case with the mistaken identification of the virial of the wall forces with the virial of the Feynman forces acting on the nuclei of a closed, isolated system. There is no rationale for doing this. The virial ν for a closed, isolated quantum system is given by

$$\nu = \int \nu(\mathbf{r})d\mathbf{r} = \langle \hat{V}_{ne} \rangle + \langle \hat{V}_{ee} \rangle + \langle \hat{V}_{nn} \rangle + \sum_{\alpha} \mathbf{X}_{\alpha} \bullet \mathbf{F}_{\alpha} = V + \sum_{\alpha} \mathbf{X}_{\alpha} \bullet \mathbf{F}_{\alpha}, \quad (13)$$

where V is the total potential energy in $E = T + V$ and $\mathbf{F}_{\alpha}$ is the Feynman force acting on nucleus α. The final term in Equation (13) is the contribution to the virial from the forces of constraint necessary to keep the nuclei in non-equilibrium configurations, and is termed the 'external virial' [14]. These forces it must be realized, act on nuclei *internal* to the system. The quantum statement of the virial theorem enables one to relate the difference between E and the kinetic energy T to the external forces of constraint, Equation (14),

$$T + E = +\sum_{\alpha} \mathbf{X}_{\alpha} \bullet \mathbf{F}_{\alpha}. \quad (14)$$

In previous calculations of pressure for a molecular system, the virial of the Feynman constraining forces acting on the nuclei given in Equation (14) as the sum $T + E$ was identified with the external virial in Equation (12) in the classical description of an ideal gas and was equated to $3pv$ [15–18]. The reader is referred to the article by Ludeña [15] for a survey of earlier work. There is no formal justification for equating the two external virials, one classical and the other quantum, one reflecting a wall constraint, the other, the forces of constraint acting on all of the nuclei internal to the system. Only by treating the problem as one requiring the physics of an open system, does one obtain a definition of pressure.

3.1. Grand ensemble approach to an open system

Schweitz was the first to recognize that the definition of pressure is a problem in the physics of an open system. He argued that the classical result based on the properties of a closed, isolated system – a petit ensemble – must be recast in terms of an open system, a system with permeable walls, replacing the petit ensemble with the grand ensemble. This he did for both a classical system [19] and a quantum system [9]. This leads Schweitz to the definition of a 'grand virial' Z for a classical system which consists of an internal virial $Z_p = \sum \mathbf{r}_i \bullet \mathbf{F}_i$ of a petit system and a 'surface flux virial' Z_s where

$$Z_s = \oint \mathrm{d}S\{\mathbf{r} \bullet (\overleftarrow{\mathbf{p}}(\mathbf{r}) - \overrightarrow{\mathbf{p}}(\mathbf{r}))\} \tag{15}$$

and here $\overrightarrow{\mathbf{p}}(\mathbf{r})$ and $\overleftarrow{\mathbf{p}}(\mathbf{r})$ are the mean flux densities of momentum (momentum/time and area) into and out of the volume of the open system. The grand virial equation is then

$$-2T = Z_p + Z_s = Z. \tag{16}$$

Schweitz proceeded to show that for a gas with an isotropic and uniform distribution of momentum, Z_s reduces to $-3pv$, and in the absence of internal forces, one obtains $2T = 3pv$, a statement of the ideal gas equation. One notes that the derivation is based on the exchange of momentum through imaginary walls, while in the case of a petit ensemble, the derivation is based on the action of forces exerted on the particles by impenetrable walls. Schweitz concludes with a discussion of the use by Slater [14], March [20] and Ross [21] of the virial theorem for a petit ensemble stated in the form

$$2T + U = -3V(\partial E/\partial V) \tag{17}$$

in the definition of pressure. ($E = T + U$, U being the internal potential energy). This is possible in the case of a homogeneous gas where the surface flux virial equals the constraint virial for the corresponding petit ensemble, a point that is rarely pointed out in the literature. One notes that $2T + U$ in Equation (17) is simply $T + E$ of Equation (14) in the quantum mechanical statement of the virial theorem and that the use of Equation (17) in the case of a quantum system is equivalent to equating the pressure to the virial of the Feynman forces acting on the nuclei. The papers by Slater, March and Ross were the origin of the improper use of Equation (14) in the calculation of the pressure.

In generalizing the virial theorem to the grand ensemble in the quantum case, Schweitz obtained the same form as obtained for a classical system. He expressed the result obtained for the 'grand virial equation' as

$$(2\langle T\rangle + Z_p + Z_s)_{\text{time average}} = 0, \tag{18}$$

where $Z_p = -\sum_i \langle \mathbf{x}_i \bullet \nabla_i \hat{V}\rangle$ is the expectation value of the petit virial, the contribution from the forces internal to the system, and Z_s is the quantum surface flux virial which, in the absence of external fields, is given by

$$Z_s = -\oint \mathrm{d}\mathbf{S} \bullet \{\mathbf{x} \bullet (\hat{\mathbf{J}}\hat{\mathbf{p}})\Gamma^{(1)}(\mathbf{x}, \mathbf{x}')\}, \tag{19}$$

where the 'flux density operator' operator is defined as $\hat{\mathbf{J}} = -(\mathrm{i}\hbar/2m)(\nabla - \nabla')$. It is important to note that in both the classical and quantum cases, Schweitz obtained clear and separate contributions from the virial of the internal forces and the virial of the forces resulting from the momentum flux through the surface of the open system.

Equation (18) is identical in form and content to the atomic statement of the virial theorem Equation (7) – the virial theorem for a proper open system – with the petit virial Z_p being the analogue of the basin virial ν_b and the surface flux virial Z_s, the analogue of the surface virial ν_s, the virial of the Ehrenfest forces acting on the surface of the open system. Schweitz's representation of the quantum stress tensor $\overleftrightarrow{\sigma}(\mathbf{r})$ in terms of the flux density operator acting on the momentum in Equation (19) makes clear its interpretation as a 'momentum flux density'. Schweitz does not, however, consider how the surface flux virial in the quantum case, Z_s or ν_s, may be related to the pv product. This, as demonstrated in the following section, has been accomplished using the atomic statement of the virial theorem [12].

3.2. Pressure and the surface flux virial: construction of the $\hat{p}\hat{v}$ operator

Marc and McMillan [22] in a demonstration of the equivalence of the quantum 'kinetic' and thermodynamic pressure have derived an operational definition of the pv product for the case of a closed system. It begins with the thermodynamic identification of p as the ensemble average of $-\partial E_i/\partial v_i$, with fixed occupation numbers. The Hellmann–Feynman theorem equates this derivative to the state average of the corresponding derivative of $\hat{H}$. The coordinates in $\hat{H}$ are then scaled with a characteristic length of the system, $v^{-(1/3)}$. For the present purpose of obtaining an open system operator, only the coordinates of a single-electron in $\hat{H} = \sum_i \hat{t}_i + \hat{V}$ are scaled, the procedure followed in the original derivation of the atomic virial theorem [23] and a necessary step in the determination of expectation values for open systems [1,2]. Differentiation of the resulting expression, after reverting to the original coordinates, yields

$$-3v\partial\hat{H}/\partial v = 2\hat{t}_i - \mathbf{r}_i \bullet \nabla_i \hat{V}. \tag{20}$$

The RHS of Equation (20) is precisely the operator expression determined by the commutator $(\mathrm{i}/\hbar)[\hat{H}, \hat{G}]$ with $\hat{G}(\mathbf{r}) = \hat{\mathbf{r}} \bullet \hat{\mathbf{p}}$, the virial operator and the averaging of this result over an open system as in Equation (2) yields the atomic statement of the virial theorem. By identifying $-\partial \hat{H}/\partial v$ with the quantum mechanical pressure operator, one is lead to the identification of $-v(\partial \hat{H}/\partial v)$ with $(pv)_{\text{op}}$ to yield the open system definition of pressure, summarized in Equation (21),

$$(N/2)\{(\mathrm{i}/\hbar)\langle\psi|[\hat{H},\mathbf{r}\bullet\mathbf{p}]|\psi\rangle_{\Omega}+cc\}=2T(\Omega)+\nu(\Omega)_b=3\langle(p\mathrm{v})_{\text{op}}\rangle_{\Omega} \quad (21)$$

and thus to the identification of a proportionality between the surface virial and $(pv)_{\text{op}}$,

$$\langle(pv)_{\text{op}}\rangle_{\Omega}=p(\Omega)v(\Omega)=-(1/3)\nu_s(\Omega). \quad (22)$$

Equation (22) properly relates the pressure to the virial of the force arising from the surface flux in the momentum density.

de Boer and Bird [24] defined $(pv)_{\text{op}}$ in analogy with the classical virial expression for a closed system of particles, in effect using the expression of Slater, Equation (16), to relate the expectation value of $(pv)_{\text{op}}$ to $-3V(\partial E/\partial v)$. As pointed out by Schweitz this is possible only in the case of a homogeneous gas and is an attempt to replace the proper contribution from the flux in the surface virial by fictitious wall forces whose energy is not included in $E = T + U$.

4. CALCULATION OF PRESSURE IN TERMS OF THE SURFACE VIRIAL

4.1. Atomic volumes and apparent origin dependence

The use of Equation (22) in the calculation of the pressure acting on a proper open system requires the definition of the volume of the open system and the fixing of an origin for the calculation of the virial of the forces acting on the system. The determination of the volume is no problem for an atom, ion or unit cell in a solid where the open system is totally enclosed by interatomic surfaces. While this in general is not the case for an atom in a molecule, there are cases where an atom is enclosed by the surfaces it shares with its bonded neighbours, examples being an atom X enclosed in an endohedral complex such X@adamantane, the central carbon atom in *neo*-pentane or the Cr atom in its hexacarbonyl complex. A standard result of the QTAIM programs is the determination of the atomic volume out to the 0.002 and 0.001 au electron density envelope, these volumes correlate well with the experimentally determined van der Waals size and shape [25]. The volume defined by the 0.001 au in general contains ~99% of the atom's density and if used consistently in an investigation, provides a suitable measure of an atom's volume.

The total virial for an open system $\nu(\Omega)$, is of course, origin independent since it equals $-2T(\Omega)$. This is not the case, however, for its expression in terms of the sum of its basin $\nu_b(\Omega)$ and surface $\nu_s(\Omega)$ contributions, whose values, according to their definitions in Equations (8) and (9), are dependent up a choice of origin. Consider a coordinate transformation denoted by $\mathbf{r} = \mathbf{r}' + \delta\mathbf{R}$ caused by a shift $\delta\mathbf{R}$ in the origin. This has the effect of changing the basin and surface virials by the same absolute amount, equal to the virial of the Ehrenfest force as given in Equation (23),

$$\delta\mathbf{R} \bullet \vec{F}(\Omega) = -\delta\mathbf{R} \bullet \oint \mathrm{d}S(\mathbf{r}, \Omega) \bullet \overleftrightarrow{\sigma}(\mathbf{r}) \tag{23}$$

but with opposite signs, leaving $\nu(\Omega)$ unchanged in value. For an open system, bounded by symmetrically equivalent surfaces, as say in a linear chain of atoms, or of course a unit cell in one-, two- or three-dimensional solids, the cancellation of the surface virials caused by the origin shift, Equation (23), occurs separately for the basin and surface contributions to the atomic virial. Thus the surface virials and the pv products they define are uniquely determined, as long as all are referenced to the same origin [12].

4.2. Surface virial as a measure of energy of formation of atomic surface

Before proceeding with the development of relating the pressure to the surface virial, it is useful to introduce a related concept, the *interatomic surface virial*. An atomic surface $S(\Omega; \mathbf{r})$ consists in general, of a number of interatomic surfaces $S(\Omega|\Omega'; \mathbf{r})$, one with each of its bonded neighbours, that is, atoms linked to it by bond paths. Correspondingly, the surface virial for the atom will consist of a sum of contributions, one for each bonded atom, and each bonded interaction will contribute to the pressure exerted on the atom. One has

$$\nu_s(\Omega) = \sum_{\Omega \neq \Omega'} \oint \mathrm{d}S(\Omega\,|\Omega'; \mathbf{r})\mathbf{r}_\Omega \bullet \overleftrightarrow{\sigma}(\mathbf{r}) \bullet \mathbf{n}_\Omega(\mathbf{r}) = \sum_{\Omega' \neq \Omega} \nu_s(\Omega\,|\Omega'). \tag{24}$$

The Ehrenfest force acting on an atom may be similarly expressed in terms a sum of interatomic contributions, one from the force exerted by each of its bonded neighbours

$$\vec{F}(\Omega) = -\sum_{\Omega' \neq \Omega} \oint \mathrm{d}S(\Omega\,|\Omega'; \mathbf{r})\, \overleftrightarrow{\sigma}(\mathbf{r}) \bullet \mathbf{n}_\Omega(\mathbf{r}) = -\sum_{\Omega \neq \Omega'} \vec{F}(\Omega\,|\Omega')\,. \tag{25}$$

While the surface and basin virials are dependent upon the choice of origin for the vector $\mathbf{r}_\Omega$, it is possible to remove the origin dependence and replace it with the measurable bond length, the vector $\mathbf{R}_{ab} = \mathbf{r}_a - \mathbf{r}_b$, where $\mathbf{R}_{ab}$ is directed from the nucleus of A to that of B and $\mathbf{r}_a$ and $\mathbf{r}_b$ are referenced to their respective nuclei. With these definitions and the property of the surface normals in Equations (24) and (25) that $\mathbf{n}_a = -\mathbf{n}_b$, one obtains the definition of the *interatomic* surface virial $\nu_s(A \parallel B)$, Equation (26).

$$\begin{aligned}\nu_s(A \parallel B) &= \nu_s(A|B) + \nu_s(B|A) \\ &= \mathbf{R_{ab}} \bullet \oint \mathrm{d}S(A|B; \mathbf{r})\, \overleftrightarrow{\sigma}(\mathbf{r}) \bullet \mathbf{n_a}(\mathbf{r}) = \mathbf{R_{ab}} \bullet \vec{F}(A|B).\end{aligned} \tag{26}$$

This expression equates the sum of the surface virials from two atoms sharing a common surface $S(A|B)$ to the scalar product of the Ehrenfest force acting on the surface with the bond length vector $\mathbf{R}_{ab}$. *The interatomic surface virial removes the dependence of the surface integrals on individual atomic origins and replaces it with the observable separation between the nuclei of the bonded atoms.* As a consequence of introducing the bond length vectors, the basin virials are necessarily nuclear referenced as well. In the case of molecule formation from separated atoms, the change in the basin virial, $\Delta\nu_b(A)$, equals the change in potential energy or twice the total energy over the basin of atom A. Thus the energy of formation of a molecule from separated atoms, ΔE_f, can be expressed in terms of contributions from the changes in the potential energy within each atomic basin plus a contribution from the virial of the Ehrenfest force $\vec{F}(A|B)$ acting in the interatomic surface between each pair of bonded atoms denoted by $\nu_s(A \parallel B)$, Equation (27),

$$\Delta E_f = 1/2 \sum_{A} \Delta\nu_b(A) + 1/2 \sum_{A|B} \nu_s(A \parallel B). \tag{27}$$

The interatomic surface virial $\nu_s(A \parallel B)$ vanishes for the separated atoms and hence $\nu_s(A \parallel B)$ is the contribution to the energy of formation of the molecule arising from the creation of an interatomic surface $S(A|B)$ between atoms A and B, a most useful result. The Ehrenfest force $\vec{F}(A|B)$ is attractive when the force exerted on the density of atom A is directed at B, the situation found for all bonded interactions. The interatomic surface virial $\nu_s(A \parallel B)$ is negative in such a case and the formation of the surface $S(A|B)$ contributes to the stability of the system.

The definition of $\nu_s(A \parallel B)$ plays a role in the assignment of the origin for the atomic virial. Consider an interatomic surface $A|B$, where the surface may be in a diatomic AB or between atoms in a polyatomic system. A shift in origin for the calculation of the basin and surface virials for atom A leaves their sum, the total virial $\nu(A)$, unchanged in value, as the contributions from the displacement of the origin to the basin and surface virials are of opposite sign, Equation (23). One may find a common origin lying between the nuclei of A and B, such that the surface integral vanishes for each atom. Shift the origin from nucleus A by the amount $\mathbf{d}_a$ so that $\mathbf{r}_a = \mathbf{d}_a + \mathbf{r}_o$, yielding the transformation

$$\nu_s(A) = \mathbf{d}_a \bullet \oint dS \overleftrightarrow{\sigma} \bullet \mathbf{n} + \oint dS\mathbf{r}_0 \bullet \overleftrightarrow{\sigma} \bullet \mathbf{n} = \mathbf{d}_a \bullet \vec{F}(A) + \oint dS\mathbf{r}_0 \bullet \overleftrightarrow{\sigma} \bullet \mathbf{n}. \quad (28)$$

The transformed surface virial vanishes if $\mathbf{d}_a = \nu_s(A)/\vec{F}(A)$. Since the basin contribution changes by $-\mathbf{d}_a \bullet \vec{F}(A)$ the remaining surface term resulting from the shift in origin is cancelled and the basin virial referenced to $\mathbf{r}_o$, $\nu_b^0(A)$, then equals the total virial $\nu(A)$. Similarly, the shift in origin for atom B, $\mathbf{d}_b$, causing the transformed surface integral to vanish is given by $\mathbf{d}_b = \nu(B)/\vec{F}(B)$, where $\mathbf{d}_b$ is directed from nucleus B towards A. Since $\vec{F}(A) = -\vec{F}(B)$, the difference between the two origins is given by $\mathbf{d}_a - \mathbf{d}_b = (\nu_s(A) + \nu_s(B))/\vec{F}(A)$. As demonstrated in Equation (26), the sum of the surface virials referenced to their respective nuclei equals $\mathbf{R}_{ab} \bullet \vec{F}(A)$ and hence $\mathbf{d}_a - \mathbf{d}_b = \mathbf{R}_{ab}$. Thus the common origin that causes both surface integrals to vanish lies between the A and B nuclei, located such that the sum of the absolute displacements from the nuclei equals the bond length $\mathbf{R}_{ab}$. The surface contributions defined by this common origin are the ones necessary to define the interatomic surface virial, $\nu_s(A \parallel B)$, that determines the contribution to the energy resulting from the formation of the surface $A|B$, Equation (27), an energy determined by the measurable bond length and free of any origin dependence, Equation (26).

4.3. Atomic pressures in diatomic molecules

An introduction to the properties of the surface virials and the pressures they determine is given in terms of the atomic contributions in diatomic molecules whose bonding covers the range from shared, to polar to ionic as represented by N_2, CO and LiF, respectively. The previously reported ground state wave functions were obtained from QCISD/SCVS/6-311++G(2df) calculations [26], where SCVS denotes 'self consistent scaling of the electronic coordinates' to ensure satisfaction of the virial theorem [27]. The energies recover over 90% of the experimental binding energies and the calculated bond lengths are in good agreement with experiment. One finds that the pressure exerted on A and B is very much determined by the nature of the interatomic surface which is characteristic for shared, polar and ionic interactions. The surfaces for the N_2, CO and LiF molecules are illustrated in Figure 1, superposed on the electron density distributions. The general nature of the conclusions obtained from the study of the above three molecules is demonstrated through the use of results obtained from near Hartree-Fock (HF) wave functions obtained from the Mulliken–Roothaan Laboratory for Molecular Spectra and Structure in Chicago (LMSS) that were

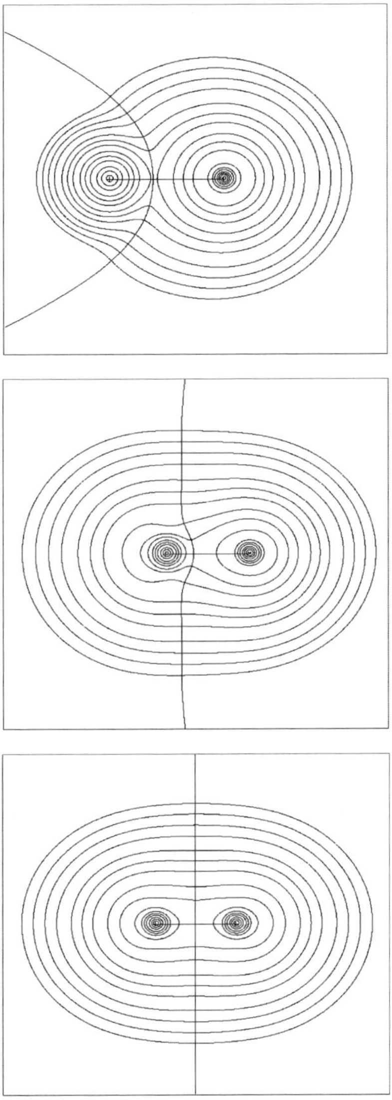

Figure 1 Ground state electron density distributions of LiF, CO and N_2 in their equilibrium geometries, superposed with bond paths and the intersection of the interatomic surfaces. The surface in LiF is characteristic of an ionic interaction, that in CO of a polar interaction and that in N_2 of a shared interaction. An interatomic surface intersects the bond path at the position of the bond critical point. The outer contour value is 0.001 au and the remaining contours increase in value in the order $2 \times 10^n, 4 \times 10^n, 8 \times 10^n$ au with n beginning at -3 and increasing in steps of unity.

generated in the 1960s using STO basis sets containing d and f basis functions with optimized exponents. The LMSS generated over 300 such functions for diatomic A_2, AH and AB molecules in ground and excited states, including ionized and electron attachment states [28–30].

The classification of the interactions as shared, polar or closed-shell (ionic in this case) is based upon the value of the density $\rho(\mathbf{r}_c)$ and its Laplacian $\nabla^2\rho(\mathbf{r}_c)$ at the bond critical point. For shared interactions, $\rho(\mathbf{r}_c) > 0.2$ au and $\nabla^2\rho(\mathbf{r}_c) < 0$, while for closed-shell interactions $\rho(\mathbf{r}_c) < 0.2$ au and $\nabla^2\rho(\mathbf{r}_c) > 0$ and is small in value as are the magnitudes of the individual contributing curvatures. The polar interactions occur in systems with significant charge transfer in which the donor atom retains nonbonded valence density. Thus $\rho(\mathbf{r}_c) > 0.2$, as for shared interactions, but because of the extensive charge transfer, the bonded radius of the donor atom approaches its core radius and, as a consequence, the positive curvature of the density at $\mathbf{r}_c$, the curvature along the bond path, is large and may exceed the magnitude of the two negative curvatures perpendicular to the bond path. (All of the curvatures of the density at the bond critical point are of large magnitude for a polar interaction, as they are for a shared interaction.) Thus $\nabla^2\rho(\mathbf{r}_c)$ may be greater or less than zero for a polar interaction. Because of the remaining valence density on the donor atom, the interatomic surface does not continue its initial parabolic shape found in the region of the donor core, as characteristic of a significant charge transfer, but instead straighten out and assume the planar nature of a shared interaction, Figure 1. In many polar interactions, the outward arms of the surface are slanted towards the acceptor atom, eg, BeH and BH discussed below. The values of $\rho(\mathbf{r}_c)$ and $\nabla^2\rho(\mathbf{r}_c)$ for the LMSS A_2, AH and AB molecules have been previously tabulated [1].

The planarity of the surface in N_2 or its near planarity, is maintained in molecules with charge transfer in cases where the extent of transfer is insufficient for the bond cp to approach the core radius of the donor atom as found in the polar molecules. Thus CF with $q(\mathrm{C}) = 0.78$e with a bonded radius $r_b(\mathrm{C}) = 0.78$ au and $\nabla^2\rho(\mathbf{r}_c) = +0.60$ au is polar with a CO type surface with arms displaced towards F, while NF with $q(N) = 0.44$e and $r_b(\mathrm{N}) = 1.02$ au and $\nabla^2\rho(\mathbf{r}_c) = -0.50$, is a shared interaction, exhibiting a surface whose surface deviates slightly from planarity with arms slightly bent towards the N atom [1].

Data for the N_2, CO and LiF molecules are given in Table 1, which lists the bond lengths R_e, the atomic charges on $A, q(A)$, and the atomic pressures $p(A)$ together with the data necessary for their determination: the surface virials $\nu_s(A)$ and the atomic volumes $v(A)$ using the relation $p(A) = -1/3\nu_s(\mathrm{A})/\mathrm{v(A)}$ obtained from Equation (22). The smallest atomic pressure is for the N atom, and this is a general feature of a planar or near planar interatomic surface that is found for molecules with relatively small or zero extent of charge transfer. The HF STO results for H_2 and the second-row homonuclear diatomic molecules Li_2 to F_2 exhibit a range of atomic pressures: from a minimum value of 3.5 GPa for H_2 rising to a maximum of 5.7 GPa for B_2 and increasing again from a minimum of 3.3 GPa for C_2

Table 1 Atomic pressures in representative molecules[a]

AB	R_e	$q(A)$	$-\nu_S(A)$	$-\nu_S(B)$	$v(A)$	$v(B)$	$p(A)$	$p(B)$
NN	2.070	0.00	0.0675		122.3		5.4	
CO	2.129	+1.22	0.5186	1.5826	112.8	138.1	45.08	112.4
LiF	2.964	+0.93	0.1520	0.1490	26.71	160.4	55.79	9.1

[a] All quantities in atomic units (au) save pressure which is reported in GPa: 1 au of pressure = 2.941×10^4 GPa.

to 8.7 GPa for F_2. (The HF value of p(N) is 6.5 GPa compared to 5.4 GPa in Table 1). The volumes also increase to a maximum at B_2 but the pressure is not diminished because of the paralleling behaviour of the Ehrenfest force and surface virial. These pressures are less than those found for ionic and polar systems.

The largest atomic pressure recorded in Table 1 is for the oxygen atom in CO. This is a general result for the electron acceptor atom in a molecule wherein the number of valence electrons on the donor atom exceeds the number of orbital vacancies on the acceptor, other examples being BeF, BF and CF and BeH and BH, for example. Thus of the four valence electrons on carbon, 1.2 are transferred to oxygen, and the remaining valence density on carbon is strongly polarized into its nonbonded region, its nonbonded radius exceeding its free atom value by 0.15 au, Figure 1. These features are reflected in the shape of the interatomic surface: because of the substantial charge transfer, the bond critical point lies close to the core of the donor atom, but as a result of the excess valence density found within the basin of the donor atom, the arms of the surface swing back towards the anion. The surface virial is of large magnitude for both C and O, resulting in large atomic pressures, being largest for the acceptor atom in spite of its larger volume. These are general results for polar molecules, as illustrated below using the HF results.

Molecules with the CO type surface possess relatively small dipole moments in spite of substantial charge transfer. Thus the near vanishing dipole moment of CO is a result of the back polarization of the nonbonded density on C essentially nullifying the charge transfer contribution to the molecular dipole. The QCISD correlated density recovers the correct sign C^-O^+: the atomic charges of $\pm 1.22e$ give a charge transfer contribution to the dipole, $q(C) \times R$, equal to 2.603 au, while the countering dipolar polarizations contribute -1.675 and -0.966 au for C and O, respectively. These values yield a dipole moment of 0.038 au compared to the experimental value of 0.048 au, (C^-O^+).

The third characteristic surface is found for molecules with closed-shell interactions, which in the present case are represented by molecules approaching the ionic limit: the acceptor atom having orbital vacancies equal

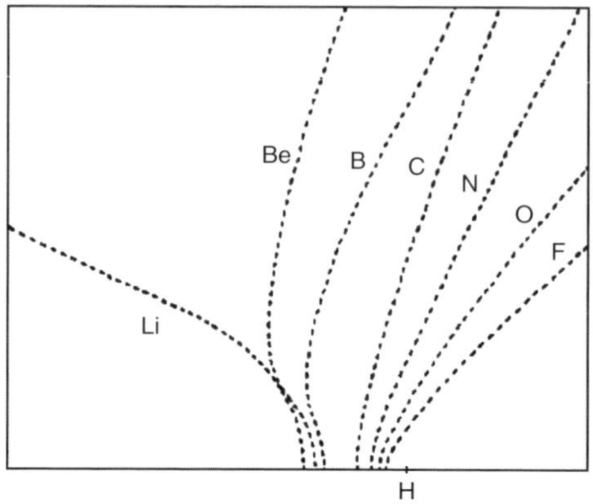

Figure 2 The zero-flux interatomic surfaces for the ground state diatomic hydrides calculated from the LMSS HF STO wave functions. The surfaces are plotted with respect to a fixed position of the proton labelled H. The bond critical point for LiH is located 1.671 au from the proton.

Table 2 Atomic pressures in AH[a]

AH	R_e	$q(A)$	$-\nu_s(A)$	$-\nu_s(\mathrm{H})$	$v(A)$	$v(\mathrm{H})$	$p(A)$	$p(\mathrm{H})$
LiH	3.015	0.91	0.1279	0.0708	30.53	193.5	41.07	3.59
BeH	2.538	0.87	0.2289	0.2871	161.0	137.7	13.94	20.44
BH	2.336	0.75	0.2734	0.5552	166.8	95.2	16.07	57.18
CH	2.124	0.03	0.0339	0.0950	169.3	54.61	2.28	17.05
NH	1.961	−0.32	0.0499	0.0457	150.1	34.70	3.26	12.91
OH	1.834	−0.59	0.1328	0.0359	144.4	21.32	9.02	16.50
FH	1.733	−0.76	0.2317	0.0387	128.7	12.45	17.65	30.43

[a] All quantities, from STO Hartree-Fock wave functions, are in au save pressure which is reported in GPa.

to or greater than the number of valence electrons on the donor. Examples are LiF, NaCl, LiH, LiO and BeO. The transfer of most of the valence density to the basin of the anion is reflected in the simple parabolic surface, Figure 1, and in the bonded radius of the cation approaching that of its core. Even in BeO, 85% of the valence density is transferred to the oxygen atom, the percentage transfers being in excess of 90% for the atoms with a single outer electron. In all cases $\rho(\mathbf{r}_c) < 0.2$ au and $\nabla^2\rho(\mathbf{r}_c) > 0$.

The second-row diatomic hydrides AH, A = Li to F, with data given in Table 2, exhibit all three types of surfaces. This is made clear in Figure 2, which displays the arms of the interatomic surfaces for the hydrides relative to a fixed position of the proton. The change in the direction of charge transfer, initially from A to H up to boron and from H to A beyond carbon (which has close to zero charge transfer) is evident in the advance of the surface towards the proton – the advance of the bond critical point – the

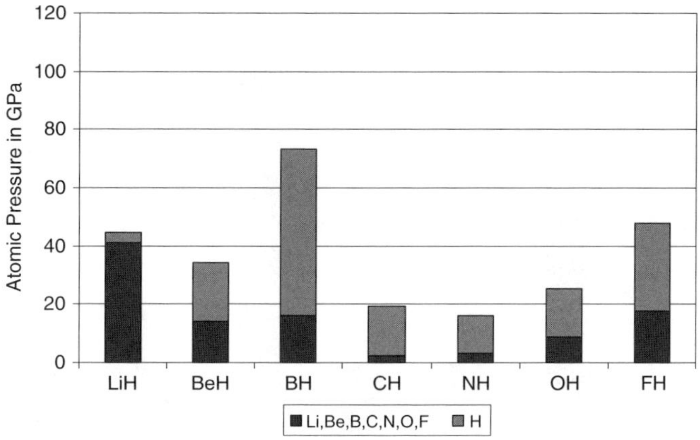

Figure 3 Bar graph of the atomic pressures in diatomic hydrides. The largest atomic pressure is for the negatively charged hydrogen atom in the polar molecule BH.

bonded radius of H changing from 1.34 au in LiH to 0.27 au in HF. The simple parabolic shape for LiH characteristic of ionic systems is transformed in BeH and BH into the surface characteristic of a polar molecule as found in CO. The arms of the remaining surfaces, while planar, are displaced towards the proton as a result of charge transfer from H to A that reaches a maximum of 0.76e in HF. Since hydrogen does not possess a core the charge transfer in NH, OH and FH does not result in ionic character, all three molecules exhibiting bond indices characteristic of shared interactions $\rho(\mathbf{r}_c) \geq 0.28$ au and $\nabla^2\rho(\mathbf{r}_c) \leq -1.6$ au.

The atomic pressures in LiH, Table 2, mirror those found in LiF with $p(\mathrm{Li}) > p(\mathrm{H})$. The value of $p(A)$ decrease for BeH and BH, as is to be anticipated for molecules that mimic the behaviour of CO, with the surface virials and pressures being largest for the acceptor atom H. The sharp decease in pressure found for the carbon atom in CH is anticipated for a shared interaction with a small extent of charge transfer. The volume of the proton decreases beyond carbon, because of the increasing degree of charge transfer leading to an accompanying increase in $p(\mathrm{H})$. The breaks in the variation in the atomic pressures $p(A)$ and $p(\mathrm{H})$ with the nature of the interaction and its accompanying surface are brought to the fore in the bar graph shown in Figure 3: $p(A)$ reaching a maximum in LiH, a minimum in CH and $p(\mathrm{H})$ reaching a maximum in BH.

A similar variation in bonding characteristics and atomic pressures is found for the fluorides, LiF to OF, Table 3. The pressure on Li in the ionic molecule LiF exceeds that on F, mimicking the behaviour of the atomic pressures found for LiH. There is a large decrease in $p(A)$ in the molecules BeF, BF and CF that exhibit the properties of the polar molecule CO [1]. As in CO, the pressure on the acceptor atom F exceeds that on the donor

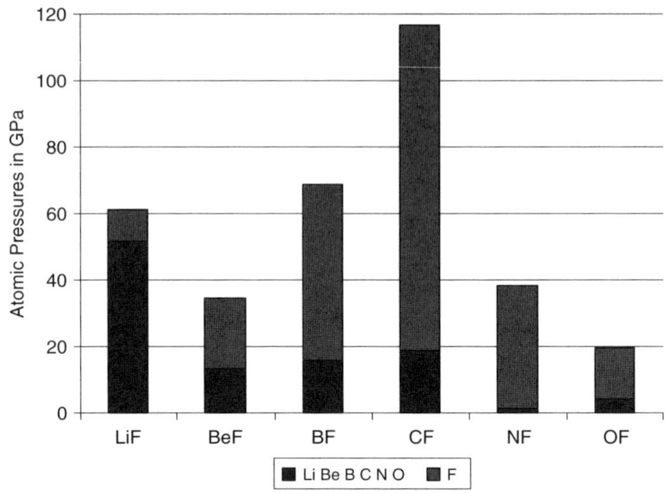

Figure 4 Bar graph of the atomic pressures in diatomic fluorides. The largest atomic pressure is for fluorine in the polar molecule CF.

Table 3 Atomic pressures in AF[a]

AF	R_e	$q(A)$	$-\nu_s(A)$	$-\nu_s(\mathrm{F})$	$v(A)$	$v(\mathrm{F})$	$p(A)$	$p(\mathrm{F})$
LiF	2.955	0.94	0.1346	0.1429	25.45	149.3	51.85	9.4
BeF	2.572	0.95	0.2153	0.3111	157.1	143.0	13.43	21.33
BF	2.391	0.93	0.2474	0.6772	155.0	125.1	15.65	53.07
CF	2.402	0.78	0.2318	1.0741	125.9	107.7	18.83	97.78
NF	2.489	0.44	0.0481	0.3719	112.6	98.9	1.42	36.87
OF	2.496	0.20	0.0457	0.1477	102.5	96.3	4.37	15.04

[a] All quantities from STO Hartree-Fock functions, are in au save pressure which is reported in GPa.

atom, the surface virial for F reaching its maximum value in CF. There is a dramatic drop in $p(A)$ for N, as the extent of charge transfer decreases and the surface assumes a nearly planar shape, features that are accentuated in OF. The surface virials in NF and OF are largest for F and its volume smallest, resulting in $p(\mathrm{F}) > p(A)$. These variations in atomic pressures are displayed in the bar graph in Figure 4, the major break from polar to shared coming at NF as compared to CH in the hydrides. Note the large atomic pressure for F associated with the polar interaction in CF.

4.4. Pressure exerted on atom confined in an adamantane cage

People have succeeded in placing an atom inside a cage molecule. The smallest such system so far synthesized was obtained by subjecting dodecahedrane $C_{20}H_{20}$ to a molecular beam of high energy He atoms [31].

Table 4 Pressure acting on atom X in $X@C_{10}H_{16}$ and associated energy change

X	$v(X)$ (au)	$\nu_s(X)$ (au)	$p(X)$ (GPa)	$\Delta\nu_b(X)$ (au)	$\Delta E(X)$ (kcal/mol)
He	15.14	−1.0177	659	−0.0268	−328
Be^{+2}	7.23	−0.5983	811	−0.6824	−402
Li^{+}	10.22	−0.6682	641	−0.1609	−261
Ne	32.09	−2.0274	620	−0.8497	−904

The resulting cage complex, referred to as $He@C_{20}H_{20}$ is predicted to possess an energy 38 kcal/mol in excess of the isolated reactants [32]. The theory of atoms in molecules has been applied to a study of the complexes formed with the adamantane molecule, $X@C_{10}H_{16}$, with X = He, Be^{+2}, Li^{+} and Ne [33]. The results were obtained using the 6-311++G(2d,2p) basis set in MP2(FULL) calculations, in conjunction with SCVS (self consistent virial scaling) to ensure satisfaction of the virial theorem.

In addition to the intrinsic interest in the pressure acting on such a confined atom, is the finding that the X atoms are totally enclosed within the adamantane cage. This is demonstrated in Figure 5, which gives the molecular graph for the adamantane molecule and its complex with He, showing that the He atom is linked by bond paths to the four methine carbon atoms, the carbon of a |C|H group, the situation found for all X. The diagram illustrates that the He atom is totally enclosed by the four interatomic surfaces it shares with the methine carbon atoms.

The atoms of the adamantane cage are not totally enclosed. Their volumes v(C) and v(H), necessary for the calculation of the atomic pressures, are determined by the intersection of the 0.001 au density envelope with the interatomic surfaces bounding the atom. This volume contains 99.991% of the density of a methine carbon atom, for example. Contributions to the surface integrals $\nu_s(X)$, beyond their intersection with the 0.001 envelope, should make negligibly small contributions to the atomic pressures.

Table 4 gives the pertinent data for the encased X atoms: the atomic volumes $v(X)$ and the values of the surface integral $\nu_s(X) = 4\nu_s(X|C1)$, where $\nu_s(X|C1)$ is the virial of the force exerted on one of the four surfaces the X atom shares with the methine carbon atoms. The pressure acting on X, $p(X)$, is given by $-1/3\nu_s(X)/v(X)$. The largest atomic pressure is for the Be^{+2} ion, the ion with the smallest volume. The He and Ne atoms possess similar pressures even though the surface virial is twice the magnitude for Ne, but this is compensated for by the volume of Ne being twice that of He. Table 5 gives the pressures acting on the atoms in an un-complexed adamantane molecule, for comparison. The pressures acting on the confined X atoms are in general six to eight times larger than that for C2, the atom with the largest pressure in free adamantane. The surface properties are characteristic for a given interaction regardless of the presence of other

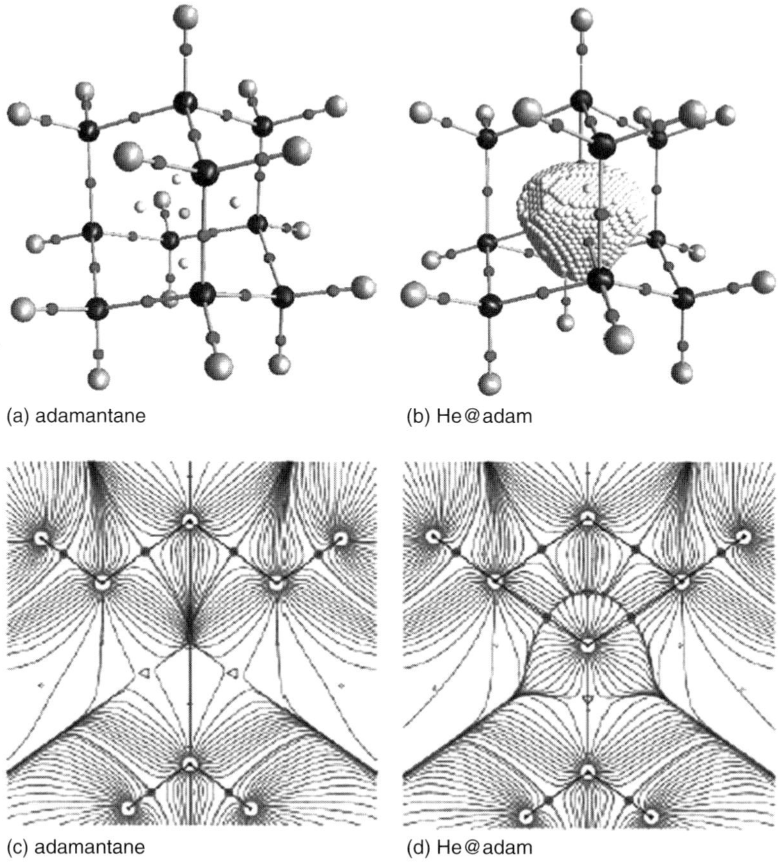

Figure 5 Molecular graphs for adamantane (LHS) and $He@C_6H_{10}$. The He atom is shown encased by its four interatomic surfaces. In the text a methine carbon is labelled C1 and its bonded H atom H1; a methylene carbon is labelled C2 and its two bonded H atoms H2. Critical points are denoted by dots; red for bond, yellow for ring and green for cage. (For interpretation of the references to colour in this figure legend, the reader is referred to the web version of this book.)

interatomic surfaces to the atoms in question. For example, the pressure exerted on the carbon and hydrogen atoms determined by the C1|H1 surface in adamantane are similar to those found in the diatomic CH and the surfaces are similar, as seen by comparing a C|H surface shown in the display of the gradient vector field in Figure 3, with that displayed in Figure 2. Thus one finds $p(\text{H})$ to exceed $p(\text{C})$ in both instances, with $p(\text{C}) = 3.9$ GPa and $p(\text{H}) = 24$ GPa in adamantane, compared to the HF results given in Table 2.

The surface virial for X, when added to the change in its basin virial $\nu_b(X)$, equals twice the *change* in its energy in forming the complex, and these changes in $\nu_b(X)$ and $E(X)$, are listed in Table 4. The surface virials,

Table 5 Atomic pressures acting in adamantane

Ω	$v(X)$ (au)	$v_S(X)$ (au)	$p(X)$ (GPa)
C1	48.47	−0.4282	87
C2	54.58	−0.5954	107
H1	50.34	−0.1236	24
H2	50.59	−0.1203	23

Table 6 Changes in adamantane atomic properties in forming $He@C_{10}H_{16}$

Ω	$\Delta N(\Omega)$	$\Delta E(\Omega)$ (kcal/mol)	$\Delta v(\Omega)$ (au)	$\Delta p(\Omega)$ (GPa)
C1	−0.02	+57.8	+2.44	−7.4
C2	+0.02	+34.1	+2.54	−46.5
H1	0.00	+4.1	+0.89	−2.6
H2	−0.01	+3.1	+0.49	−1.5

Table 7 Changes in adamantane atomic properties in forming $Be^{2+}@C_{10}H_{16}$

Ω	$\Delta N(\Omega)$	$\Delta E(\Omega)$ (kcal/mol)	$\Delta v(\Omega)$ (au)	$\Delta p(\Omega)$ (GPa)
C1	+0.33	−88.7	+8.88	+11.6
C2	+0.15	−19.4	+5.49	−13.9
H1	−0.21	+57.5	−10.86	−9.6
H2	−0.14	+35.1	−7.85	−5.4

which determine $|3pv|$, dominate the energy changes for He, Ne and Li^+, with the basin and surface contributions being of similar magnitude for Be^{+2}. Only the formation of the complex with Be^{+2} is exothermic, by 223 kcal/mol. While the formation of the He, Li^+ and Ne complexes are endothermic, by 156, 70 and 395 kcal/mol respectively, all the complexes reside in potential wells and are structurally stable [33,34]. The single largest stabilizing energy change caused by the insertion of X arises from the attractive interaction of the density of the X atom with the nuclei of the cage [33], the same mode of stabilization operative in metallic carbonyl complexes such as $Cr(CO)_6$ and the metallocenes [26].

Tables 6–9 give the changes in the populations, energies, volumes and pressures for the adamantane atoms following the insertion of the X atom. These results, as discussed below, make clear that the pressure exerted on an atom in a molecule is determined in concert with its interactions with the confining atoms that comprise its surface.

Table 8 Changes in adamantane atomic properties in forming $Li^{+}@C_{10}H_{16}$

Ω	$\Delta N(\Omega)$	$\Delta E(\Omega)$ (kcal/mol)	$\Delta v(\Omega)$ (au)	$\Delta p(\Omega)$ (GPa)
C1	+0.11	−10.1	+6.16	−15.0
C2	+0.09	+8.8	+4.60	−16.4
H1	−0.11	+28.5	−6.30	−5.5
H2	−0.07	+17.1	−4.25	−3.6

Table 9 Changes in adamantane atomic properties in forming $Ne@C_{10}H_{16}$

Ω	$\Delta N(\Omega)$	$\Delta E(\Omega)$ (kcal/mol)	$\Delta v(\Omega)$ (au)	$\Delta p(\Omega)$ (GPa)
C1	0.00	+149.8	+12.06	−38.1
C2	+0.03	+94.1	+9.61	−41.3
H1	−0.01	+9.8	+2.09	−5.5
H2	−0.02	+8.0	+1.35	−3.6

4.4.1. Charge transfer and stability

There is a negligible transfer of electronic charge, ~0.1e, to the He and Ne atoms from C1 and the hydrogen atoms. By far the most significant charge transfer is from the hydrogen atoms to the interior atoms of the cage in the case of the charged atoms, Be^{+2} and Li^{+}, 2.52 and 1.28e, respectively. It is this transfer of charge from the outer atoms to the interior that accounts for the stabilization of the Be^{+2} complex and both C1 and C2 are stabilized along with Be. The reduced extent of charger transfer in the case of the Li^{+} complex is insufficient to override the attendant destabilization resulting from the larger volume of Li. One would predict an even larger exothermicity for the insertion of B^{+3} and this is indeed the case [33]. The formation of $B^{+3}@C_{10}H_{16}$ is predicted to be exothermic by 826 kcal/mol, compared to the 223 kcal/mol for the insertion of Be^{+2}, with the a transfer of 3.7e from the hydrogens to the cage interior, with 0.79e being transferred to B. This increase in charge transfer causes an increase in the exothermicity in spite of the volume of B being greater than that of Be in the complex, 9.6 compared to 7.2 au.

4.4.2. Volume changes of confining atoms and their effect upon the pressure

The volumes of all of the adamantane atoms increase with the insertion of He or Ne and they undergo a corresponding decrease in pressure, the effect being largest for the Ne complex, with the methylene carbons undergoing the largest decrease. The C1–C2 separations increase upon insertion, being largest for insertion of He or Ne. As a consequence of the distortion of the adamantane cage, the energies of the cage atoms increase as a result of the

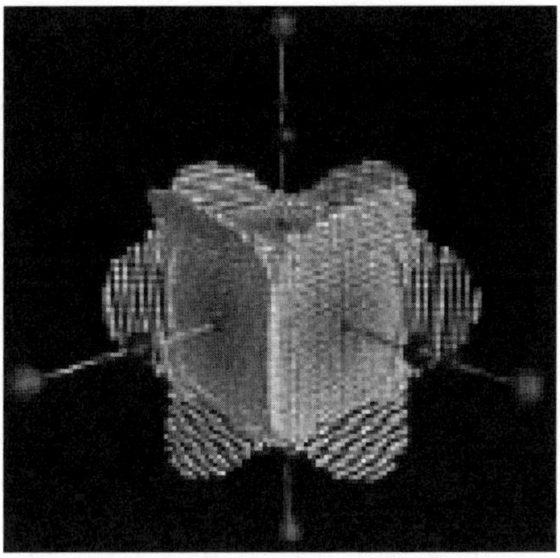

Figure 6 Three of the six interatomic surfaces bounding the Cr atom in $Cr(CO)_6$. The red dot in the centre of each surface denotes the bond critical point, the origin of the bond path that links the Cr to a carbonyl carbon, denoted by a black ball, that in turn is linked to an oxygen, denoted by a red ball. Note the planarity of the Cr|C surfaces. (For interpretation of the references to colour in this figure legend, the reader is referred to the web version of this book.)

decreased attractive interactions of the density in their atomic basins with the external nuclei.

The volumes of C1 and C2 increase with the insertion of the ions, while the volumes of the hydrogen atoms decrease because of their transfer of electron density to the interior atoms. The pressure on C1 increases in the case of Be^{+2} and decreases for the remaining atoms. The increase for C1 in the Be complex is unsurprising since it undergoes the largest decrease in volume. Because of the transfer of charge from the hydrogens and their attendant increase in energy, the energies of both C1 and C2 decrease with the insertion of Be^{+2}, a decrease that is restricted to C1 in the case of Li^{+}.

The metal atoms Cr, Fe and Ni in their carbonyl complexes are enclosed by the interatomic surfaces they share with the carbon atoms of the carbonyl groups [26]. This is evident in Figure 6 which shows the six interatomic surfaces that define the volume occupied by the Cr atom in $Cr(CO)_6$. In contrast to the curved $X|C$ interatomic surfaces bounding an X atom confined within an adamantane cage Figure 5, the Cr|C surfaces are essentially planar, mimicking the surface characteristic of A_2 molecules, in spite of a transfer of 1.2e to the six carbonyl groups. Because of the lack of surface curvature the surface virial for the Cr|C interaction is of the same order of magnitude as those found for the A_2 molecules, with $\nu_s(\mathrm{Cr}|\mathrm{C}) = 0.0357$ au giving a contribution of 6.2 GPa to the pressure acting on the Cr atom, to be compared with 5.4 GPa N in N_2. Thus the total pressure acting on the Cr atom, $p(\mathrm{Cr}) = 37.0$ GPa, is much less than the pressure exerted on an X atom confined within an adamantane cage, the smaller pressure reflecting the lack of curvature in the Cr|C surfaces compared to that found for the X|C surfaces whose surface virials are considerably greater.

The Cr atom is tightly bound in the complex. Its energy decreases by 620 kcal/mol [26], the same order of magnitude as that found for the X atoms in the adamantane complexes, and it is stabilized in the same manner as are the X atoms in their complexes: by the interaction of the density of the confined atom with the nuclei of the host. The atoms in the carbonyl complexes, while totally confined and tightly bound, are not compressed, the Cr atom for example, having an atomic volume of 57 au. Thus they are not subjected to the high pressures found for the X atoms in $X@C_{10}H_{16}$ molecules. Haaland et al. [35] have argued that the He|C1 interaction is 'antibonding', in spite of the finding that the atoms are linked by a bond path, a line, that by definition, is one of maximum density linking the nuclei of neighbouring atoms in a geometry where no repulsive Feynman forces act on the nuclei. One should also recall that $He@C_{10}H_{16}$ is a stable molecule, residing in an energy minimum. The Ehrenfest forces acting on the atoms are, as discussed above, attractive and the resulting surface virials are negative [33]. Thus the surface virials contribute stabilizing contributions of the energy of formation of the complexes. The forces responsible for the 'antibonding' are never defined, nor discussed other than in arguments 'that enclosing an atom in a cage must lead to the operation of repulsive forces'. There are only two forces acting in a molecular system [26,36,37]: the Feynman force exerted on the nuclei that vanish in an equilibrium geometry and the Ehrenfest force acting on the electron density that is attractive in the neighbourhood of an equilibrium geometry. Arguments based on classical notions of chemistry, that undefined 'repulsive forces' must be present even in situations where atoms are linked by a bond path, are readily refuted by physics [38]. People proposing such arguments should instead think in terms of *force per unit area experienced by an atom*, that is the *pressure acting on it*, as opposed to the force exerted on its nuclei or on its density. An atom closely confined in an adamantane cage experiences a much greater pressure than one bounded by an open network of neighbouring molecules as found in the metal carbonyl complexes, for example. There are no 'repulsive forces' acting on atoms confined in an admantane cage, but they are subject to large forces per unit area of surface, that is, to large pressures.

4.5. Compression of hydrogen molecules in a neon vise

The system first used to introduce the quantum definition of pressure consists of a linear chain of hydrogen molecules compressed between a pair of Ne atoms [12]. The vise is depicted in Figure 7 for the compression of five H_2 molecules, the system Ne|H_{10}|Ne. The calculations were performed at the SCF/SCVS level using the 6-311G** basis. The effect of electron correlation on the calculated pressure and on the variation of properties with pressure is found to be minimal, amounting to a few per cent, as determined in

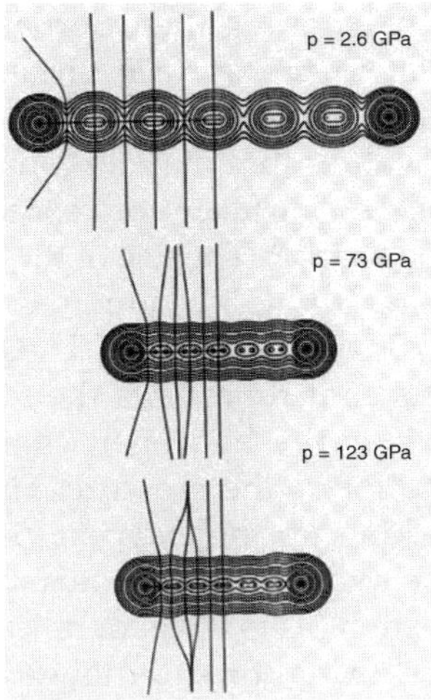

Figure 7 Contour maps of the electron density for the Ne$|H_{10}|$Ne vise. The intersections of the interatomic surfaces with the plane of the diagram are shown for the left half of each molecule, as are the atomic interaction lines. There is a $(3, -1)$ critical point in the density at each intersection of an interaction line with an interatomic surface. The indicated structures are invariant to an increase in pressure.

MP2 calculations at selected pressures. The volumes employed were those determined by the 0.001 au density envelope and shown to contain up 99% of the electronic charge and significantly more than 98% of the atomic virials and kinetic energies. It was found that upon compression, the chain undergoes a significant contraction perpendicular to the chain length as evident in Figure 7, increasing the percentage of the density contained within the 0.001 envelope. Thus the width of the central $|H_2|$ in the $|H_{10}|$ system was reduced from the value of 6.0 au for an isolated molecule to 5.6 au. The volume of an isolated H_2 equals 120 au, which is reduced to ~60 au for the central $|H_2|$ group at the highest pressure. The geometry was optimized for each Ne–Ne separation and thus only the Ne atoms experience a non-vanishing Feynman force on their nuclei, a force that is equal and opposite to the external force applied to a neon atom. These external forces contribute a term $\Sigma_{Ne}\mathbf{X}_{Ne} \bullet \mathbf{F}_{Ne}$ to the total virial, but do not directly determine the pressure acting on the hydrogen chain. Table 1 of Ref. [12] lists all of the important properties of the total system, including the external and total virials, whose values demonstrate that the virial theorem $2 + (\nu/T)$ is satisfied to within small errors, of the order of 1×10^{-9} to 1×10^{-7} au. The external virial $-\Sigma_{Ne}\mathbf{X}_{Ne} \bullet \mathbf{F}_{Ne}$, for example, equals $T + E$ for each Ne–Ne separation as required for satisfaction of Equation (14), and its positive value indicates that T exceeds $|E|$, showing that repulsive Feynman forces act on the system. The use of the classical expression Equation (14) to calculate the pressure, by equating the external virial to $3pv$, is in effect equating the external virial to the surface virial. Table 1 of the Ne vise paper [12] lists the values of the external virial acting in the $|H_{10}|$ system for all Ne–Ne separations, values that may be compared to the values given in Table 2 for $\nu_s(|H_{10}|)$ [12]. The surface virial exceeds the external virial for large R and is less than it for short R. The use of the external virial rather than $\nu_s(|H_{10}|)$ to calculate p yields 0.19 GPa compared to 1.31 GPa at $R = 20$ Å and 153 GPa rather than 123 GPa at $R = 8$ Å. The external virial necessarily increases with the pressure, as it represents the force that must be applied on the Ne atoms to compress the hydrogen chain, but there is no direct relationship of this contribution to the pressure. The present case, where the only force of repulsion making up the external virial is from the Ne atoms exerting the pressure on the hydrogen molecules with no contributions from the nuclei interior to the system (as is the general case), should maximize the possibility of modelling $3pv$ using the external virial rather than the surface virial.

Pressures were calculated for the $|H_{10}|$, and for the internal $|H_6|$ and $|H_2|$ groups, for Ne–Ne separations ranging from 20 to 8 Å to give pressures ranging from 1.3 to 123 GPa. The finite chain mimics an infinite chain by exhibiting approximately the same pressures for inner groups. Thus the pressures exerted on the central $|H_2|$ and $|H_6|$ groups were found to be essentially equal at moderate pressures and differ by only 4 GPa at the highest pressure of 123 GPa. It was not possible to obtain pressures in excess of 160 GPa for the $|H_{10}|$ system because of convergence problems with both the SCF routine and the attainment of equilibrium geometries. The highest pressure attained was 158 GPa for the $Ne|H_6|Ne$ system, which is within the pressure range where solid hydrogen undergoes a phase transition [39–41].

4.5.1. Effect of pressure on geometrical and topological parameters

The properties of the linear chain of atoms fall into three sets: those for the intramolecular, the intermolecular and the closed-shell Ne–H interactions. The intramolecular bond lengths vary from $R_e = 1.39$ au, the equilibrium value to 1.04 au obtained for the central molecule in the $Ne|H_6|Ne$ vise at a pressure of 158 GPa. Only at the highest pressures do the intermolecular separations approach R_e. Unlike the LDA calculations on solid hcp hydrogen, no increase in separation over R_e is found for high pressures. The Ne–H separations decrease at a slower rate.

The presence of a $(3, -1)$ critical point in the electron density between neighbouring atoms in an equilibrium geometry signifies that the atoms are linked by a line of maximum density, a bond path, and that the atoms are bonded to one another. The bond path is defined by the unique pair of trajectories of the gradient vector field of the density $\nabla\rho(\mathbf{r})$ that terminate, one each at the nuclei. The set of trajectories of $\nabla\rho(\mathbf{r})$ that terminate at a $(3, -1)$ critical point defines the interatomic surface that separates the neighbouring atomic basins. This basic topological property of the density is found to be invariant to increases in the pressure, as evident from Figure 7, which illustrates that the nuclei of neighbouring hydrogen atoms in the chain are linked by bond paths and their basins separated by interatomic surfaces. Repulsive forces act on the Ne nuclei, and the line of maximum density linking a Ne atom to the chain is termed an atomic interaction line. The Ne atoms are not bound to the chain. If the external force of constraint is removed, the Ne atoms will separate from the chain.

As previously discussed, bond critical point properties serve to classify the bonding they define. One extreme is a shared interaction where electron density is accumulated between the nuclei, the other is the interaction found between closed-shell atoms or molecules wherein density is removed from the region of the bond critical point and accumulated in the basins of the neighbouring atoms. The discussion here focuses on the role of $\nabla^2\rho(\mathbf{r}_c)$ in the bonding classification. The formation of an interatomic surface and a bond path is a result of a competition between perpendicular contractions and parallel expansion of the ρ, effects that are summarized by the Laplacian of the density at the bond critical point, $\nabla^2\rho(\mathbf{r}_c)$. The contractions are determined by the duo of negative eigenvalues (downward curvatures of the density) of the Hessian of ρ, which leads to a concentration of electronic charge along the bond path, while the expansion is measured by the positive eigenvalue (upward curvature of the density) leading to a depletion of charge in the surface and to its separate concentration in the neighbouring basins. The sign of $\nabla^2\rho(\mathbf{r}_c)$ determines which of these two competing effects is dominant in any given interaction.

The isolated hydrogen molecule is an example of a shared interaction with $\rho(\mathbf{r}_c) = 0.27$ au and $\nabla^2\rho(\mathbf{r}_c) = -1.07$ au. The value of $\rho(\mathbf{r}_c)$ undergoes a continual increase while that of $\nabla^2\rho(\mathbf{r}_c)$ becomes increasingly more negative for the intramolecular interactions as the pressure is increased. The magnitude of $\nabla^2\rho(\mathbf{r}_c)$ more than doubles at high compression. Thus density is accumulated at these critical points by an enhancement of the mechanism that leads to bonding, that is, by an increasing accumulation in and contraction of ρ towards the bond path.

The Ne|H interaction is closed-shell with $\rho(\mathbf{r}_c)$ initially less than 0.05 au and $\nabla^2\rho(\mathbf{r}_c) > 0$. While $\rho(\mathbf{r}_c)$ does increase with an increase in pressure, attaining a value of $\sim$0.18 au, the interaction is dominated by the increasingly positive value for $\nabla^2\rho(\mathbf{r}_c)$ which attains a value of $\sim$1 au for the highest pressures. The Ne|H interactions accumulate less density at the bond critical point than do the intermolecular interactions between the hydrogen molecules. The intermolecular interactions exhibit the hallmarks of closed-shell interactions at low pressures, but $\rho(\mathbf{r}_c)$ increases with increasing pressure and in the case of Ne|H_6|Ne the value of $\rho(\mathbf{r}_c)$ slightly exceeds the value for an isolated hydrogen molecule at the highest pressure of 158 GPa. The positive curvature dominates $\nabla^2\rho(\mathbf{r}_c)$, but the increase is slight and at the highest pressures undergoes a decrease towards negative values, a situation attained in the case of the Ne|H_6|Ne system at the highest pressure of 158 GPa. Thus the intermolecular interactions are transformed from closed-shell to shared interactions with an increase in pressure, the high pressure limit presumably being an ultimate loss of distinction between the inter- and intramolecular interactions.

The Laplacian of the density appears in the local expression of the virial theorem given in Equation (10). The kinetic energy density $G(\mathbf{r}) > 0$, while the total potential energy density of the electrons, the virial field $\nu(\mathbf{r}) < 0$. Thus closed-shell interactions with $\nabla^2\rho(\mathbf{r}_c) > 0$ are dominated by $G(\mathbf{r})$, while shared interactions with $\nabla^2\rho(\mathbf{r}_c) < 0$ are dominated by $\nu(\mathbf{r})$, that is, the decrease in the potential energy density resulting from the accumulation of density along the bond path. $G(\mathbf{r})$ reduces to the Weizsäcker correction to the kinetic energy density where it is determined by the gradients of the density, a situation that also occurs for a system described by a single molecular orbital at the Hartree-Fock level. Thus $G(\mathbf{r}_c) = 0$ for an isolated hydrogen molecule at H–F, and its value remains small for the intramolecular bonding, even for high pressures where it attains a maximum value of 0.073 au for the central |H_2| in Ne|H_6|Ne at the maximum pressure of 158 GPa. $G(\mathbf{r})$ becomes increasingly larger for the closed-shell Ne|H interactions as the pressure is increased, attaining a value of 0.38 for the highest pressure. The increase in $G(\mathbf{r})$ with increasing pressure for the intermolecular interactions is intermediate between that for the shared and closed-shell interactions. In summary, $G(\mathbf{r}_c)$ is largest and increases most rapidly for the closed-shell interactions and $\nu(\mathbf{r}_c)$ is most negative and undergoes the greatest decrease for the shared interactions, all as anticipated on the basis of the local virial theorem and the characteristics of closed-shell and shared bonding characteristics.

The analysis of the critical point properties allows for the following conclusions concerning the nature of the intra- and intermolecular interactions; (a) the intermolecular bonded interactions become progressively stronger and shorter as the pressure is increased, their potential energy exhibiting the largest decrease for the smallest accompanying increase in kinetic energy. (b) The Ne|H wall interaction exhibits all of the hallmarks of a closed-shell repulsive interaction and the rapid rise in $G(\mathbf{r}_c)$, that far outstrips the moderate decrease in $\nu(\mathbf{r}_c)$, indicates that it becomes increasingly more repulsive with increasing compression. The change in sign of $\nabla^2\rho(\mathbf{r}_c)$ for the H_2|H_2 interactions at high pressures indicates that they are approaching a change-over from closed-shell to shared interactions. The increase in their $\rho(\mathbf{r}_c)$ values is greater than for the Ne|H interactions, while the increase in their $G(\mathbf{r}_c)$ values is more moderate.

The double integration of the exchange density over a single atomic basin or over two basins determines the degree to which the Fermi correlation is localized within a given atom or shared between two atoms [42].

It therefore provides a physical measure of the localization/delocalization of the electrons as determined by the pair density. The delocalization index for a pair of bonded atoms provides a measure of bond order, equalling one, two and three for the equal sharing of a corresponding number of electron pairs [40]. Thus the delocalization index for an isolated H_2 molecule equals unity. The bond orders of the H_2 molecules decrease with increasing pressure, being reduced by one-third at the maximum pressure of 123 GPa. The bond orders for the intermolecular interactions – those between H atoms in neighbouring H_2 molecules on the other hand, increase from zero with increasing pressure to 0.25 at the highest pressure. Thus the effect of increasing pressure on the delocalization of the electrons is in the direction anticipated for attaining metallic behaviour, the bond orders of the intramolecular and intermolecular interactions decreasing and increasing respectively, changing in the direction of attaining a common, limiting value of one-half.

For an isolated hydrogen molecule, twice the electronic kinetic energy equals the virial, which for an isolated molecule equals the basin virial $\nu_b(H_2)$. Thus for a molecule under pressure, the surface virial $\nu_s(H_2)$ measures the disparity between the increase in $T(H_2)$ and the change in the internal virial caused by the creation of the zero-flux surface that results from the interaction of the molecule with its neighbours. For example, twice the kinetic energy for the central $|H_2|$ in the $Ne|H_{10}|Ne$ system is in excess of the magnitude of its basin virial for all finite pressures. This difference determines the $3pv$ contribution to the energy arising from the force exerted on the group by the neighbouring $Ne|H_4|$ groups, the term $\nu_s(H_2)$. For a separation of 10 Å, $\nu_s(H_2)$ equals -0.35 au or -191 kcal/mol. The magnitude of this surface term is a measure of the excess kinetic energy engendered in the electrons of $|H_2|$ in response to the pressure acting on its surface.

4.5.2. More recent use of the 'classical approach' in the definition of pressure

A number of papers appeared in the 1980s by Nielsen and Martin [47–49] and one in 2002 by Pendas [50] that employ the classical approach in the definition of pressure as explored by Slater [14] and others and embodied in Equation (17). This approach identifies the pv product with the virial of the external forces acting on the nuclei relating the pressure, in "analogy to classical thinking" [50], to the trace of a stress tensor, Equation (29)

$$p(\mathbf{r}) = -(1/3)\mathrm{Tr}\ \overleftrightarrow{\sigma}(\mathbf{r}) \tag{29}$$

the density $p(\mathbf{r})$ integrating to the pv product. If instead, one considers the definition of pressure as requiring the physics of an open system, then as shown by Schweitz, the identification of pressure with the virial of the external forces is possible only for the case of a homogeneous gas where the surface flux virial equals the constraint virial for the corresponding petit ensemble. Correspondingly, Equation (29) has been shown to hold true in the quantum case of an open system only for a homogeneous gas of uniform density in which case the quantum stress tensor is constant and isotropic [12]. Only the paper by Pendas [50], which reviews and builds on the work of Nielsen and Martin will be considered in detail, since he applies his approach to the definition of the pressure acting on an atom in a molecule using the atomic statement of the virial theorem Equation (7). None of the above authors make any appeal to the physics of an open system in the development of their approaches, although they refer to applications to unit cells in a solid and to atoms in molecules. Pendas however, uses the expressions for the virial and virial theorem provided by the physics of an open system.

Pendas states that the stress tensor appearing in Equation (29) must consist of an electronic part, as given here in Equation (6) and generating the Ehrenfest force density on the electrons, Equation (5), plus a nuclear part generating the Feynman force density on the nuclei. Thus he defines a 'total stress field' $\overleftrightarrow{\sigma}(\mathbf{r}) = \overleftrightarrow{\sigma}_k(\mathbf{r}) + \overleftrightarrow{\sigma}_p(\mathbf{r})$, consisting respectively, of an electronic (kinetic) part and a potential part. He bases his criticism of the approach presented in this treatise to its neglect of the nuclear contribution, claiming that "pressures include both kinetic and potential terms". Pendas introduces the potential part through the use of the 'classical expression' applied to the total stress field in Equation (29), that yields Equation (30) for the pv product for an open system A

$$3p(A)v(A) = \sum_{\alpha \in A} \mathbf{R}_\alpha \bullet \mathbf{F}_\alpha - \oint \mathrm{d}\mathbf{S} \bullet \mathbf{r} \bullet \overleftrightarrow{\sigma} = 2T(A) + V(A). \tag{30}$$

Thus the classical approach predicts that the pv product for an open system vanishes for a system in electrostatic equilibrium, since in that case $2T(A) = -V(A)$. This result is of course, at odds with the quantum definition of pressure that equates the pv product for an open system to $-(1/3)\mathcal{V}_s(A)$, Equation (22). The surface virial $\mathcal{V}_s(A)$ vanishes only for a free, isolated atom. According to the quantum definition, any bound atom experiences a pressure acting on its surface whether or not external forces act on its nucleus. This makes chemical sense in that one appreciates that even in an equilibrium geometry, an atom confined to a cage will experience a greater pressure than one bound in an open system. As discussed in 4.4.2, the pressure acting on an atom provides an important facet towards the understanding of its bonding, whether or not the Feynman force on its nucleus vanishes.

Since the appearance of the paper by Pendas [50], QTAIM has progressed and the atomic statement of the virial theorem can now be applied to define the energy of an atom in a molecule in non-equilibrium situations [51].

The definition of the atomic contribution to the virial of the external forces of constraint is nontrivial. Keith has developed the procedure for the atomic partitioning of 'null' properties; properties such as the sum of the Feynman forces on the nuclei, that sum to zero for the entire molecule [51]. Thus in analogy with the expression for a total system, the energy of atom A is given by

$$E(A) = -T(A) + W(A), \tag{31}$$

where $W(A)$ is the atom's share of the external forces of constraint, its share of the virial of the Feynman forces acting on the nuclei, a term that vanishes for an equilibrium geometry. The use of Equation (31) requires that the wave function satisfies both the virial and Feynman theorems. Thus, as found for the total system, when the forces acting on the nuclei are attractive, $W(A) < 0$ and $T(A) < -E(A)$, and when repulsive, $W(A) > 0$ and $T(A) > -E(A)$. Exploratory work is under way to determine how closely the quantum definition of pressure corresponds to the thermodynamic expression

$$p = -(\mathrm{d}E/\mathrm{d}v). \tag{32}$$

This is a severe test, as the pressure acting on an individual atom is necessarily far from the 'thermodynamic limit' making the comparison more interesting. It is clear that applying this relation to the definition of an atom in a molecule relates p, not to just changes in the external work $W(A)$ as done in the Pendas approach, but to changes in the atom's kinetic energy as well. It is after all, the kinetic energy that determines the pv product in the ideal gas law.

As an initial example, the derivative $-(\mathrm{d}E(A)/\mathrm{d}v(A))$ evaluated about the equilibrium separation for the H atom in H_2 at Hartree-Fock, yields $p(\mathrm{H}) = 5.6$ GPa, a result of the same order of magnitude as the value 3.4 GPa determined by the use of $(-1/3)\ \nu_s(H)/v(\mathrm{H})$. No such approximate agreement is obtained by the use of the change in $W(A)$ alone, as demanded by the Pendas approach. $T(A)$ and $W(A)$ change in opposite directions for a displacement from equilibrium, resulting in a small net change in $E(A)$. Equating $p(A)$ to changes in $W(A)$ alone would generate extremely large pressures.

While Pendas makes no attempt to disprove the quantum definition of pressure obtained through the scaling procedure of Marc and McMillan [22] as presented here, he does state that "the use of electron-only scaling to study stressed situations, where the virials due to the nuclear system cannot be neglected, is not a very consistent procedure." In reality, the physics of an open system does not neglect 'the virials due to the nuclear system', they are included in the total virial that is defined by taking the virial of the Ehrenfest force, see for example Equation (13). The virial theorem illustrates the beauty of quantum mechanics. There are only two forces operative in a field-free molecular system: the Ehrenfest force acting on the electrons and the Feynman force acting on the nuclei, and both are united in the virial theorem. The virial theorem relates the virial of the Ehrenfest force acting on the electrons to their kinetic energy, the molecular virial *including a contribution from the virial of the Feynman forces acting on the nuclei* [14]. Thus through the Ehrenfest and Feynman theorems, one has the tools that are needed to describe the forces acting in a molecule and through the virial theorem, to relate these forces to the molecule's energy and its kinetic and potential contributions in the manner promulgated by Slater. Thus while one can talk about a nuclear contribution as supposedly existing separately from that defined by quantum mechanics in the kinetic stress, the definition of a system's virial, including its nuclear contribution, is totally determined by the quantum stress tensor Equation (6). One is faced with a simple choice: a definition of pressure based on a classical analogy, but using the physics of an open system, or one determined entirely by the quantum mechanics of an open system as presented here.

5. DISCUSSION AND CONCLUSIONS

Previous discussions of the metallization of hydrogen relate its onset to the 'breaking of the bonds in the H_2 molecules' [39,40]. Such a view is incompatible with the topological definition of structure. It has been demonstrated that the topology of the density is invariant to increases in pressure and the bond paths linking neighbouring hydrogen atoms persist at all pressures. The presence of a bond path meets the sufficient and necessary conditions for two atoms to be bonded to one another [41]. It has been demonstrated that the electron density in metals exhibits the same bonding topology as that found in isolated molecules [42], the properties of the density at the bond critical points forming the basis for structure-property relationships for metals and alloys [43]. Thus one anticipates that each hydrogen atom in the solid phase of hydrogen, which may be bcc or hcp, will necessarily be linked by bond paths to 14 or 12 other atoms respectively, just as a sodium atom is linked by bond paths to 14 other atoms in bcc metallic sodium [44]. Gibbs et al. [45] have provided a detailed analysis of the bond critical point properties for the silica polymorph cosite for pressures extending up to 17 GPa that were obtained from first principles calculations. More relevant, is the paper by Gibbs et al. [46] wherein it is demonstrated that in the conducting NiS mineral heazlewoodite, the Ni atoms are linked to one another by bond paths forming a continuous network throughout the crystal, with bond critical point properties similar to those found in Ni metal. Another NiS mineral, vaesite, on the other hand, which is an insulator, exhibits no bond paths linking the Ni atoms. Thus bond paths appear as the 'wires' between the Ni atoms necessary for electron transport, and bond paths must link the atoms in any conductor.

An interatomic surface is a result of the balance achieved in the interaction of two atomic basins. Just as an atom and its properties are transferable between molecules, so is its atomic surface, its form and properties being characteristic for a given interaction [43]. A property of particular importance is its curvature, that is found to be characteristic of the nature of the atom's interactions with its neighbours. The degree of surface curvature determines the magnitude of the Ehrenfest force and its associated virial, whose value determines the atomic pressure. Since an atomic surface is composed of a set of interatomic surfaces – one for each bonded neighbour – the pressure acting on an atom is the sum of contributions, one from each of its neighbouring atoms. Atoms on the boundary of a system share an interatomic surface with the atoms that serve to confine it and this interaction makes a contribution to the system's pressure. If the system of interest is a single confined atom, as in the inclusion complexes of adamantane, then the interactions with the confining atoms determine the total pressure and one may relate the pressure acting on the atom to its interactions with the atoms of the confining cage. By relating the pressure to an interatomic surface virial, the physics of an open system makes possible the determination of the pressure acting on some total system or on any of its individual components, as demonstrated in the discussion of the neon vise.

ACKNOWLEDGEMENTS

I wish to thank Dr. Fernando Cortés-Guzmán of the Universidad Nacional Autonoma de México, for performing the QCISD calculations and generating the related figure. I also wish to thank Dr. Todd Keith for providing wave functions satisfying both the virial and Feynman theorems required for the calculation of atomic pressure in terms of the change in $E(A)$ with respect to $v(A)$.

REFERENCES

[1] R.F.W. Bader, Atoms in Molecules: A Quantum Theory, Oxford University Press, Oxford, UK, 1990.
[2] R.F.W. Bader, Phys. Rev. B49 (1994) 13348.
[3] R.F.W. Bader, T.T. Nguyen-Dang, Y. Tal, Rep. Prog. Phys. 44 (1981) 893.
[4] R.F.W. Bader, T.T. Nguyen-Dang, Ad. Quantum Chem. 14 (1981) 63.
[5] J. Schwinger, Phys. Rev. 82 (1951) 914.
[6] E. Schrödinger, Ann. D. Physik 81 (1926) 109.
[7] J. Schwinger, Lect. Part. Field Theory, Brandies Summer Inst. Theor. Phys. 2 (1964) 157.
[8] R.F.W. Bader, J. Chem. Phys. 73 (1980) 2871.
[9] J.-A. Schweitz, J. Phys. A: Math. Gen. 10 (1977) 517.
[10] T. Takabayasi, Prog. Theor. Phys. 8 (1952) 143.
[11] R.F.W. Bader, P.M. Beddall, J. Chem. Phys. 56 (1972) 3320.

[12] R.F.W. Bader, M.A. Austen, J. Chem. Phys. 107 (1997) 4271.
[13] J.O. Hirschfelder, C.F. Curtiss, R.B. Bird, Molecular Theory of Gases and Liquids, Wiley, NY, 1967, pp. 134–135.
[14] J.C. Slater, J. Chem. Phys. 1 (1933) 687.
[15] E.V. Ludeña, J. Chem. Phys. 66 (1977) 468.
[16] E.V. Ludeña, J. Chem. Phys. 69 (1978) 1770.
[17] E. Ley-Koo, S. Rubinstein, J. Chem. Phys. 71 (1979) 351.
[18] N. Aquino, Int. J. Quantum Chem. 54 (1995) 107.
[19] J.-A. Schweitz, J. Phys. A: Math. Gen. 10 (1977) 507.
[20] N.H. March, Phys. Rev. 110 (1958) 604.
[21] M. Ross, Phys. Rev. 179 (1969) 612.
[22] G. Marc, W.G. Mcmillan, Adv. Chem. Phys. 58 (1985) 209.
[23] S. Srebrenik, R.F.W. Bader, J. Chem. Phys. 63 (1975) 3945.
[24] J. de Boer, R.B. Bird, in: J.O. Hirschfelder, C.F. Curtiss, R.B. Bird (Eds.), Molecular Theory of Gases and Liquids, Wiley, New York, 1967, pp. 398–400.
[25] R.F.W. Bader, M.T. Carroll, J.R. Cheeseman, C. Chang, J. Am. Chem. Soc. 109 (1987) 7968.
[26] F. Cortés-Guzmán, R.F.W. Bader, Coordination Chem. Rev. 249 (2005) 633.
[27] T. Keith (Ed.), AIMALL, 2008, aim.tkgristmill.com.
[28] P.E. Cade, W.M. Huo, At. Data Nucl. Data Tables 12 (1973) 415.
[29] P.E. Cade, A.C. Wahl, At. Data Nucl. Data Tables 13 (1974) 339.
[30] P.E. Cade, W.M. Huo, At. Data Nucl. Data Tables 15 (1975) 1.
[31] R.J. Cross, M. Saunders, H. Prinzbach, Org. Lett. 1 (1999) 1479.
[32] D. Moran, F. Stahl, E.D. Jemmis, H.F.I. Schaefer, P.v.R. Schleyer, J. Phys. Chem. A 106 (2002) 5144.
[33] R.F.W. Bader, F. De-Cai, J. Chem. Theory and Comp. 1 (2005) 403.
[34] D. Moran, H.L. Woodcock, Z. Chen, H.F. Schaefer III, P.V.R. Schleyer, J. Am. Chem. Soc. 125 (2003) 11442.
[35] A. Haaland, D.J. Shorokhov, N.V. Tverdova, Chem. Eur. J. 10 (2004) 4416.
[36] R.F.W. Bader, J. Hernández-Trujillo, F. Cortés-Guzmán, J. Comput. Chem., (2006) (submitted for publication).
[37] J. Hernández-Trujillo, F. Cortés-Guzmán, D.C. Fang, R.F.W. Bader, Faraday Discuss. 135 (2007) 79.
[38] R.F.W. Bader, Chem. Eur. J. 12 (2006) 2896.
[39] H.K. Mao, R.J. Hemley, Science 244 (1989) 1462.
[40] E. Kaxiras, J. Bruoughton, R.J. Hemley, Phys. Rev. Lett. 67 (1991) 1138.
[41] L. Cui, N.H. Chen, I.F. Silvera, Phys. Rev. B 51 (1995) 14987.
[42] R.F.W. Bader, M.E. Stephens, J. Am. Chem. Soc. 97 (1975) 7391.
[43] R.F.W. Bader, F.J. Martin, Can. J. Chem. 76 (1998) 284.
[44] P.F. Zou, R.F.W. Bader, Acta Cryst. A 50 (1994) 714.
[45] G.V. Gibbs, M.B. Boisen, K.M. Rossi, D.M. Teter, S.T. Bukowinski, J. Phys. Chem. B 104 (2000) 10534.
[46] G.V. Gibbs, R.T. Downs, C.T.R.K.M. Prewitt, N.L. Ross, D.F. Cox, J. Phys. Chem. B 109 (2005) 21788.
[47] O.H. Nielsen, R.M. Martin, Phys. Rev. Lett. 50 (1983) 697.
[48] O.H. Nielsen, R.M. Martin, Phys. Rev. B 32 (1985) 3792.
[49] O.H. Nielsen, R.M. Martin, Phys. Rev. B 32 (1985) 3780.
[50] A.M. Pendás, J. Chem. Phys. 117 (2002) 965.
[51] T.A. Keith, in: C.F. Matta, R.J. Boyd (Eds.), Atomic Response Properties, in: Quantum Theory of Atoms In Molecules From Solid State to DNA and Drug Design edn., Wiley-VCH, Weinheim, 2007, pp. 61–94 (chapter 3).

SUBJECT INDEX